BOLESLAW LIMANOWSKI (1835-1935)

A STUDY IN SOCIALISM AND NATIONALISM

BY
KAZIMIERA JANINA COTTAM

EAST EUROPEAN QUARTERLY, BOULDER
DISTRIBUTED BY COLUMBIA UNIVERSITY PRESS
NEW YORK

1978

EAST EUROPEAN MONOGRAPHS, NO. XLI

Winner of the Kosciuszko Foundation Doctoral Dissertation Award
for 1972

*This book has been published with the help of a grant from
the Social Science Federation of Canada, using funds provided by
The Canada Council*

Printed in the United States of America

IN MEMORY OF MY PARENTS

Kazimiera J. Cottam—a brief biographical sketch

Born in Czestochowa, Poland, in the late forties, Kazimiera J.
Cottam attended a secondary school named after Boleslaw Limanowski, located in the Warsaw suburb of Zoliborz. Hence her early
interest in his career. In 1949, she emigrated to Canada, graduating
from the Montreal High School for Girls. However, it proved impossible for her immediately to pursue her academic studies, due to
lack of funds.

Eventually, in 1964, she obtained her Bachelor of Arts degree
from Sir George Williams University in Montreal (now the Sir George
Williams Campus of Concordia University), graduating as the highest
ranking student in Arts who also received the Lieutenant Governor
of Quebec Silver Medal for History. She obtained her Master of Arts
degree from the University of Toronto in 1966. To pursue her graduate studies, she was granted a number of fellowships and scholarships
including, among others, the Canada Council Fellowship, the Poland-
Unesco Fellowship, the Province of Ontario Graduate Fellowship
and the Adam Mickiewicz Scholarship.

In October 1967, she went to Poland to research her doctoral dissertation on Limanowski, who had appealed to her, first of all, as a
patriot and feminist. On returning to Canada in December 1968, she
continued working on her thesis and was employed as a teaching
assistant in the Department of History at the University of Toronto.
The successful defense of her dissertation took place in November
1970. This book represents a revised version of her dissertation.
Dr. Cottam published four articles based on her thesis, as listed in
the Preface. She is now employed as a Quality Controller in the
Translation Bureau of the Department of the Secretary of State in
Ottawa.

This book represents a study in an aspect of the nationality problem in Tsarist Russia (i.e., the "Polish Question") in the second half of the nineteenth and the beginning of the twentieth centuries. It hopes, too, to contribute to the history of ideas. Its main theme is an attempt by a scholar, whose busy and politically eventful life spanned an entire century, to construct a unique socio-political theory.

What was his contemporaries' opinion of Limanowski, who was so "old-fashioned" and so modern at the same time? When he died, his political friends in the Polish Socialist Party (PPS) paid the following tribute to his memory, as engraved on his tombstone: "To the Senator of the Polish Republic: thinker, great scholar, and student of the history of the Polish revolution; an indomitable champion of the common people and of the cause of independence of the Polish Republic; veteran of the uprising of 1863, in the ranks of Polish socialism till the very last days of his great life."

The steadfastness of Limanowski's political convictions was most unusual. During his lifetime, a number of Polish generations forsook their youthful idealism: the generation which struggled in the uprising of 1863; the generation formerly in the ranks of the party Proletariat (1878-1892); the generation which took part in the Revolution of 1905; and, finally, the generation of veterans who fought in the Polish Legions (1914-1918). But Limanowski continued striving, almost till the end of his very long life, for a Poland of social and political justice. Many of those who suspected his patriotism, because he was a socialist, eventually themselves became loyalist toward the partitioning powers—Austria, Prussia and Russia; and many of those who doubted his socialism, because he was a patriot, in the long run left the socialist camp. Yet Limanowski remained as he always was—both socialist and patriot. According to his admirer, the PPS publicist and historian of the interwar (1919-1939) period, Adam Prochnik: "[Limanowski inspired] respect and great admiration in all [who met him]. There were more prominent activists, greater scholars and writers than Limanowski, but there never was a finer personality. [Moreover], his life represented a long period in Polish history, the entire story of Polish socialism, and important chapters in Polish historiography as well as sociology."[1]

His political opponent on the right and former friend, W. Studnicki-Gizbert, in 1912 characterized Limanowski thus: "Fifty years of toil; a pile of books; hundreds of articles; clarity of thought; broad knowledge; topics of the greatest timeliness, discussed conscientiously; great love for the entire Polish people; special sympathy with the underprivileged strata; fifty years of wandering in exile, many years of ideological isolation and lack of recognition—such briefly was the story of Limanowski. No fictional character was ever as worthy of study as this living human being. There was so much in him of the soul of the Polish nation."[2]

On the other hand, his enemy on the left—Rosa Luxemburg—who, according to her most recent biographer, "all along confronted idealism and beauty with the pessimism of historical necessity," in 1908 ridiculed Limanowski as follows: "The mystic sentimental socialism which ran wild in Germany in the 1830s, represented by Karl Grün and Moses Hess, emerged in a suitably messy version, after 40 years, in the ideas of Limanowski. . . . Mr. Limanowski, the later leader of the Polish Socialist Party, united Polish socialism on the basis that socialism is undoubtedly a beautiful idea and patriotism a no less beautiful idea; hence 'why should not two such beautiful ideas unite together?' "[3] And, finally, a Soviet author in the second edition of *Bol'shaia Sovietskaia Entsiklopediia* attributed to Limanowski the sellout of the proletariat to the Polish bourgeoisie.

I shall attempt to present the real man, rather than Limanowski the "myth," as well as to analyze the significance of the "myth" and its impact on his compatriots. Prior to 1918, in spite of appearances to the contrary, Limanowski had never been a front-line political activist—partly due to his long exile. However, in his writings Limanowski discussed vital problems. Already by the time of the Polish uprising of 1863 his basic ideology was formed; this he shared with many *émigré* Polish democrats, notably the poet Mickiewicz in his radical years of 1848-1849. He envisioned no regime of justice in Europe until after the Poles had shaken off their foreign yoke, thus facilitating the destruction of despotism and militarism and, hence, the victory of the triple slogan of the French Revolution—Liberty, Equality and Fraternity. And he saw no real freedom for the Poles without social justice in their own country, and without equality of status for all nationalities within and without the future Polish commonwealth. This quiet and modest man, who always kept his own person in the background, never compromised on these basic ethical and political issues.

Between the Franco-Prussian war and the Revolution of 1905, during the years of so-called "stabilization" in Europe, which was an era characterized by development of both nationalism and socialism, the Polish question no longer appeared on the international agenda. Nevertheless, there were educated Poles—scholars, novelists or political activists—who never came to terms with the *status quo*. They have been labelled the "epigones" of the radical ideology associated with the uprising of 1863. Limanowski was one of them. In Polish history textbooks he is usually presented , above all, as the founder in 1881 of the ephemeral socialist organization, the Polish People, and as chairman at the founding congress of the PPS in 1892, thus becoming a prominent figure in the history of Polish democratic socialism. But Limanowski was not among the sponsors of this congress and was reluctant to attend it, even after receiving an invitation to do so. Although when he died, the obituaries in the organs of the PPS and in the allied press referred to him as "leader of Polish socialism," strictly speaking he was not one of its leaders.

However, Limanowski did play an important political role both outside of, and within, the Polish socialist camp. He was among its early activists, and he became the first serious Polish socialist publicist. His social philosophy was to inspire all non-Marxist socialist groups in his home country at large, and he especially influenced the founders of nationalistic socialism in Russian Poland (i.e., the formerly autonomous Congress Kingdom created at the Congress of Vienna, in 1815, under the rule of Alexander I.) Limanowski became a symbol, as it were, of Polish democratic socialism, being venerated as a special veteran of the PPS, whose opinion was always considered by this party. It was his commitment to both patriotism and socialism which justifies reference to him as the "father" of the patriotic PPS, even if Polish democratic socialism might well have come into being independently of Limanowski. True, his insurrectionary "epigonism," that is, his endorsement of the old methods of struggle for independence, led him for tactical reasons even to put his socialism aside until after Poland became liberated. Hence his alliance with the future dictator of Poland, Jozef Pilsudski (one of the leaders of the PPS), and every insurrectionary Polish element. On the eve of independence, Limanowski by virtue of his prestige facilitated the realization of

Pilsudski's patriotic but non-socialist aims. Yet, after 1918, Limanowski stressed the need for rallying the entire Polish proletariat to the PPS which he considered the best instrument of social justice in independent Poland. And he never lost faith in the ultimate victory of evolutionary and democratic socialism everywhere.

Limanowski, a versatile writer, distinguished himself as historian-scholar and historian-politician; as "sociologist" (social scientist and political theoretician); and as ideologist of just "world-society," that is, socialist and prophet of a united Europe. In this study, I shall attempt to narrow the discussion of Limanowski as sociologist and historian to those issues which were most relevant to his politics, with only brief references to Limanowski as the "pure" scholar. He was influenced, as we shall see later, by French and British thinkers rooted in the Enlightenment like Condorcet, Saint-Simon, Comte, Augustin Thierry, Guizot, Grote, Buckle and John Stuart Mill, as well as by naturalism of three varieties (i.e., organicism represented, for instance, by Albert Schäffle and René Worms; social Darwinism of Bagehot and Gumplowicz; and Spencer's evolutionism). Like many contemporary thinkers, Limanowski absorbed all intellectual trends to which he was exposed.[4]

Almost invariably every sociological problem Limanowski tackled led him to consider the problem of nationality, which limited the scope of his inquiry. Limanowski's sociology as a social science was outdated by the time his writings went to press; it proved insufficiently original to interest the Polish community of scholars centered around the universities, and to influence Polish scholarship in sociology.[5] Except as a contribution to the history of ideas, his sociological theory scarcely calls for further examination.

There are pitfalls in studying an aspect of Limanowski's ideology rather than the whole man, partly because his sociology, history and political ideology were so intertwined—even sometimes in a single work—and partly because his theory usually served him as a means of analysis and reflection and was not really a dogma.

I have encountered two basic difficulties in preparing this study, apart from those due to the indirectness of Limanowski's political role—namely, the complexity of the subject and the need for much background material. I did not dwell too much on the economic setting, in view of the nature of Limanowski's ideology which was essentially an ethical rather than an economic creed.

Limanowski's writings comprise: 1) lectures, articles and books on history; 2) semi-theoretical atrticles and pamphlets connected with contemporary events; and 3) theoretical treatises which, however, were seldom an exposition of "pure" theory. A single work of Limanowski might well contain elements of two or even of all three aspects of his writing. Nevertheless, for the sake of coherent organization, it proved necessary to attempt to differentiate between them.

Chronology as such presented me with no problems; I have given all dates in the New Style. All translations, except where otherwise stated and those from German (see Acknowledgements) are my own. In transliterating Russian words and proper nouns, I generally followed the *Slavic Review* system. And, for the sake of uniformity, I used the modern Polish orthography throughout. In my notes I attempted to list the sources in the order in which I utilized them, except where the final passage in a paragraph is a direct quotation; in such instances, I gave the source of the direct quotation first.

Confronted with the problem of overabundance of sources, I have attempted to examine, if not utilize, all available manuscripts and printed sources in Poland bearing upon Limanowski's social and political activity, while omitting the less important of his writings, such as reviews of current Polish historiography. For practical reasons, I have confined my research to Canada and Poland. Therefore, as a rule I have not been consulting Limanowski's articles in non-Polish periodicals, unless they were available in Canada or Poland, or government documents in the archives of Paris, Geneva, Lvov, Moscow and Vienna. However, many of the Austrian documents dealing with Limanowski were made available to me—either as copies or in the original—in Cracow and Warsaw, as well as through a secondary source which cited a number of them *in extenso*. Some Austrian documents. deposited in the State Archives in Cracow, have been appended to the third volume of Limanowski's *Memoirs (Pamietniki)*. I have utilized government and public documents; private correspondence; Limanowski's books, pamphlets and various bibliographical material. However, I have made no attempt to list here standard bibliographies or reference works.

I have studied Limanowski's outgoing private correspondence, consisting of several hundred letters, in the National Library in Warsaw and in the Ossolinski Institute of the Polish Academy of Sciences in Wroclaw (Breslau). Of these two collections, the most important are those letters which Limanowski wrote to his brother and friend, Lucjan, and to his distant relatives, Boleslaw and Maria Wyslouch, respectively; apart from purely personal matters, they contain much material pertaining to Limanowski's public life. His incoming correspondence is deposited chiefly in the library of the Jagiellonian

University of Cracow and in the archives of the Institute of Party History at the Central Committee of the United Polish Workers' Party. The latter repository contains over one hundred of Limanowski's letters written to the Centralization, the executive of the Union of Polish Socialists Abroad, which from about 1900 was known as the Foreign Department of the PPS. They deal with a variety of topics: his relations with a number of individuals; socialist and other publications, both Polish and foreign; his contributions to party publications and the publication of his pamphlets and monographs by the party; his administration of the "Red Cross" (an originally nonpartisan charitable organization, whose aim was to aid Polish political prisoners and exiles); organizational and other party affairs—especially those of the Paris Branch of the PPS—and of other Polish political parties based both at home and abroad, as well as of some non-Polish (notably Russian) parties; matters concerning Polish emigration in general; and, finally, international affairs.

I have usually read the latest revised editions of Limanowski's works, after having compared them with the original. He wrote a number of popular pamphlets for the purpose of propaganda and agitation. These proved especially valuable for this study, because they contain Limanowski's ideas in "concentrated" form. His book criticisms—omitting historiographical reviews of books by Polish authors (as mentioned above)—are also interesting for in them Limanowski theorized on subjects only indirectly related to the themes of the books or articles which he was reviewing.

Usually, his numerous articles present no identification problems; he signed most of his writings, whether using full name, initials, or an easily identified pseudonym. He disliked secrecy and wished to be able to proclaim his political convictions openly. It is only fair to add that, writing mostly in emigration, Limanowski could more readily identify himself than the publicist at home. Limanowski made numerous speeches (many of which, of course, were not recorded and their general import could be derived only from his *Memoirs*).

Turning now to the biographical and autobiographical data, I should note that there are many brief biographical sketches of Limanowski, usually written by his contemporaries, most of the authors being under the spell of Limanowski's "myth." However, to my knowledge, there is as yet no comprehensive political biography written in the West, synthesizing all these views and attempting a critical analysis of Limanowski's activity from a historical perspective.

(Two monographs on Limanowski appeared recently in Poland, in 1971 and 1973.)

Parts of the contents of this book were published as follows:

"Boleslaw Limanowski and the Nationality Problems of the Polish Eastern Borderlands," *Polish Review*, New York, No. 2, Spring 1972.

"Boleslaw Limanowski and the Rise of Student Activism at the University of Moscow, 1857-1858," *Canadian Contributions to the Seventh International Congress of Slavists*, Mouton, 1973.

"Boleslaw Limanowski: A Polish Theoretician of Agrarian Socialism," *Slavonic and East European Review*, University of London, England, No. 122, January 1973.

"Boleslaw Limanowski and the Polish Labor Movement, 1876-1884," *Canadian Slavonic Papers*, Carleton University, Ottawa, No. 4, 1972.

The basic source for the study of Limanowski are his *Memoirs*, the first volume of which appeared two years after his death, in 1937, because of public demand, and was reissued in 1957. Volumes II and III came out in 1958 and 1961, respectively. Also, the final instalment of Limanowski's reminiscences, a diary covering the period from 1919 to 1928, was being prepared for the press at the time of my research in Poland, in 1968. It appeared in 1973.

Limanowski's *Memoirs* constitute a precious document in Polish history—covering the period of the insurrection of 1863 and subsequent Russian repressions, formation of new socio-political trends in both emigration and at home after 1863, development of Polish socialism, the struggle for independence and the reconstruction of the Polish state after 1918.[6] Writing late in life (i.e., between his seventy-sixth and ninety-fourth year) Limanowski had the advantage of a historical perspective; on the other hand, he was handicapped by failing memory. However, by using correspondence and other documents to verify his recollections, he produced a record of high credibility which, as such, constitutes a valuable source for the historian. Appearing in his *Memoirs* as both activist and keen observer, in youth, maturity and advanced old age, with a life spanning an entire century, Limanowski evaluated the events of a long by-gone era with a modern outlook and, conversely, saw modern events through the eyes of a man born very long ago. His *Memoirs* were not history; he did not attempt to answer the question "what happened?" but, rather, to tell what he subjectively perceived. This approach was perhaps due to the fact that he himself was a practicing historian, as had been pointed out by Prochnik.

Having destroyed his original draft composed in Moscow, in his student days, as too revealing, Limanowski omitted all intimate and unpleasant personal matters from the final verison of his *Memoirs*. Rejecting Rousseau's "psychological" approach, he instead adopted a "sociological" method, concentrating on the environment imping-ing on his personal and public life. (This enhanced the value of the *Memoirs* as a historical source, limiting its usefulness for the literary biographer.) Limanowski believed that his method was appropriate for him as a historian. He wrote: "In using the sociological approach.... the author need not confess his shortcomings and errors. His own person recedes into the background. Perhaps such a method is the only valid and honest one... [and the one] most useful to the his-torian."[7] And in an unpublished version of his memoirs he says:

> Memoirs written for posterity might serve two alternative purposes: either they become source material for psychological research or pro-vide. . . . the future historian with a keener insight into the spirit of the epoch described. I chose the second course. . . . I could perhaps with reference to myself repeat Rousseau's words which he wrote in the introduction to his *Confessions:* "Etre éternel, rassemble autour de moi l'innombrable foule de mes semblables; qu'ils écoutent mes confessions, qu'ils gemissent de mes indignités, qu'ils rougissent de mes misères. Que chacun d'eux découvre à son tour son cœur aux pieds de ton trône avec la même sincerité; et puis qu'un seul te dise, s'il ose: 'Je fus meilleur que cet homme-la."[8]

Ottawa, Canada   K. J. COTTAM
May 1, 1976

ACKNOWLEDGEMENTS

I wish to thank both Unesco and the Canada Council for financing my research in Poland, as well as to express my gratitude to the supervisor of my dissertation, Professor Peter Brock, for his assistance. The following Polish scholars have aided me in my task: Professors S. Kieniewicz, J. Durko, J. Buszko and the late Professor M. Zychowski, as well as Drs. J. Skrowronek, J. Zdrada and J. Kancewicz. They are in no way responsible for any opinion expressed in this study, except where specific acknowledgement is made.

I am very grateful to Mmes Jankowska, Bobrowska and Zielinska for carrying out for me several translations from the German, as well as to all the librarians and archivists who helped me at the following institutions in Poland: University of Warsaw, the Institute for Party History at the Central Committee of the Polish United Workers' Party (PZPR), the National Library, the Public Library, and the Central State Archives for Old Documents in Warsaw; the Jagiellonian University, the State Archives of Cracow, and the Library and Archives of the Polish Academy of Sciences in Cracow; and the Ossolinski Institute of the Polish Academy of Sciences in Wroclaw. I am also much indebted to Dr. Z. Zechowski for his recent study on Limanowski as a sociologist.

Finally, I wish to express my appreciation to both the Social Science Federation of Canada and the Kosciuszko Foundation in New York for their subsidies, without which the publication of this book would not have been possible.

LIST OF ABBREVIATIONS

Add: Additional (MS), i.e., recently acquired MS (Przybytki).

AGAD: Archiwum Glowne Akt Dawnych (The Central State Archives for Old Documents, Warsaw).

AM: Archiwum Mikrofilmow (Microfilmed Archives).

APKr.: Archiwum Panstwowe Krakowa (The State Archives of Cracow).

APPS: Archiwum Polskiej Partii Socjalistycznej (Archives of the Polish Socialist Party).

ASEER: *The American Slavic and East European Review.*

AZHP: Archiwum Zakladu Historii Partii przy KC PZPR (Archives of the Institute for Party History at the Central Committee of the Polish United Workers' Party).

BBWR: Bezpartyjny Blok Wspolpracy z Rzadem (Non-party Bloc of Cooperation with the Government).

BJ: Biblioteka Jagiellonska (The Library of the Jagiellonian University of Cracow).

BN: Biblioteka Narodowa (The National Library, Warsaw).

BP: Biblioteka Publiczna (The Public Library, Warsaw).

BPAN: Biblioteka Polskiej Akademii Nauk (The Library of the Polish Academy of Sciences, Cracow).

JCEA: *Journal of Central European Affairs.*

KH: *Kwartalnik Historyczny (The Historical Quarterly).*

KZ PPS: Komitet Zagraniczny Polskiej Partii Socjalistycznej (The Foreign Committee of the Polish Socialist Party).

Mfm: Microfilm.

NKN: Naczelny Komitet Narodowy (The Supreme National Committee).

Ossol: Biblioteka Polskiej Akademii Nauk, Zaklad im. Ossolinskich (The Ossolinski Institute of the Polish Academy of Science in Wroclaw).

OZ PPS: Oddzial Zagraniczny Polskiej Partii Socjalistycznej (The Foreign Department of the Polish Socialist Party).

PH: *Przeglad Historyczny (The Historical Review).*

PPS: Polska Partia Socjalistyczna (The Polish Socialist Party).

PPSD: Polska Partia Socjal-Demokratyczna Galicji i Slaska Cieszynskiego (The Polish Social Democratic Party of Galicia and Teschen Silesia).

PSB: *Polski Slownik Biograficzny (The Polish Biographical Dictionary)*

SKDP: Socjaldemokracja Krolestwa Polskiego (Social Democracy of the Polish Kingdom).

SDKPiL: Socjaldemokracja Krolestwa Polskiego i LItwy (Social Democracy of the Polish Kingdom and Lithuania).

SEER: *The Slavonic and East European Review.*

SR: Stronnictwo Ludowe (Peasant Party).

St.GKr: Starostwo Grodzkie Krakowa (Sub-prefecture of the City of Cracow).

TKSSN: Tymczasowa Komisja Skonfederowanych Stronnictw Niepodleglosciowych (The Temporary Commission of Confederated Independence Parties).

ZHP: Zaklad Historii Partii (The Institute for Party History; see AZHP, above).

ZZSP: Zwiazek Zagraniczny Socjalistow Polskich (The Union of Polish Socialists Abroad).

TABLE OF CONTENTS

Limanowski's ideology was rooted in the ideology of Polish democracy which, in turn, was rooted in the idealism of the French Revolution. The Revolution left a dual legacy of social and nationalist movements to nineteenth-century Europe. In the countries remaining under foreign rule, the ideas of the Revolution inspired above all struggles for national independence. The key issues during the first half of the century in East-Central Europe, in the countries ruled by Russia and the Habsburg Monarchy, were the agrarian problem and the problem of nationality.

The final partitition of Poland between Tsarist Russia, Hohenzollern Prussia and Habsburg Austria, in 1795, affected the balance of power in Europe. Herself partly owing her ascendancy to Russia, in turn Prussia helped Russia to establish her influence in Poland during the eighteenth century, which was facilitated by the reactionary elements among the Polish aristocracy. However, Prussia became fearful of Russia's absorbing the whole of Poland and, on the other hand, she disliked the rise of the Polish anti-aristocratic movement, being alarmed by the prospect of Poland's recovery and fearful of the potential spread of Polish social-revolutionary ideas to neighboring states. Therefore, Prussia's aim became to exploit Russia's difficulties in Poland by claiming her share (in accordance with the traditional expansion of the Hohenzollerns, at the cost of Poland, of which they were once vassals for East Prussia). The partitions enhanced Prussia's status in Germany, and united the three powers for the sake of maintaining the status quo in Poland. For Russia this alliance had dual significance: it enabled her to retain her share of Poland, as well as the status of the arbiter of East-Central Europe.[1]

Revised and sanctioned by the Congress of Vienna in 1815, the partitions delayed the process of industrialization in Poland, which began toward the end of the eighteenth century. Nevertheless modern industrial development took place, especially in Russian Poland. However, each region developed separately. Unlike Austrian and

Prussian Poland which were incorporated into states more advanced economically, remaining backward hinterlands of Austria and Prussia, Russian Poland was subjected to an extensive socio-economic transformation and became one of the most industrialized regions of the Russian Empire. On the other hand, as the result of peasant emancipation by the Prussian authorities in 1823 and their oppressive policies vis-à-vis the Polish element, in Prussian Poland modernization of agriculture proceeded at a quicker pace than elsewhere in the lands of the former Polish-Lithuanian Commonwealth.

While the most important problem in Germany toward the middle of the nineteenth century was unification, and in Russia—peasant emancipation, the situation in Poland demanded a coordination of the struggle for national independence with agrarian reform in Russian and Austrian Poland. (In Galicia, serfdom was abolished by the partitioning power, Austria, in 1848). Polish patriots began to consider the masses to be the instrument of liberation; conversely, social reform was deemed by them impossible without the struggle for independence against the reactionary Monarchies allied with native conservatism. Because the three partitioning powers were until the middle of the nineteenth century the mainstay of reaction in Europe, the independence of Poland became the cause of European democracy. European democrats supported the Polish cause and on the other hand, the Poles participated in every revolution in Europe.[2]

In addition to being influenced by the French Revolution, Polish democratic ideology was formed in conjunction with contemporary events in Poland—the Constitution of 1791 and the insurrection of 1794. In 1791 the five major social strata in Poland numbered: nobility (725,000), clergy (50,000), "Christian" bourgeoisie (500,000), Jews (900,000), and peasants (6,365,000). About one million of the latter belonged to clergy, one million to the Crown and to the public estates *(starostwa)*, while 3,500,000 were serfs on private estates. The Christian bourgeoisie in Poland constituted two-thirds of the noble population, and their weakness vis-à-vis the nobility was not so much due to their numbers, as to their predominantly foreign origin and inferior economic status. The Polish-Lithuanian Commonwealth was principally a noble society, in which the middle class could play only a subordinate role.[3]

The franchise defined by the Constitution of 1791—which was an attempt to strengthen the Commonwealth after the first partition of 1772 by the reformist Four-Year Diet (1788-1792)—excluded only the poorest nobility while including only the richest bourgeosie.

Twenty-four middle class deputies were to represent the royal towns in the Diet—out of the total number of 500—on matters concerning trade and commerce. Only the Christian bourghers in the royal towns were to enjoy civil rights equal to those of the nobility; the private towns were unaffected by the Constitution. The Constitution theoretically provided legal protection to the peasants, but this concession proved meaningless in practice, because Polish law remained obscure and uncodified. The Constitution in effect reaffirmed the political preponderance of the nobility.[4]

These reforms were inspired by the ideology of the priest Stanislaw Staszic, adopted by an activist group of propagandists, "The Forge *(Kuznica)*," led by another priest Hugo Kollataj. The so-called "Kollataj school" of political liberalism deemed the people entitled to natural rights, but attributed political rights to the propertied classes only and avoided economic reforms. The vital problem of security of tenure, the burdens of the corvée, and patrimonial jurisdiction were not discussed by the reformers. However, influenced by contemporary criticism to this effect, the reformers issued a decree on April 24, 1792 recognizing the hereditary rights of royal and state peasants, but providing no relief to serfs on private estates. In the course of the insurrection of 1794, its leader Tadeusz Kosciuszko in his Manifesto of Polaniec proclaimed on May 7 the right to leave their landlords, applicable to all peasants, as well as limitation of their corvée among other concessions, for the duration of the uprising. He defined the machinery to make the reforms effective, promising their implementation upon the insurgents' victory. The legislation was drafted in the spirit of reform of Joseph II of Austria; serf economy remained intact. However, there was provision for security of tenure and alleviation of services. Therefore, peasants ralled in greater numbers to Kosciuszko than in subsequent uprisings.[5]

The insurrection represented a vehement protest against the second partition in 1793. The opponents of the Constitution and of agrarian reform confederated in the town of Tagowica; they secured Russia's aid and thereby sparked a civil war with foreign intervention, which resulted in the second partition. All that remained of the old Commonwealth were parts of central Poland (Mazovia, Podlachia and a section of southern Poland), and the western half of historical Lithuania (Courland, Samogitia, Lithuania proper and Western Belorussia as well as Western Volhynia). The lesser gentry who benefited from the moderate Constitution of 1791, and represented the restless element in Polish politics, precipitated the insurection and for the first

time emerged as an independent political force. Therefore, the Confederation of Targowica in effect, though not in intent, destroyed
the monopoly of aristocracy, and allegiance to the national cause replaced allegiance to the great aristocratic families. Also, for the first
time in Polish modern history the peasants were affected by the rising tide of Polish patriotism. The Constitution itself was voted in the
Diet while the majority of the deputies were absent from Warsaw
during the Easter recess of 1791; when the other members returned,
they were faced with a *fait accompli*. In time, the uncertain reception of the reforms of 1791 was forgotten, and "Targowica" became
the most derogatory word in Polish political terminology.[6]

The legacy of the events of 1791 and 1794 was ambivalent. Later
a consensus developed in Poland that the Constitution represented
the beginning of moral and political recovery—stopped by native reaction allied with foreign intervention. The second event left the
memory of violence and passions displayed by the Warsaw mob, and
of lynchings by the lesser gentry of their political opponents, which
inspired the upper classes with fear of revolution.[7]

Meanwhile, the military reforms effected by the Four-Year Diet
benefited the medium gentry. A small but efficient professional
army was created wtih officer cadres drawn from this social stratum.
Thus modern Polish militarism was born in 1791. Polish patriotism
and militarism, evolved in the internal struggles between the men of
the reformist Four-Year Diet (1788-1792) and the "men of Targowica," and in the external struggles with the partitioning powers at
the end of the eighteenth century, were strengthened in the course
of the Napoleonic campaigns. After the final partition in 1795,
many Poles centered their hopes on Napoleon, and envisioned independent Poland being restored by the legionaries marching back to
Poland from Italy. The legacy of these wars was a romantic approach
to politics, that is, solidarity with revolutionary Europe and, on the
other hand, the quest for an alliance with every real or potential
enemy state of the partititioning powers.[8]

The Napoleonic cult in Poland was associated with fond memories
of the Duchy of Warsaw (1807-1815) created under Napoleon's auspices and comprising the territories taken by Prussia in the second
and third partition, and from Austria in the third partition (i.e., Poznania, Mazovia, Podlachia, and parts of southern Poland). Napoleon
granted the Duchy a constitution which was less liberal than that of
1791, but the regime of the Duchy was more democratic, with more
democratic representation and staffing of the bureaucracy and army.

Moreover, he endowed the Duchy with two milestones in modernization: his Civil Code and formal emancipation of the peasantry. In the latter instance, his legislation was anticipated by Kosciuszko. Yet, emancipation without land did not really serve the peasant's interest, who was subject to eviction by the landlord. With its modernized government and social relations, the Duchy was significant in that it laid a basis for independence in the future.

In 1815, due to Napoleon's defeat, and following the Congress of Vienna, the Duchy ceased to exist. Prussia regained Poznania; the ancient Polish capital of Cracow with its nearby rural districts survived as a quasi-independent Republic of Cracow until 1846, at which time it was absorbed by the Habsburg Monarchy; and the remainder of the Duchy became the Congress Kingdom (Russian Poland). This Kingdom, created by the Congress of Vienna, was united to Russia in the person of the Tsar, Alexander I, but had an autonomous administration and army. It comprised most of the territories taken by Prussia and Austria in the final partition (i.e., Mazovia, Podlachia and parts of southern Poland), but the eastern borderlands remained incorporated into Russia (i.e., Courland, Polish Livonia, Samogitia, Lithuania proper, Belorussia, Volhynia, Right-Bank Ukraine, and Podolia). Therefore, in the 1820s the landed gentry conspired to regain these territories in the East. (On the other hand, the liberal opposition in the Warsaw Diet—led by the brothers Bonawentura and Wincenty Niemojowski and based in west-central Poland—confined their demands to the maintenance of the Constitution granted by the Tsar, blocking legislation which they deemed contrary to its letter and spirit).

Organized by young people of gentry origin, the conspiracies of the 1820s were connected with Romanticism as the great cultural trend. Romanticism viewed the national society as a spiritual organism evolving in the course of history, and tended to idealize the common people. On the other hand, the Enlightenment considered society as a group of individuals endowed with equal natural rights. Both these views were implicit in the development of the idea of social justice and democracy in Poland. However, the legacy of Romanticism was ambiguous. Romanticism also inspired a conservative social philosophy idealizing the past, stressing tradition rather than change, and opposing democratization which social reformers postulated as an instrument of national unity as well as of establishing a regime of social justice. The Polish conspirators of the 1820s were only vaguely conscious of the need for social reform in Poland. They revolted

against the political status quo that was being justified by their elders
by narrow "empiricism" and appeal to "reason." Thus, the great
Romantic Polish literature was born, an expression of protest against
the loyalist older generation associated with Classicist literature and
the philosophy of the Polish Enlightenment. Perhaps at no other
time in Polish history was literature and philosophy so directly committed to political struggle as on the eve of the insurrection of 1830.[9]

Since 1815 there was a cleavage within the educated class in Poland; the propertied element was usually loyal to the partitioning
powers, while the professional *declassé* radicals refused to condone
indefinite dismemberment of their country.[10]

Hence, the radicals forced the upper classes to take part in the
November insurrection of 1830, by confronting them with a *fait accompli*. Being opposed to social reform at this time, the upper classes
transformed the insurrection into a "conservative revolution." During the insurrection, the radical lawyer and deputy to the Diet Jan
Olrych Szaniecki proposed peasant emancipation with land and compensation to the landlords; he began voicing this demand just prior
to the uprising, receiving only marginal support from the educated
classes. However, his "economic" approach to the peasant problem
represented a revolution in thinking, as compared to the earlier ideology of the "Kollataj school" of political liberalism. (See above,
p. 3.)[11]

After 1830, the conservatives split on the national and social issues;
One section remained in the home country; fearful of the masses,
they abandoned the struggle for independence. The *émigré* conservatives, led by Prince Adam Czartoryski, played a major role in the independence movement. They produced a moderate program of social
reform, eventually including emancipation of the peasantry with
land and compensation to landlords; they abandoned the stipulation
for compensation in 1848, demanding immediate emancipation.
However, they were not prepared to coerce the gentry.[12]

The military defeat of 1831 shocked the educated Poles, already
humiliated by the partitions. It resulted in the "Great Emigration"
(1831-1863) composed of much of the elite of the nation, which
may have numbered as many as 10,000, with a majority concentrated in France,[13] that was motivated by the common aim of preparing
another insurrection. Some of the patriots now began to scrutinize
the national tradition in order to discover the causes of the defeat
and the guidelines for the future. They speculated that the fate of
the uprising might have been different if the masses had taken an

active part. At this time two-thirds of the population in Poland were illiterate, without protection of the law, and burdened out of proportion to their capacity to pay taxes. Unsympathetic to the national cause, they were potentially a dangerous opponent of the movement for independence. All the *émigrés*—from conservatives to extreme radicals—admitted the need for agrarian reform. The Prussian government had already initiated peasant emancipation with land in the region it occupied, and most of the enlightened Poles hoped that elsewhere in Poland the reform would not have to come from above, that is, from the Russian and Austrian governments.

Notwithstanding their common aims of insurrection and agrarian reform, the *émigrés* differed as to the means of achieving independence. The conservatives solicited aid from European governments, the democrats—from "peoples." The conservatives were usually constitutional monarchists, the democrats—republicans. They also differed in their approach to social reform. While the conservatives tended to separate the issues of social reform and national liberation, the democrats assumed their close interdependence and agitated for an unconditional emancipation with land.[14]

The idea of democracy in Europe, rooted in the Enlightenment and in the idealism of the French Revolution, passed through the crucible of Romanticism, emerging as a synthesis of both these great cultural and philosophic trends. Originally, the idea of democracy was embodied in the demand for equal privileges for the three Estates (i.e., civil rights afor all and political rights for propertied classes). Eventually, it was embodied in the demand for abolition of the Estates and equal civil and political rights for all, regardless of origin and wealth. The development of the Polish idea of democracy was affected by the foreign rule. The defeat of the 1830 uprising and the loss of autonomy in Russian Poland precluded educated Poles from taking part in practical politics, and it also made some of them aware of the connection between Poland's faulty social relations and her political decline in the eighteenth century. Taking stock of the past and visualizing the future, they scrutinized the basic slogans of the French Revolution: "Liberty, Equality and Fraternity." While some of them adopted extreme positions, serious thinkers attempted to arrive at a balanced synthesis between the rights of the individual and the welfare of society as a whole. Polish democratic thought, developed in exile apart from practical politics, was a voluntaristic type of ideology, neither fully realistic nor fully Utopian, stressing individual commitment to the armed struggle for independence and democracy.[15]

Unlike the *émigré* conservatives who remained united, the democrats soon split into factions, the most important of which was the
Democratic Society, founded in 1832; which played an important
role in conspiracies of the 1830s and 1840s. According to the
Democratic Society, "the Polish cause ought to become the people's
cause." The intent was to enlist the masses in the struggle for independence, without pushing away the nobility, whose estates were to
remain intact, and to reconcile the landlord with the peasant. (However, the Manifesto of 1836 contemplated force being used against
the gentry, should they resist unconditional peasant emancipation
with land.) The goal of the Democratic Society was restoration of
a politically democratic Poland with privately owned property.[16]

There was a great deal of discussion in Poland on the peasant
issue, which led to delving into the controversial origins of the Polish
state and the original status of the peasant. The priest F. S. Jezierski,
who was a member of the Reform party which prepared the Constitution of 1791, argued that the peace loving Sarmatians, the alleged
original inhabitants of Poland, succumbed to the warlike Slavonians,
a kindred tribe. Hence serfdom in Poland was based on nothing but
sheer conquest. Some protagonists of the Norse theory of conquest—
put forward by Polish historians in the early nineteenth century—
insisted that peasants received their land from the gentry. And those
who rejected this theory could nevertheless argue that peasants were
descendants of slaves or prisoners of war, whom the gentry or their
predecessors settled on land. On the other hand, the followers of the
historian Joachim Lelewel (1786-1861) were convinced that peasants
were originally freeholders. Yet, we may add, conceptions of private
land ownership are comparatively recent in Poland. The gentry was
not concerned with the legal status of land, of which there was plenty
in Poland until the nineteenth century, but rather with their rights
over the persons of peasants, because labor was in short supply. According to twentieth century historians, there is no conclusive evidence as to the origins of the Polish state or the original status of
the Polish peasant.[17]

The chief protagonist of the theory that the peasants were originally freeholders, Lelewel, concluded on the basis of his researches
that Poles like the other Slavs had originally lived under a system of
collective landownership, which was similar to the modern Russian
mir. According to Lelewel, already in the prehistoric pagan times
there emerged, as the result of petty tribal wars, two social groups in
Poland: the peasant Kmiecie and the later privileged Lechici. How-

ever, both groups enjoyed equal civil and political rights, performing similar military duties, and differed only as to their system of land-ownership. The Kmiecie enjoyed equality of status within their communal system of landownership, whereas the Lechici had private hereditary holdings of unequal size. The ruling dynasty of Piasts, themselves of Kmiec origin, based themselves on the Lechici. Moreover, since Poland's baptism in 966, under the influence of Western civilization (i.e., Christianity and feudalism), the Kmiecie were gradually subjugated by the Lechici. After the death of their royal ally, Boleslaw the Bold, toward the end of the eleventh century, the people no longer held offices of state and were confined to the more menial duties of warfare. Precluded from intermarriage, two completely different social classes came into being in Poland by the fourteenth century: the well-born gentry and the enserfed peasants. Thus, according to Lelewel, the rise of the Polish gentry was basically a spontaneous internal development.

Although the people had been subjugated by the gentry, he felt that their democratic traditions survived in the gentry democracy of the old Commonwealth. Therefore, the task of Polish democracy was to liberate the people, and return to them the land which had been taken away from them by the gentry. (This was the standpoint of the Polish Democratic Society.) Lelewel was not a socialist, and his theory could well apply to programs for political democracy and decentralization. It proved very popular among the *émigrés*, as well as in the home country, among literary men, philosophers and politicians. Many Polish works on agrarian reform drew their inspiration from Lelewel's ancient Slav commune (*gmina*). For Polish socialists his theory provided a native ideological root, akin in significance to that which the Russian *mir* had for Herzen. Although future researchers were to show the falseness of Lelewel's theory, his role in the shaping of Polish socialist ideology was important, and for many resurrected commune remained the basis for common ownership of land and a regime of liberty and equality.[18]

Lelewel argued that independent Poland never knew individual property ownership as such. That which the individual held in land he deemed "possession" and not "ownership." However, he admitted that the people held land indirectly, leasing it from the landlords for an indeterminable period. Thus peasants were not unconditional "possessors" of their allotments. It was easy for conservatives to contend that, although undoubtedly the peasant "possessed" his land, it was the landlord who actually "owned" it.[19]

The basic inconsistency in Lelewel's historiography was due to his combining elements of the Enlightenment and Romanticism. He derived the process of political and social change in Poland from the interaction of the native principle of democracy embodied in the people—struggling with the alien principle of aristocracy embodied in the gentry. He felt that primitive Slav equality was destroyed by an alien way of life—Christianity and feudalism. On the other hand, he was influenced by the Western idea of progress. Having overcome the idea of non-Slav origin of the gentry in Poland, Lelewel replaced the theory of "ethnic conquest" by the notion of a "cultural conquest" as explanation of the modern Polish class society.[20]

The most popular solution of the peasant problem in Poland was that propounded by the Democratic Society, which postulated individual landownership by peasant families without compensation to the gentry. A small minority of the *émigrés* advocated collective ownership of land and other resources, being influenced by the French Utopians, the radical Catholicism of the priest Lamennais, as well as by the *babouviste* element among the *carbonari*—due to their egalitarian and revolutionary appeal. These early Polish socialists drew their inspiration from two main ideological sources: Christianity and the concept of the ancient Slav commune.

The first Polish socialist organization, the Polish People (*Lud Polski*), was a break-away group from the Democratic Society, which came into being in Portsmouth, England, on October 30, 1835. It comprised 212 soldiers and non-commissioned officers of peasant and artisan origin, veterans of the uprising of 1830 who—having refused the amnesty granted them by the Tsar—had been interned in Prussian prisons for two years prior to their arrival in Portsmouth.[21] During the first phase of its existence the leaders of the Polish People were the middle class "Jacobin" Tadeusz Krepowiecki and Count Stanislaw Worcell, former members of the Democratic Society's branch on the island of Jersey. The intellectuals and the rank and file of the Polish People were all united by their faith in the common people, being conscious of the gulf separating the peasant from the gentry and the bourgeoisie. However, hostility toward the gentry was not equally shared by all members. Some condemned the gentry-dominated Polish past, others rejected the historical role of the nobility in general, but this criticism diminished with reference to the contemporary era for the sake of struggle for independence requiring national unity.[22]

The Polish People were bitterly opposed to the rival Democratic Society with their program of private peasant proprietorship. They felt that a program of this kind would bring the evils of capitalism to Poland and result only in a facade of democracy. Peasants would elect gentry as their representatives; the gentry would remain the dominant class. For the landless, there would be a new system of serfdom, i.e., economic dependence on the landlords as well as on the new class of rich peasants. The Polish People opposed all privilege and inequality, including inherited wealth. According to their program, work was to be compulsory for all, the tools being supplied by the state through local communes—conceived in the nature of the Russian repartitional communes—granting allotments to the individual families and reassigning them for the sake of maintaining equality.[23]

Worcell in his treatise *On Property* (1836) regarded it as a historical rather than natural category which, as such, was subject to change, evolving to perfection through consecutive epochs, each with specific needs as defined by the ruling Idea (Revelation), the "Revelation" of the contemporary era being the voice of Christ preaching brotherhood. A protagonist of the Saint-Simonian idea of socialism as the final outcome of progress, Worcell considered socialism as the fulfilment of Christian idealism.[24]

Rejecting the concept of human nature postulated by the Enlightenment as a non-historical principle producing an atomized society, the Polish People opposed the Democratic Society's justification of private property in terms of natural law. Worcell differed from the Society in his interpretation of the basic slogans of the French Revolution: "Liberty, Equality and Fraternity." According to Worcell, equality had as its premise economic equality; freedom was defined by the obligation to work and had as its premise the economic independence of one individual from another (which independence Worcell expect to result from abolition of private ownership of the means of production); and brotherhood had as its premise collective ownership and cooperation, because real brotherhood was precluded in a society based on the principle of egoism and on the anarchism of private property. The leadership of the Polish People argued in their first Manifesto of November 6, 1835 that the Democratic Society had failed in its attempt to come to grips with basic social problems: in its theory it replaced the rule of the gentry with the rule of the bourgeoisie, thus substituting one type of tyranny for another.[25]

The Democratic Society defined history in terms of a struggle between the principles of individualism and the principle of socialism. While postulating the ideal of harmony between the individual's rights and his duties to society, and between equality and freedom, they condemned "extreme socialism." On the other hand, the Polish People viewed history as a struggle of "spiritualism" with material- ism. They considered socialism to be "spiritualism" applied to social relations. While idealizing the early Christian society, they tended to deny the egalitarian political implications of the pagan commune; and were in favor of a hierarchical structure of government and army.[26] They opposed, on one hand, the "atheism" of the Democratic Soci- ety and, on the other hand, its policy of "half-measures" on the social question.

However, as was to be the case with the entire non-Marxist Polish left, they considered social revolution impossible prior to regaining independence. Moreover, under the influence of Mazzini's national mystique, already in the 1840s they sought reconciliation with the gentry in the hope that it would rally to the cause of the "Universal Church" (i.e., socialist World Commonwealth).[27]

Although the Polish People remained a small "sect"—until their dissolution in 1846, in order to unite with the rest of the demo- cratic *émigrés*—they provided an inspiration to future socialist groups in Poland. In their stressing interdependence of the nationalist and socialist goals, the Polish people were forerunners of Polish national- istic socialism.[28]

By 1850 Worcell was back in the Democratic Society; he had left the Polish People in 1839 due to their sectarianism, eventually be- coming a member of the Centralization (i.e., executive) of the Soci- ety, which was driven from Paris to London in 1849. Worcell again emerged as a leading figure in Polish *émigré* politics. His socialism evolved to become like that of Ledru-Rollin and his fellow exiles from France, with whom he and the other Polish democrats collabor- ated at this time. To Worcell, as to Ledru-Rollin, socialism represent- ed the continuation of democracy.

This moderate and reformist character of Worcell's socialism re- pelled some of the Polish exiles in London. In 1855, they founded a new organization, the Revolutionary Commune of the Polish People, modelling itself directly on the second phase of the Polish People in the 1840s—after Worcell's departure—when it was led by Zeno Swietoslawski, who became the author of the program adopted by the Revolutionary Commune. He postulated a regime of com- plete political and economic equality, preceded by a transitional dictatorship of the people, and Slav federation within a world-wide socialist republic. However, he now laid greater emphasis than earlier on industry and labor, which was the result of his exposure to Chartism.[29]

Both leaders of the two Polish organizations in London—Worcell and Swietoslawski—were friends of Herzen. The latter maintained close ties with the other Russian exiles in London, too. Like the Russians, he wished to build a new society in both Russia and Poland that would be based on the idea of the ancient Slav village commune, bypassing the capitalist stage of economic development.[30]

Until the outbreak of the brief Cracow insurrection in March, 1846, Polish Utopian Socialism was based on British soil. However, in France, which was the center of the "Great Emigration," there were from the beginning small ephemeral groups of Polish converts to socialism. The most prominent among these socialists was the Romantic poet Adam Mickiewiecz (1798-1855), a student of Lelewel when the latter was professor of history at the University of Wilno. During the 1830s and 1840s Mickiewicz was not formally associated with any party. However, writing in *La Tribune des Peuples* (founded in March 1849), Mickiewicz declared himself a socialist. His socialism was an attitude of mind, rather than a concrete political and economic doctrine. He was a political mystic, naive as to practical politics, alternating between conservatism, nationalism and socialism. Eventually, the controversies in regard to his true political role died down; his enemies were silenced; and he became known as a hallowed national figure above partisan politics.

In 1848 Mickiewicz wrote a political program known as "Collection of Principles *(Sklad Zasad)*" for the Polish legion that he was then organizing in Rome. The clause on agrarian socialism in this program read: "To each family [to be granted individually possessed] land under the supervision of the commune; to each commune [to be granted collectively owned] land under the supervision of the state." However, in another point in the same program he considered private property inviolable. Mickiewicz might well have contemplated leaving the gentry estates intact, along with peasants' land held in common. This was not really a socialist measure, but one that was similar to the schemes of the Russian Slavophiles.[31]

Polish Utopian socialism combined the historicism of Saint-Simon with *babouvisme*'s revolutionary egalitarianism and was related to the teachings of Fourier, Cabet and Proudhon. Yet in some of its features it differed from Utopian socialism in the West.[32] First, it was an agrarian ideology. Secondly, it anticipated the idea of the historical mission of the proletariat, put forward by Marx and Lassalle, in stressing the creative role of the common people in history. Thirdly, it deemed the mission of the Polish people to be a socialist mission, believing that the moral rebirth of mankind would come from Slavdom led by Poland.[33]

However, non-socialist Polish ideologists believed in Poland's "mission," too. During the peak of Polish Romanticism from 1830 to 1863, the prevailing *Weltanschauung* in Poland was a native version of Hegelianism. The Polish philosophers—Cieszkowski, Libelt, Trentowski and others—each in his own way arrived at a similar world outlook, rooted in emotion and will, stressing the importance of nationality and community, rather than of the individual, and accepting Lelewel's theory of the ancient Slav commune. They envisaged a new era when mankind would be morally reborn by means of Slavdom, specifically by Poland. This messianism of the "national philosophy" was popularized by the Romantic poets Adam Mickiewicz and Zygmunt Krasinski.[34]

In what way did the messianism of the philosophers differ from that of the Utopian socialists and of the Polish democrats? They all considered nationality to be determined by the mystical "national spirit." However, unlike the moderate reformers such as Trentowski, both socialists and democrats understood Polish messianism as a slogan rallying the Poles to armed struggle for independence, in alliance with revolutionary Europe. They envisaged the masses as the real nation; for instance, Lelewel idealized the common people as the embodiment of the native element in Polish history, struggling against the alien Latin principle represented by the gentry.[35]

By the early 1840s, when socialism appeared in the home country, perhaps the most important role in the national movement was played by two radical Hegelians, cousins and friends, Henryk Kamienski and Edward Dembowski, as well as the philosopher Karol Libelt. Although hampered by the backwardness of the peasants, the weakness of the middle class, as well as by having to compete against the influence of the Centralization, Dembowski was busy preparing another insurrection. In Prussian Poland—since 1840 the center of Polish cultural life and conspiracy—Libelt headed a patriotic committee planning an uprising based on gentry support. On the other hand, Dembowski was allied with the radical Union of Plebeians organized by the Poznan bookseller Walenty Stefanski, which competed with Libelt. (However, arrests soon immobilized the plotters in Prussian Poland.) Active consecutively in Russian, Prussian and Austrian Poland, Dembowski was killed during the brief uprising in Cracow, in

February 1846, which was put down by the Austrians. Meanwhile, his cousin Kamienski inspired a patriotic organization, the Union of the Polish Nation, which had ties with the peasant conspiracy in Russian Poland, led by the priest Piotr Sciegienny and suppressed by the Russians in 1844.[36]

This conspirator envisaged an independent socialist federation in Poland, comprising theocratic villages inhabited by prosperous farmers, with democratic representation at the county, provincial and central levels. Sciegienny's program was a moderate version of agrarian socialism. Like the Slav commune it combined individual possession and communal ownership of land.[37]

On the other hand, Kamienski—who was a convert to socialism as a distant goal—considered contemporary Poland as "unripe" for collective landownership. He was best known for his propaganda for a mass "people's war." Hoping that social reform in Poland would be accomplished bloodlessly, like the Manifesto of the Democratic Society of 1836 he was prepared to force the gentry to cooperate by threatening it with terrorism, in order to effect the necessary unity to wage a war for national liberation.[38]

Unlike Kamienski, his brilliant young cousin Dembowski was primarily a man of action rather than a theorist. Contemporary Polish historians consider him more radical than Kamienski, and his philosophy to have been akin to Marxian materialism, notwithstanding his agreement with the idealistic interpretation of Polish history according to Lelewel.[39] However, in practice, Dembowski's vision of an ideal society became a distant goal for him, too; in practical politics he differed little from Kamienski. The realization of socialism in Poland depended on securing independence. Since both gentry and peasant support was needed to struggle for independence, and both usually objected to socialism, even such radicals as Dembowski had to temporarily abandon their socialism for tactical reasons. In order to win independence, the Polish socialists were compelled to come to terms for the time being with the principle of private ownership.[40]

The defeat of the uprising of 1863 resulted in a new emigration to the West of about 5,000, again centered in Paris. It must be stressed that this group, constituting about half of the "Great Emigration," was much less important, less illustrious and less self-sufficient economically than its predecessor had been. Hence it was more susceptible to social radicalism, but its preoccupation with earning the daily living precluded political activity on the scale of the earlier emigration. Most of the new *émigrés* soon returned home, attracted by the favorable situation in Galicia, while others became assimilated abroad. Altogether, the post-1863 Polish exiles in Western Europe were active for only eleven years as opposed to the thirty-two years of *émigré* politics between 1831 and 1863.

Although the revolutionary left in exile after 1863 had no influence on contemporary events in Poland, it anticipated future ideological developments in the home country.[41] Concentrating in Paris, London and Geneva, this group was subject to both foreign and native influences (e.g., the ideology of Herzen, Ogarev, Dobroliubov, the brothers Serno-Solovievich, Bakunin, Mazzini, Sismondi, the French Utopians, Lelewel, the Democratic Society, and the Polish People). However, their revolutionary zeal was directed toward fighting for independence rather than toward social reform. Like the radical exiles after 1831, they were allied with the revolutionary movement in Europe.[42]

These Poles participated in the activities of the First International founded in 1864 under the slogan of aid to Poland, and in the League for Peace and Freedom created in 1867 and active during the late 1860s and 1870s. Unlike the International, the latter was a liberal international organization of the Mazzini variety, with pacifistic aims. As such, it was combatted by the International. Utilizing both organizations as mouthpieces for their patriotic speeches, the Poles stressed that national self-determination must come before peace in Europe. The socialist congresses usually took place in Geneva, which was becoming the most important revolutionary center in Europe. The problem of private property was subject to fervid discussions in both organizations. And in 1868 the Brussels Congress of the First International for the first time considered agrarian socialism on its agenda, which was an issue of great interest to the Poles.[43]

The first post-1863 *émigré* political organization came into being in 1863, created by the leader of the unsuccessful rebellion in Poznania in 1848, General Ludwik Mieroslawski (1814-1878), who modelled his program on that of the now defunct Democratic Society. Subsequently, a group of the General's supporters published a radical paper *Le Peuple Polonais* in Geneva (1868-69) with a nationalistic socialist program, voicing confidence in the General's revolutionary dictatorship, as well as solidarity with the labor movement in Europe. Influenced by the French Utopians and Lelewel's idea of the ancient Slav commune, this organization envisioned Poland as a decentralized republic and a member of a Slav federation.[44]

At first the entire emigration was united on the platform of the Manifesto of January 22, 1863, issued by the insurgent National

"Government." This document had granted to the peasants the land which they tilled for themselves unconditionally, as well as small allotments to the landless insurgents, and stipulated equal civil and political rights for all citizens. (The Manifesto was regarded as a revolutionary document at the time; for instance, by Herzen and Engels.) However, by 1866 the revolutionary left moved beyond the Manifesto. In 1867 Jozef Hauke-Bosak (1834-1871) and Ludwik Bulewski (1824-1883) published in Geneva a program for a new organization, the Polish Repyblican Center *(Ognisko Republikanskie Polskie)*, which was a branch of Mazzini's Alliance Republicaine Universelle. They announced that land in Poland should be held as property of the nation and not of the commune. (Bulewski, having just returned from a trip to the United States undertaken on behalf of Mazzini, was impressed by land distribution under the Homestead Act of 1862.) The smallest of the four main *émigré* organizations, the Center was important due to the caliber of its leaders as well as its influential and comprehensive program.[45]

Also departing from communitarian socialism in the strict sense of the word, another political group of Polish exiles came into being in Switzerland. Led by the intimate friend of Bakunin Walerian Mrockowski and by Jozef Cwierciakiewicz, the ideologist of the co-operative commonwealth, who was opposed to Bakunin's anarchism, this group drew its inspiration from such diverse sources as Lelewel, Herzen and Ogarev, Owen, Fourier, Proudhon and Blanc.[46]

However, it was the appearance of the slavophile *Commune* *(Gmina)* which was the first symptom of ideological polarization among the post-1863 Polish exiles. The *Commune*, appearing from September 20, 1866 to July 15, 1867, was published in Geneva by Jozef Tokarzewicz (1841-1919) and Jozef Brzezinski, becoming Tokarzewicz's mouthpiece for propagation of his brand of agrarian socialism that was based on the idea of the ancient village commune adapted to modern conditions. Although he considered himself a disciple of Herzen, Tokarzewicz was also influenced by Polish authors and ideologies: the Polish People, Lelewel and the historian Karol Szajnocha (1818-1868), who had espoused the theory of Norse conquest. (He identified the Sarmatians with the Normans and alleged Norman origins of the Polish state and of the Polish gentry.) Tokarzewicz condemned the Manifesto of 1863 for postulating an alien reform based on the alien principle of private property as represented by the gentry, rather than on the communal principle as represented by the people. His concrete criticicism of the Manifesto was important; he pointed out that it appealed only to the richer peasant, because the allotment it offered to the landless was insufficient for bare subsistence.[47]

In 1868 Tokarzewicz joined the largest *émigré* organization at the time, the Union of Polish Democracy *(Zjednoczenie Demokracji Polskiej)*, and served on its executive along with the future Communard Walery Wroblewski (1836-1908), during the balance of its existence (1869-1870). In that period, he replaced the novelist and former member of the Democratic Society Zygmunt Milkowski (1824-1915) as editor of the Union's organ *Independence (Niepodleglosc)*—published from July 10, 1866 to January 22, 1870—using it as mouthpiece for his propaganda of socialism.[48]

One of the characteristics which usually distinguished the revolutionary left from the rest of the Polish exiles was its belief in the possibility of reviving the ancient Slav eommune (as well as its belief in Lelewel's conception of Polish history as the hsitory of the Polish people struggling with the gentry for their rights, the native principle struggling against the alien principle).[49]

Almost all of the *émigrés* aimed at Poland with boundaries of 1772, idealizing to this end the Union of Lublin (1569) which cemented the Polish-Lithuanian Commonwealth. However, unlike most "Whites," the "Reds" advocated autonomy for the constituent nationalities in the former Commonwealth, proposing federation of Poland with Ukraine and Lithuania, as a step toward a united Europe. While some sections of public opinion in the West favored the revival of the old historical Poland for the sake of weakening Russia, the Russian anarchist Bakunin opposed the boundaries of 1772. Tokarzewicz clashed with him on this issue of boundaries in Zurich, in 1872, while discussing the proposed program for the ephemeral Polish Social Democratic Association. The demand for historical Poland put forth by the "Reds" was also motivated by domestic considerations; it was directed against the loyalism of those conservatives who like Marquess Wielopolski were reconciled to the Congress Poland created by the Congress of Vienna in 1815. The federations planned by the *émigrés* were also inspired by the example of Switzerland and that of the United States.[50]

Usually, these plans excluded Russia. Some of the *émigrés* accepted the theory of Duchinski alleging the "Turanian origin of Muscovy." However, the "Reds" opposed Tsarist Russia as a "reactionary" power hostile to national self-determination, declaring their alliance with "peoples," rather than "governments," in the cause of progress.[51]

Moreover, the "Reds," especially the left-wing "Reds," were protagonists of an armed struggle envisaged as a "people's war," which they deemed inevitable. This also distinguished them from the "Whites," who usually mistrusted the masses.[52]

The post-1863 Polish exiles contributed to the maintenance of the tradition of struggles for independence in all three regions of Poland, but they failed in their attempt to produce a coherent and clear program of social reform. Their communal socialism was by then even more anachronistic with reference to Poland than was populism in Russia, in the 1870s. However, they were conscious of their role as an intermediary between the Utopian and modern Polish socialism.[53]

Because their primary goal was the struggle for independence, these exiles formed strategic alliances; they proved willing to come to terms with the gentry and, on the other hand, participated in the activities of the International and in the Paris Commune. They considered Marx first of all as a friend of Poland, their ally against Proudhon and Bakunin. Moreover, they had connections with such prominent European socialists as W. Liebknecht and J. F. Becker. Generally speaking, as in the past, these Polish radicals were more revolutionary in Western Europe than at home and more radical in reference to Western European problems than in reference to Poland. *Emigré* Poles fought in the Paris Commune either out of sympathy with the Parisian people, or because they felt that the Commune represented the beginning of a revolution in Europe, which would result in an independent Poland.[54]

However, even some of the latter eventually became socialists as well, due to having participated in the Commune. For instance, the veteran of the 1863 uprising, Walery Wroblewski, who was made general in the course of the struggle, subsequently settled in London. He became a close friend of Marx, a member of the General Council and secretary for Poland, and with Marx represented the Council at the Hague Congress in 1872. The Polish branch of the International included thirty ex-Communards. To commemorate the centennial anniversary of the first partition, in 1872 Wroblewski and associates transformed this group into the Union of Polish People (*Zwiazek Ludu Polskiego*), which became an ideological successor to the Revolutionary Commune of the 1850s. This group, numbering less than one hundred people, sponsored special meetings and regular anniversary commemorations, and contributed to the British press, too. It collaborated with the populist association of Lavrov grouped around his paper *Forward (Vpered)*, published in London from 1874 to 1877. However, the Union's belief in the imminence of another uprising in Poland had more and more isolated it from opinion at home.[55]

Already on the eve of the insurrection of 1863, a new trend of so-called "pre-positivism" began replacing Hegelian philosophy in Poland, as evidenced by the appearance of Volume I of a treatise by Jozef Supinski, entitled *The Polish School of Political Economy (Szkola polska gospodarstwa spolecznego)*, published in 1862. The new philosophy of positivism was to justify in ideological terms, the "organic work" in Russian Poland after 1863. (This term originated following the defeat of 1831. It denoted the patriotic non-political, non-conspiratorial activity of the liberal intelligentsia, aimed at overcoming the Polish cultural and socio-economic backwardness. "Organic work" comprised such activities as promotion of the arts, scholarship and popular education; charitable and social work among the masses; fostering of industrialization; and development of various forms of association. Due to the unfavorable domestic and international situation following the defeat of the Polish insurrection of 1863 and the humiliation of France in 1871, the liberal intelligentsia in Russian Poland temporarily denied the need for an armed struggle, devoting its energies to economic and cultural tasks, in an attempt to strengthen Polish nationality in this manner. A similar situation also prevailed in Austrian and Prussian Poland since the abortive uprisings of 1846 and 1848.)

It was Boleslaw Limanowski who, by summarizing J. S. Mill's appraisal of Comte's philosophy in Warsaw's *Weekly Review (Przeglad Tygodniowy)* in 1869, became an initiator of the so-called Warsaw Positivism which would be led from 1871 by Aleksander Swietochowski (1849-1938). Adapting to the Polish situation Comte's philosophy, Mill's economics, Spencer's sociology, Buckle's interpretation of history and Darwin's biology, the Warsaw Positivists attacked Polish gentry privilege, superstition, subjection of women, and economic backwardness. Aiming at universal primary education and professing faith in progress and "positivistic" science (science that was shorn of metaphysics and applied the methodology of the natural sciences to the study of social problems), unlike Limanowski they were protagonists of the economic doctrine of "laissez faire." Even the revolutionary left in exile embraced much of the outlook of positivism, although it was opposed to "organic work," arguing against it on the basis of both Lelewel's ideology and that of the Polish People of the 1830s and 1840s.[56]

Prompted by the outbreak of the Russo-Turkish war, in 1877 Wroblewski moved from London to Geneva, to be closer to home and at the center of revolutionary activity. The Union of Polish People ceased to meet. Meanwhile, new Polish socialist exiles began arriving in France and Switzerland from Russian Poland and Galicia, and in 1878 Limanowski himself came to Geneva. General Wroblewski became for some time the intermediary between the older exiles influenced by Western socialism and the new arrivals. Thus the banner of the "irredenta" passed to the Polish socialist movement via the "epigones" of the left "Red" ideology, such as Wroblewski, Mroczkowski and Limanowski. Although it must be stressed again that the post-1863 emigration was much less important and active than the "Great Emigration," it was indeed significant for the ideological development of Limanowski.[57]

CHAPTER II

FROM PATRIOTISM TO SOCIALISM
AN EXILE FOR THE NATIONAL CAUSE
(THE EARLY CAREER OF BOLESLAW LIMANOWSKI, 1835-1870)

Limanowski was perhaps the most prominent among the handful of Polish left revolutionaries of the 1860s and 1870s who survived the First World War. His native estate Podgorz (Zamkolny—in Latvian) was located in the county of Dünaburg—named Daugavpils in Latvian and Dzwinsk in Polish—of the Vitebsk *guberniia*, in the formerly Polish section of Livonia (which was annexed to Russia in the first partition). Here he was born on October 30, 1835, just four years following the defeat of the November insurrection (1830-1831), on the same day that the first Polish socialist group, the Polish People, was founded on British soil. His family milieu was that of lesser gentry settled eastward from ethnic Poland on lands which for several centuries represented the bone of contention between Poland and Russia. Devoted to the military traditions of the old Polish-Lithuanian Commonwealth, the borderland gentry was surrounded by a potentially hostile peasantry of Latvian or Lithuanian descent in the north-east, Belorussian toward the south, and Ukrainian in the south-east. (The borderland town element, however, was predominantly Jewish.) During the period following the partitions, it was this lesser gentry of Polonized ethnic stock or of mixed origins which produced many outstanding Polish politicians, and played a significant role in all Polish uprisings.[1]

Probably Limanowski owed to such ancestry both his moral courage and his steadfastness. On his maternal side, he was descended from an old Lithuanian-Ruthenian family, the Wyslouchs, who were possibly related distantly to the Jagiellonian dynasty which ruled Poland from 1386 to 1572. Originally from Bohemia-Moravia or Silesia, the Wyslouchs lived during the early Middle Ages in the province of Red Ruthenia in the south-western corner of the Kievan State. By the end of the fifteenth century, they moved via Novgorod to Lithuania and finally settled in the area which became Grodno *guberniia* of the Russian Empire. Related by marriage to prominent Polish-

Lithuanian aristocratic families, nevertheless the Wyslouchs failed to secure the higher offices of state and did not accumulate any large fortunes. They participated in the expeditions against Moscow under Stefan Batory and Zybmunt III in the sixteenth and seventeenth centuries, and in the Vienna campaign against the Turks under Jan Sobieski in 1683. In the eighteenth century some of them became associated with the Reform movement in the era of Polish Enlightenment. The great-grandfather of Limanowski, Zenon Wyslouch, was a personal friend of the reformist King Stanislaw August. As a member of the Four-Year Diet and protagonist of the Constitution of 1791, he was forced to flee Poland by the "Men of Targowica."[2]

On his paternal side, Limanowski's genealogy was less illustrious. According to the family traditions, he was descended from a Cossack who lived beside one of the *limany* (lagoons) along the Dnieper river and was ennobled by Prince Vitovt of Lithuania (1340-1420). Rejecting this legend, Limanowski traced his origins to the milieu of petty Polish gentry from Galicia, eventually settled in the northeastern borderlands.[3] Like the Wyslouchs, the Limanowskis held public offices at the country level. Boleslaw's paternal grandfather was a justice of the court in the county of Dryzien, which was adjacent to Dünaburg. Boleslaw's father, an officer in the Russian army during the reign of Paul, also held for a time a similar post in his native county of Dünaburg. The father's brother married a daughter of the Russian general in command of his division; from him descended the Russian branch of the family, which was sympathetic to the Polish branch.[4]

Limanowski's native Podgorz was located on the border line between Livonia and Belorussia. Before his birth his father purchased this landed estate from the aristocratic Plater family. Comprising several villages, Podgorz was territorially a large manor when compared with the nearby estates of the lesser gentry, but there seems to have been a shortage of serfs to work the land, which might have accounted for their severe exploitation. However, the Limanowskis enjoyed a very modest standard of living. Their two-storey manor house stood on a stone foundation, but was built of wood; it was surrounded by a flower garden and large orchard. The family was large, comprising one sister and five brothers, one of whom died in infancy. As was customary in his milieu at this time, Boleslaw was educated at home at the primary level, by a number of tutors. His favorite childhood game was a make-believe war against "Muscovy," which he played with his two younger brothers, Lucjan and Jozef, and the servants' children. The brothers were determined to fight against Russia upon reaching their adulthood.

However, their parents appeared to be reconciled to Russian rule. The father, who had served in the Russian army, intended to enroll Boleslaw in the Russian Imperial Cadet Corps. Instead, the boys acquired their patriotism from books, under the influence of their two elder brothers, Aleksander and Anicety. In his early childhood, Boleslaw was especially impressed by the *Historical Songs (Spiewy historyczne)*, written by the well-known Polish author Julian Niemcewicz (1787-1841) and the *Pilgrim in Dobromil (Pielgrzym w Dobromilu)* by Princess Izabela Czartoryska (1746-1835, the mother of the famous Prince Adam. Both books were popularized histories with special appeal to very young children or to uneducated people. The boys' personal library included novels about the Napoleonic wars and patriotic Polish Romantic poetry. Boleslaw knew by heart some of this poetry, including Adam Mickiewicz's "Alpuhara" from his epic *Konrad Wallenrod*.[5]

The oldest brother, Aleksander, Boleslaw's senior by ten years, a law student at the University of Moscow, spent his summer vacations at home. Here he discussed with Boleslaw such topics as the partitions of Poland, the uprising of 1794, and the cruelties of Grand Duke Constantine (the commander-in-chief of the Polish army during the period of the Congress Kingdom). It was Aleksander who introduced Limanowski to Lelewel's historical writings. However, Boleslaw shared his room with the next brother, Anicety, who was nine years older. Originally a pupil of the gymnasium in Dünaburg, Anicety remained at home as the result of a chronic illness (probably tuberculosis). Anicety awakened Boleslaw's interest in books by their joint readings. He told the younger brother of the uprising of 1830 and of subsequent Russian repressions ordered by Nicholas I, particularly of the martyrdom of Szymon Konarski in 1839, who was the emissary to Lithuania of "Young Poland" allied with Mazzini's "Young Italy." Thereupon, Boleslaw swore to become an avenger of injustice, especially yearning to struggle against national oppression. Of the books which he read jointly with Anicety, he was most impressed by the fantastic novel by the French author Paul Féval (known under the pseudonym of Sir Francis Trolopp), *Les mystères de Londres* (1844). This was a story of the scion of a wealthy Irish family who, allied with the London underworld in order to avenge his fatherland, master-minded a conspiracy to rob and destroy the Bank of England as one of the bastions of the British Empire.[6]

Limanowski's grandmother had brought him up as a devout Roman Catholic (though in his childhood he eventually became more familiar with Greek mythology than with teachings of his own Church). At bedtime, he prayed for the restitution of his fatherland, as well as for the abolition of serfdom. His father, though indulgent to his children, was a cruel master, and Limanowski rebelled against the maltreatment of the serfs on his father's estate. Spending much time in the company of servants and their children, Limanowski was aware of their plight; and the moans and screams of the flogged peasants evoked his indignation and intense compassion.

The fate of the peasants in the immediate area was indeed far from enviable. They dwelled in cabins without chimneys, with tiny windows—where the sole items of furniture were tables and benches. Heavily overworked, existing on a diet of black bread and potatoes, these people were compelled to feed on grass in the spring. Due to poor harvests, the last two years of Limanowski's stay at home—1845 and 1846—were especially hungry for the peasants.

In the north-eastern borderlands the attitude of the gentry toward their peasants was patriarchal. Originally, there was the strong bond of a common religion between the peasant and his lord, but this relationship deteriorated after 1848. Henceforth, on the one hand the Belorussian peasants hated the gentry and, on the other hand, felt hostile toward the Russian state for persecuting their religion. At the time of the massacre of landlords in Galicia in 1846, there were individual cases of retribution by the peasants in Belorussia, as Limanowski notes in his *Memoirs*. During the spring of 1847 the peasants in the Belorussian counties of Vitebsk *guberniia* became excited and abandoned their homes, in order to put themselves under the rule of the "White Tsar," assuming perhaps that the one presently ruling over them was completely "Black." According to Limanowski, around 10,000 peasants left their homes and advanced toward Riga. In the vicinity of Podgorz they were either scattered or captured by a detachment of troops from Dünaburg.[7]

After his father's sudden death on April 4, 1847, it was decided that Boleslaw would accompany Aleksander to Moscow in the fall. He regretted this family decision for the rest of his life, for he felt a sense of shame in having been educated in Russia.[8]

Limanowski spent eleven years in Moscow; seven years in secondary schools (1847-1854) and four years at the University of Moscow (1854-1858). During his second year in gymnasium, his new stepfather Kazimierz Sawicki, accompanied by Boleslaw's mother and sister, arrived in Moscow to complete his legal studies. He then temporarily came under the influence of the ardent Catholicism of Sawicki, who introduced him to Mickiewicz's drama *Forefathers (Dziady)*, the third part of which described the martyrdom of young people exiled to Russia in the Polish cause.[9] However, upon the return home of his mother and sister, which almost coincided with the untimely deaths of Aleksander and Anicety, he virtually lost touch with his family and home country. For four years he associated almost entirely with Russians, while acting as resident tutor to the sons of the Russian merchant Sireishchikov, and being treated as a member of the family.[10]

During the reign of Nicholas I, Russian gymnasia were of two categories from the fourth year on: the classical and the "realist"—specializing in the natural sciences. Having decided to pursue the study of history and literature at the University of Moscow, Limanowski chose the classical program. In his fourth year one of his teachers, Vasilii Grivtsev, introduced Limanowski to Augustin Thierry's *Récits des temps merovingiens* in which the statesman-historian Guizot (1787-1874) had convincingly associated civilization with progress.[11]

Limanowski's final high school year was marked by the outbreak of the Crimean War. At first he sympathized with the Balkan Slavs; but because of Napoleon III, who was allied with Turkey, Limanowski soon came down on the side of the Turks. Recollecting his childhood interest in the Napoleonic era, he idealized Napoleon III as the nephew of the famous uncle. However, the object of his special cult was the defeated Napoleon I, the Napoleon of the "one hundred days," rather than the victor, for he always sympathized with the underdog.[12]

Having passed his final examination which enabled him to enroll at the University, Limanowski felt liberated from the vague fear which he experienced during his gymnasium years. Although throughout this period he himself was well treated, every day when he entered his classroom he anticipated a humiliation. He hated any form of corporal punishment and was disgusted by the screams of the flogged boys. Having abandoned his school uniform for civilian clothes, Limanowski felt grown up and independent. However, he was frustrated in his plans to pursue history and literature at the University of Moscow.[13]

Frightened by the events of 1848-1849, Nicholas I considered universities a seedbed of revolutionary propaganda. Therefore, he limited each academic department at Moscow's University to 300 students, with the exception of medicine. Thus Limanowski was faced with a dilemma; either to delay for a year or two his enrolment in the Department of History and Philology, to which wealthy graduates of a lyceum enjoyed priority of admission, or to study medicine. He reluctantly decided to follow the latter course in order to enter the University immediately.

However, during his first year he paid little attention to his medical studies, and instead attended lectures on history and archeology. Around this time, the faculty of the University of Moscow included two famous historians, both freethinkers and humanitarians: Timofei Granovsky (1813-1855) and Petr Kudriavtsev (1816-1858). Unfortunately, Granovsky died soon after Limanowski's arrival at the University. However, he was given the opportunity to attend the lectures of Kudriavtsev on ancient history, and was impressed by this professor's sympathy for conquered peoples struggling for their independence.[14]

Indifferent to the defeat in the war, the students at the University responded to the passing of the Tsar with great joy. Limanowski himself hated him for ordering repressions in Poland, and for his despotism in general. The successor to the throne, Alexander II, was popular with Moscow's intelligentsia, due to his unassuming behavior in society. Soon Limanowski was able to take a close look at the new Tsar when the latter came to inspect the University.[15]

During his second year Limanowski became interested in physiology. He contemplated devising a new science which he termed "sociology," dealing with society rather than with the individual organism. Limanowski then compared the family to a cell in the biological organism, independently of such scholars of the nineteenth century "organicist" school as the sociologist and Austrian statesman, Albert Schäffle (1831-1903).[16]

However, his interest in social studies was short-lived. During his third year of University, the student "revolution" of October 1857 changed the course of his life. This rise of student activism was provoked by an arbitrary police raid of an innocent Polish student party. Thereupon, a vehement student protest was submitted to the district curator, who authorized a commission of inquiry composed of student deputies and university administrators. The latter concurred with the students. As a result, city police were reprimanded and students became conscious of their collective power. In the course of the week following the raid, they held meetings to discuss their needs, framed their resolutions, and sent a petition to the University authorities demanding the punishment of the guilty police officials, as well as the right of association and assembly. Limanowski was among the student leaders.[17]

Previously he had rebelled against the despotism of landlords, against the Russian autocracy, and against the despotic regime in schools. He longed to become another William Tell. Now, at the University, his rebellious mood became more intense, under the influence of Herzeń and Dobroliubov; he read the latter's literary reviews in the *Contemporary (Sovremennik)* in the home of the merchant Sireishchikov. Moreover, the students who were victimized by the police happened to be Poles, as mentioned above. Thus, as a result of the raid, Limanowski identified himself with the activist Polish students. Overcoming his superficial assimilation with the milieu of the Russian intelligentsia, he recovered his ardent Polish patriotism.[18]

There were about 600 Poles at the University of Moscow at this time, mainly from the borderlands. They had founded a separate student organization for the purpose of mutual aid, which was also a school of Polish patriotism. The overall organization consisted of separate associations from each of the academic departments; it had its diet and dietines as well as a handwritten daily paper. Each association employed a cashier and a librarian, the latter being entrusted with the distribution of forbidden books to the members, who studied Polish *émigré* political literature, Polish history and the works of Polish poets associated with Romanticism in Poland. The majority of Polish students at this time adopted the political program of the Polish Democratic Society.[19]

On joining an inner secret "circle" which—like the socialist circles in Russian Poland, formed after the 1863 uprising—combined self-education with propaganda activity, Limanowski lectured on Polish history for the benefit of his fellow members in the circle. He based these lectures on writings by Lelewel and by his disciples Henryk Schmitt and Karol Szajnocha. The group also edited a students' paper, to which Limanowski contributed translations from *Gegenwart*, a German periodical sympathetic toward Polish struggles for independence.[20]

The semi-legal organization, with its secret inner circles, became standard among Polish students in Russia. Moreover, by bringing Russian and Polish student activities together, the student "revolution" of 1857 gave the impetus to the Russian student movement, which was organized on the Polish model. The Russians advocated a joint political front with the Poles against their common enemy, the Russian autocracy. At the University of Moscow, they issued a proclamation to this effect, to which they appended a copy of a moving obituary in honor of the Polish *émigré* socialist Worcell, written by Herzen in his *Polar Star (Poliarnaia Zvezda)*. Limanowski was much affected by this homage to Worcell, and henceforth he attempted to learn everything he could about this early Polish socialist and patriot.[21]

The Poles had remained calm during the Crimean War. On the arrival of Alexander II in Warsaw, in 1856, there was an expectation of major changes, partly as the reward for this loyalty. Immediately, the Tsar advised the Poles not to expect too much. He conceded an amnesty to the Polish political exiles in Siberia (the "Siberians"); the opening of an Academy of Medicine in Warsaw; and a charter to an important gentry organization known as the Agricultural Society, to provide a forum for discussion of the imminent agrarian reform. Polish students in Moscow (and elsewhere) considered these concessions insufficient; they were convinced that an armed uprising was inevitable. They put pressure on the gentry to anticipate the agrarian reform from above, in order to gain peasant allegiance for the Polish cause.[22]

Around this time, the Lithuanian nobility, being prompted by Governor Nazimov, had asked the Tsar to end serfdom, but on terms favorable to the gentry. Thereupon he appointed *guberniia* committees of gentry in 1857, to discuss the impending reform. This became a factor in forcing the nobility in the former Kingdom of Poland, too, to face the issue of peasant emancipation with land.[23]

In 1858 there was a re-awakening of aspirations to independence in Russian Poland. Polish students in Moscow at this time were intensely interested in the peasant question. They entertained the "Siberians" on their way home from exile. Affected by the atmosphere in Poland, by the growing popularity of Napoleon III among the Poles, and the student "revolution" of 1857, the Polish student activists in Moscow attempted to contact their counterparts elsewhere in Russia, as well as student radicals in Poland; they maintained ties with Polish *émigré* leaders, too. The effect of all this on Limanowski was to cause him to transfer from Moscow to the German university at Dorpat, for the sake of removing himself from the milieu of a Russian university, which he now hated.[24]

In Dorpat Polish youth was organized in two groups—"Majority and "Minority," the latter comprising mainly students from the north-

eastern borderlands. Limanowski joined with the "Majority"; however, he hoped to unite these two organizations. He attended lectures in psychology by the German philosopher Ludwig Strumpell.
This encouraged Limanowski to give up medicine and to study
philosophy instead. (He could never get used to watching surgical
operations, and his pride was hurt by each minor mistake which he
committed as a medical student.) Strumpell's lectures on social
philosophy awakened his interest in the history of ideas. However,
in the spring of 1859 he interrupted his studies in Dorpat and returned home, in order to convert serf labor to rent on his family
estate, in accordance with the agreement entered into with his
brothers.[25]

It proved impossible for him to accomplish this task, due to lack
of cash, workers, horses and implements. Feeling guilty about his
inability to do away with the *corvée* entirely, he limited free services
to the minimum required for the smooth functioning of the farm,
and reduced the length of the working day. Nevertheless, the necessary work was completed earlier than on the other estates in the
vicinity, where farming proceeded by the traditional methods.[26]
Notwithstanding his attachment to rural life, Limanowski returned
to Dorpat in the fall. This was made possible by the arrival of his
brother Lucjan, who had been expelled from the Medical Academy
in Warsaw and now relieved Limanowski from the management of
the estate.

Meanwhile, abroad Italy was being gradually liberated from foreign
rule. In Poland the defeat of one of the partitioning powers, Austria,
encouraged the lesser gentry to contemplate another insurrection.
For Limanowski the academic year of 1859-1860 was a year of re-
awakened hopes. Writing in Dorpat in a student paper at this time, he
advocated union of Poland and Lithuania, influenced by Szajnocha's
idealization of the gentry democracy in the old Commonwealth, the
benefits of which Limanowski—like Lelewel—wished to extend to the
common people of Poland.[27]

He was not yet attracted to socialism; his program was that of the
Democratic Society. During the recent five months' stay on his family
estate, he discovered from talking to his peasants that they aspired
to individual ownership of the land which they tilled for themselves.
This confirmed Limanowski's belief at this time that, before peasants
could be persuaded to adopt the idea of collective ownership, it was
necessary to emancipate them from the tutelage of their landlords and
give them a taste of owning their own land. In any case, Limanowski

was then primarily concerned with the armed struggle for independence; he had no patience for socialist propaganda.[28]

Since his arrival in Dorpat in the fall of 1859, he had effected a merger of the two student groups and was now chosen to preside over the new organization. Nevertheless, Polish students in Dorpat were generally indifferent to politics. Disgusted with this situation, Limanowski abandoned the University in September 1860, on hearing a rumor that a Polish legion was being formed in Western Europe.[29]

On his way to the West, Limanowski for the first time came in contact with industrial workers in the city of Riga, where he was given the opportunity to inspect a few factories. In a somewhat anarchistic manner, he evinced no interest in the working conditions as such, but was impressed with the workers' sense of dignity, pride, and absence of servility toward their masters.[30]

On arrival in Liège he learned that Garibaldi had appointed Mieroslawski, the commander-in-chief of the proposed Polish, Hungarian, German and French legions. Limanowski then decided to attend Mieroslawski's Polish Military School scheduled to open in Paris on December 1, 1860. Arriving in Paris in November, Limanowski remained there until the following March. In Paris he attended lectures at the Sorbonne and the College de France as well, studying mathematics, political economy and international law. He also spent long hours in the Polish Library on the Quai d'Orléans, expanding his knowledge of Polish *émigré* politics. He identified himself ideologically with the group associated with the *Review of Polish Affairs* (*Przeglad rzeczy polskich*), a periodical published in Paris between 1857 and 1863 by an active member of the Democratic Society, Seweryn Elzanowski. Limanowski began his career as a writer by contributing an article to the December issue of this paper, entitled "Correspondence from Livonia." In it he advocated a voluntary entry of an autonomous Latvia into the future independent federation, being aware of the national movement among the Latvians. Suspecting Limanowski—albeit unjustly—of chauvinism, Elzanowski omitted the offensive passage. At this time, Limowski researched his first biographical sketch—on the heroine of the 1830 uprising Emilia Plater—which was published in Lvov's *Literary Daily (Dziennik Literacki)* in 1861.[31]

His aim now was to become a military instructor. To this end he attended Mieroslawski's lectures on strategy and strategic geography, given in the Polish Military School, and drilled in a large indoor arena. While still in Dorpat, like other Polish students, Limanowski had

come under the spell of the General's charisma; the latter was considered by the young people as the natural leader of the next insurrection, due to his alleged social radicalism, his criticism of the
leadership of the 1830 uprising, and of his active role in the revolutions of 1848-1849 in Poznania, Sicily and Baden. Yet Mieroslawski's agnosticism repelled those members of the young generation
who accepted the need for religion as a way to commune with the
peasants, and who doubted the efficacy of religious appeals to the
masses by leaders devoid of a religious faith. In Paris Limanowski
was disgusted with those compatriots of his who considered the
peasants and their religion solely as a means of enlisting them in the
struggle for independence. Thus he began to concern himself with
social justice as an end in itself. On the other hand, under the spell
of Mickiewicz's mystical patriotism, Limanowski was above all a protagonist of national unity, and hence deplored squabbling among the
émigré democrats.[32]

Meanwhile, since November 1860, when Warsaw students formed
an alliance with the artisan class in Warsaw, the capital of Russian
Poland assumed a special role in the liberation movement. The fatal
shooting by police on February 27, 1861 of several demonstrators
in Warsaw and the subsequent rise in activism—patriotic-religious
manifestations and counter-terror against police and spies—convinced
Limanowski that insurrection was at hand. He decided to return
home.[33]

Just prior to his departure from Paris in March, Limanowski met
with several compatriots to form a revolutionary committee which
would organize a patriotic-religious manifestation in Lithuania, in
solidarity with the nationalist movement in Russian Poland. To this
end he travelled to Dorpat in order to rally his former fellow students
to a meeting in Wilno (Vilnius—in Lithuanian). As a result, about
twenty young men from various parts of the north-eastern borderlands gathered in Wilno in the middle of April. There was strong opposition to the proposed demonstration by the local gentry, who objected to the intrusion of outsiders, ostensibly on the grounds that
such intrusion would spoil an amicable settlement of the agrarian
question on the basis of the Emancipation Edict of March 1861.
Actually, a patriotic demonstration was not in the gentry's interest.
The argument of Jakub Gieysztor, who was spokesman for the opposition and future leader of the "Whites" in Lithuania, was that
patriotic activism would result in the interference of the bureaucracy
in relations between the gentry and the people, thereby harming the
national cause in the long run. Gieysztor was a liberal landowner and

protagonist of emancipation with land, who opposed a demonstration at a time when the government was choosing Arbiters of the Peace *(mirovye posredniki)* from the local gentry in each district to implement the reform, and when the Lithuanian gentry's proposal for the establishment of a Land Credit Association (such as existed in Russian Poland) was being considered by the government.

Limanowski contended that a manifestation of this kind was necessary for the sake of public opinion in the West, that an armed uprising was the only means to freedom, and that the success of such an uprising being contingent on peasant support, a patriotic manifestation was the only way to secure it. Although Gieysztor's view was understood, it did not prevail; most of the young people voted to go ahead with the plan, because they felt bound by their word of honor to do so. The manifestation—the singing of a patriotic humn *(Boze cos Polske)*—took place as scheduled on May 20, 1861, in the Wilno Cathedral. As a result, Limanowski and his associates were arrested, and many such church manifestations subsequently occurred in Lithuania.[34]

The administrative verdict with regard to Limanowski was exile to distant Archangel *guberniia*. On arrival under escort in Archangel, Limanowski was assured by the Governor that upon good behavior he would be transferred back to the capital from the town of Mezen which was his final destination for the time being. Mezen was a miserable settlement of some 1500 inhabitants, located on the shore of the White Sea. It consisted of two rows of decrepit wooden buildings on either side of an unpaved road, and was adjacent to the tundra inhabited by the Samoed nomads with their herds of reindeer. At Mezen Limanowski was met by six compatriots. With some exceptions, the local elite was exactly like that described by Gogol and Saltykov-Shchedrin in their satires of Russian provincial society. However, several families proved very kind and hospitable to Limanowski. In Mezen he experienced a pleasant social life with frequent parties in the winter and outdoor amusement in the summer.[35]

Here he was assigned to clerical duties at the local land court. His supervisor, the town's chief administrator *(ispravnik)* Porfirii Shvetsov was a friend of the Governor, Nikolai Arandarenko. A despotic but scrupulously honest official, Shvetsov helped Limanowski to achieve in Mezen the lowest rank in the Russian civil service, that of "collegiate registrar." On the basis of an official assignment, Limanowski wrote here two articles in Russian: "A report on foreign trade in the White Sea region in the reign of Anna Ivanovna," and another one

"On the administrative measures to improve the breeds of cattle and horses under Catherine II." A contribution to Russian economic history, the first work was based on the customs receipts (*Kabatskie Knigi*) regarding official documents. Both articles were published in Archangel's *Guberniia News (Gubernskie Vedomosti)*.[36]

During his stay in Mezen from September 1861 to May 1862, Limanowski read voraciously. He obtained from the local school library the monumental *History of Greece* by the enlightened English banker George Grote, in a Russian translation. (Writing in the 1840s, as a representative of the rising and progressive British middle class, Grote embodied in his history an ideal picture of the Athenian democracy, in which Plato resembled a Benthamite reformer.) From the *Contemporary* Limanowski learned of the existence of Auguste Comte, whose name remained unknown to him while studying in the three university cities of Moscow, Dorpat and even Paris—the birthplace of the "father of sociology!" His interest in social studies revived, and he began to arrange systematically the ideas which had germinated earlier in his mind. By the time he was leaving Mezen, he had about ten sheets of his first manuscript on "sociology." Limanowski was aware that this work had no scientific value as such, but the benefit which he derived from his essay was two-fold; it aided him in the crystallization of his ideology and accustomed him to think and write systematically.[37]

His thinking about the family as the primary social unit persuaded him that women should be accorded complete equality, for the sake of society as a whole. He was convinced that the arguments against their emancipation applied to the majority of men as well. He also felt that the women of the Polish borderland gentry, whom he knew personally, were not less endowed with patriotism, the sense of social justice, and intellect than men of their class; these were the qualities which he valued most in both men and women. His conversion to the ideal of equality for women conditioned him to admit the need for abolition of all privilege.[38]

When thinking about the next social unit, the community, he questioned the right to inheritance. Limanowski knew that the "Men of Targowica" acquired enormous fortunes. Thus, was not property often the result of violence and theft? Yet he felt that inheritance was sanctified by tradition, and was a by-product of heredity. He solved the dilemma in his mind for the time being by urging the limitation of inheritance to the direct heirs, subject to a special estate duty.[39]

While contemplating the agrarian problem, being influenced by Chernyshevsky, Limanowski nevertheless felt that the Russian *mir* was not justified even on economic grounds; in this respect he preferred the agrarian communities of the Moravian Brethren. In 1862, in the period, that is, prior to peasant emancipation with land in Russian Poland, he rejected agrarian collectivism outright; he nevertheless envisioned—but in the distant future after emancipation—model communities like those of the Brethren being set up on public estates in Poland. It was to university graduates that he assigned the task of bringing innovations to the village, such as cooperative farming and setting up of peasant self-government. However, categorically opposed to any form of compulsory collectivization, Limanowski was not yet fully reconciled to agrarian socialism for the Polish village, even in a remote future.[40]

The news which reached Mezen from Poland was depressing. Both Russian Poland and Lithuania were subject to martial law. The wish to escape was always present in Limanowski's mind. Meanwhile, the famous novel by Victor Hugo—*Les Misérables*—reached him from Poland in instalments in the *Polish Gazette (Gazeta Polska)*, edited by the writer Ignacy Kraszewski. This novel, which told the story of a convict and a prostitute, in effect awakened the social consciousness of Europe. In conjunction with his social studies, and against the background of the dreariness and monotony of northern winter days, the novel with its vivid presentation of injustice made a strong impression on Limanowski.[41]

In the spring of 1862 he was transferred to Archangel. On his arrival there in the capital of the *guberniia*, he was employed in the Governor's office, and was given access to confidential files on the Old Believers and on political exiles, who were mainly Poles. In this connection he handled instructions on potential escapes. Where possible, he ignored these instructions or, if unable to do so, he changed names and identifying descriptions. He did not come under suspicion. On the basis of an article entitled "On the Population in Archangel," he was recommended for a confidential secretarial post in that committee. However, being fearful of some humiliation in the course of frequent contacts with the higher authority—this fear was perhaps due to his high school experience in Moscow—he refused the position, which was later granted to the Ukrainian ethnographer and Limanowski's friend, Chubinsky. Instead, Limanowski was nominated to the post of department head (*stolonachal'nik*) of the *guberniia* administration under the Lieutenant Governor. Yet he had no ambition to pursue a Russian civil service career.[42]

His whole attention was directed toward the events in Poland; he anticipated an uprising there in the near future. The agrarian crisis in Poland, the revolutionary situation in Russia, the victories of Garibaldi in Italy—all were factors enhancing the chances of a Polish victory and encouraging the Poles to rise. Actually, the situation was not as favorable as might appear. Napoleon had no desire to revolutionize Europe, the strength of the Russian revolutionary movement was overrated by the Poles, and the Polish radicals were hampered by the legacy of the peasant revolt in Galicia in 1846.

The fatal shooting of several demonstrators in Warsaw on February 27, 1861, resulted in the initial victory for the moderates, the newly formed City Delegation and the Agricultural Society. Marquess Aleksander Wielopolski—the protagonist of the revival of Congress Poland based on the Constitution of 1815—became Minister of Education and Religious Denominations. Russian Poland also secured an appointed consultative Council of State and elected city and county councils. However, the long-term program of the coalition of gentry and middle class deputies of the City Delegation included full independence and restitution of the borderlands. Therefore, they pressed for further concessions while holding back the radicals. The result was the defeat of the moderates—the abolition of the Delegation and the Agricultural Society by the Tsar. Several casualties among the populace in Warsaw on April 8, 1861, due to another confrontation between the Russian troops and a patriotic demonstration, widened the gulf between the moderates and the authorities. Wielopolski became isolated from everyone. The populace in Warsaw continued to be rebellious. Barred from the streets, it sang patriotic hymns in churches. On December 14, 1861, martial law was proclaimed in Warsaw. The next day police, upon breaking into two churches, arrested 1500 men. A polarization took place in the patriotic move-"Reds" gained the upper hand; and in the spring of 1862 they formed the National Central Committee which was to become the future insurgent government.

In 1862 Russian Poland secured further reforms: 1) the conversion of peasants' *corvée* to the payment of rent; 2) civil rights for Jews; and 3) Polonization of the educational system. These reforms produced a Polonized administration; an evolutionary trend toward a liberal regime was apparent, which would have safeguarded the interest of the gentry. However, there was no prospect of full independence. Therefore, patriotic opinion remained unmollified.

While negotiating with Herzen and Ogarev, who represented the Russian revolutionary *Zemlia i Volia*, the Central Committee was prepared to give the borderland peoples an option to secede from Poland. This Mieroslawski protested as treason to Polish national interest. The General now established a rival Revolutionary Committee in Warsaw, which created diversion and confusion in the radical ranks. The Poles were handicapped by lack of trained officer cadres and shortage of arms. Nevertheless, the Central Committee slated the uprising for the spring of 1863. However, the military draft of potential insurgents, ordered by Wielopolski—which took place in Warsaw in the middle of January 1863, and was expected elsewhere by the end of the month—precipitated the insurrection.[43]

Planning to escape, Limanowski followed these events in Poland in *Le Temps* and the London *Times*. When the uprising finally broke out on January 22, he was confident that both the West and the Russian revolutionaries would come to the insurgents' aid. In the spring of 1863 he attempted to board a British ship in the port of Archangel, but was apprehended by customs officials and placed under arrest. Having lost his civil service post, Limanowski on his release from prison two months later was compelled to give lessons and engage in part-time clerical work in the office of vice-admiral Fleischer, the Norwegian consul in Archangel. He could not touch the income derived from his landed estate; Podgorz was sequestrated by the authorities, and his share was slated for compulsory sale.[44]

Meanwhile, the autocracy struggled with the native revolutionary movement; and Russian and Ukrainian radicals began arriving in Archangel. Among them was the well-known Ukrainian ethnographer Pavel Chubinsky (1839-1884). At the time of his arrival in 1862, like Limanowski Chubinsky was a liberal and patriot with vaguely populist views. Limanowski joined his political circle consisting of several Ukrainian and Russian populists. This group, influenced by Shevchenko and Herzen, envisioned an alliance of the Poles, Ukrainians and Russians in a common revolutionary struggle for "freedom." However, by 1864 the Polish insurrection was petering out; Russian liberals influenced by the nationalist Katkov were now alienated from the Polish cause. Even Limanowski himself temporarily lost faith in the efficacy of Polish national uprisings. Instead, overrating the revolutionary movement in Russia (until it was suppressed, due to Karakozov's attempt on the life of the Tsar in 1866), he began to believe in the imminence of revolution in Russia as a means of securing Polish independence.[45]

Political calm prompted him to return to his studies. Having read
the first volume of Buckle's influential *History of Civilization in
England* (imported from Lvov), Limanowski prepared an article on
the significance of morality in history. In it he opposed Buckle's
view—popularizing the ideas of Comte and Mill—that progress was
determined solely by intellect, that the emotional and moral factors
were static, and that religion was not a decisive influence in the onward movement of progress. Limanowski submitted his first "sociological" work intended for publication to the *Warsaw Library (Biblioteka Warszawska)*. Its editor rejected Limanowski's treatise, ostensibly because of its vagueness.[46]

Among the books which Chubinsky imported in 1863, in his official capacity as secretary of the Statistical Committee, was John
Stuart Mill's *Principles of Political Economy* (originally published in
1848) in a French translation. After reading it, Limanowski acquired
a Polish translation of this treatise, published in St. Petersburg in
1859. When comparing the two editions, he was intrigued to note
that the Russian censor deleted the chapter in which Mill discussed
socialism and communism. Therefore, he copied the forbidden text.[47]

Buckle and Mill revolutionized Limanowski's thinking. Buckle
convinced him of the power of knowledge, the advancement of
which Buckle equated with progress, and undermined Limanowski's
already shaky religious belief. On the other hand, Mill made him
aware of the role of the economic factor in social relationships. By
1865 the insurrection was completely suppressed; and Limanowski
seems to have lost his faith in gentry patriotism, in the Polish clergy's
patriotic influence on the peasants, and in the aid of European governments, as factors working for Polish independence. By 1866 he
also temporarily lost his faith in revolution in Russia. Henceforth he
counted on the outbreak of a social revolution in Europe. When he
read the *Workingman's Programme (Arbeiter-programm)* by Lassalle
in a Russian translation in the *Contemporary*, it represented for him
an intensely emotional experience. He described his feelings thus:
"Sun rays pierced dark and heavy clouds of despair . . . . From the
high heaven, shining like a lodestar, the idea of socialism descended
to earth becoming embodied in the hearts and minds of the working
class. Socialism represented for me the political power which would
save us from falling into the abyss of hell, into which we were being
pushed by the Muravievs and Bergs."[48]

Previously socialism had appealed to him solely on ethical grounds,
as a fulfilment of the idea of Fraternity. Mill introduced him to the

economic aspect of socialism and Lassalle to the idea of the political mission of the working class. In the North, Limanowski also came in contact with a group of Polish workers, about a thousand artisans exiled to Archangel for their part in the uprising of 1863. Limanowski felt that Polish artisans—whom he equated with "proletariat"— were the mainstay of this insurrection. Under the spell of *Les Misérables* which he re-read in the original, and of *Le compagnon d'un tour de France* by Georges Sand, as well as of his contacts with the Polish artisans in Archangel, he began to idealize the industrial proletariat.[49]

Limanowski's conversion to socialism was a gradual process. His ideology was grounded in the ideas of the French Revolution, of the "Spring of Nations" of 1848, and of the left "Reds" of the 1860s. It was also grounded in his rebellion against serfdom, in Russian revolutionary propaganda and Western social literature and social thought. There were five stages in his intellectual development. His years in Moscow and Dorpat represented the beginning of the populist phase of his ideology, formed under the influence of the ideas of Herzen, Dobroliubov and Chernyshevsky, as well as Polish *émigré* politics. During the second period of his intellectual development, in Paris, Limanowski broadened his knowledge of the latter and was introduced to the disciplines of political economy and international relations. Thirdly, in exile he discovered Comte, Mill and Lassalle, while under the spell of Western literature of social protest (e.g., Hugo and Sand—a disciple herself of the humanitarian Communist Pierre Leroux), as well as having come into contact with Polish artisans and Ukrainian patriots. I shall discuss the two final stages in his development in the two subsequent chapters.

In the spring of 1866 Limanowski secured a transfer to south Russia, due to his deteriorating eyesight. Meanwhile, the final verdict regarding his escape attempt arrived from the capital. It recommended sending him to prison for a period of six weeks, for having on his person at the time of his apprehension the poem entitled "Ukraine still lives (*Shche ne vmerla Ukraina*)" written by Chubinsky, which became the Ukrainian national hymn. In privileged quarters of the prison, he received many visitors and packages of food, in some instances from people he hardly knew. The authorities in Archangel proved on the whole very lenient in his case, and Limanowski retained fond memories of his exile in the North—where he was happy, except for his yearning to play an active role in politics at home. Generously endowed with money and food, he departed from Archangel under escort in April 1866.[50]

However, travelling by stages, he experienced a tiring journey which lasted nearly five months. Sleeping in stale air, among vermin of all kinds, he was exasperated by the slowness of movement with frequent and lengthy rests at the staging points arranged in prisons. Also, he deplored his complete dependence on his escorts—Russian soldiers with limited education. Yet the trip proved beneficial to his ailing eyesight, and he learned first hand about the hidden aspects of the Russian life, the life and opinions of the common people.[51]

During the trip he often met parties of Polish political prisoners, whose revolutionary propaganda, he suspected, affected many peasants and soldiers in Russia. Himself engaging in a similar acticity, in Archangel he had written a letter to the Polish colony in Vologda, advocating a common front against Tsarist despotism with Russian radicals, as a countermeasure against Russification of Poland.[52]

During these months of travelling, Limanowski contemplated organizing a Polish workers' party as an independent political force, after his return to Poland from exile. He now felt that Poland must embrace the socialist creed, in order to keep up with progressive development in the West. He identified the mission of the proletariat with the historical mission of Poland, and he considered the industrial workers the elite of the working class as a whole. He was aware that some *émigré* Poles collaborated with the First International, and he idealized this organization as a means of rejuvenating Europe. However, at this time he did not make himself clear to what extent he agreed with its principles.

Upon his arrival at his destination in Pavlovsk-on-the-Don in the Voronezh *guberniia*, Limanowski attempted to propagate his ideas among fellow exiles, but without success. He also contributed an article to a liberal publication in Warsaw, advocating "mutual assistance" (workers' self-organization) in opposition to the then fashionable in Poland liberal slogan of "self-reliance" (private enterprise). For the first time, in this article Limanowski revealed himself as a protagonist of cooperation; he cited the contemporary trend in the West, for instance the Rochdale experiment in Britain in 1844, and based himself, too, on the tradition of the apostolic early Christian communitarianism. He argued that cooperation was beneficial on both ethical and material grounds; the decline in poverty resulting from the movement led to a better educated and more humanitarian society, and produced an informed and influential public opinion.[53]

In Pavlovsk Limanowski supported himself by giving lessons in private homes. He pursued his social studies, becoming convinced that socialism was the fulfilment of the triple slogan associated with the ideology of the French Revolution. For Limanowski socialism now became not merely an attitude of mind and an outcome of his emotions, but a practical guide to a better ordering of economic and political relationships.[54] He longed to play an active role in line with his new social creed and his positivist scientific outlook. In the fall of 1867, as a result of the general amnesty granted by the Tsar, Limanowski was permitted to settle in Russian Poland, but he was barred from his native Livonia. A week before receiving his passport, he learned that Podgorz would likely be returned to its owners.[55]

On arrival in Warsaw, Limanowski found employment as a manual worker in a factory owned by an acquaintance. His work day began at 8:00 a.m. and ended at 7:00 p.m. His wages did not cover the cost of his bare subsistence, but he was privileged in that the owner invited him to his home for dinner. While becoming accustomed somewhat to physical labor and succeeding in gaining the confidence of his fellow workers, he nevertheless suffered from the monotony of routine work and missed his intellectual pursuits, from which he was precluded during the week by long hours of labor and resultant fatigue. Although in Warsaw Limanowski lived among compatriots, he felt more oppressed by the Russian regime than when in exile. His material situation deteriorated, in spite of assistance from an uncle. About a month after commencing employment at the factory, Limanowski became a private tutor in Warsaw. It seems that his brief acquaintance with physical labor was not so much motivated by his desire for "going to the people *(khozhdenie v narod)*," as has been suggested by a number of his admirers, but rather by the sheer economic need.[56]

He soon moved to a landed estate near Lublin, Galezow. Here was an opportunity for him to investigate the conditions of rural life in the unfamiliar parts of Russian Poland. The year 1868 and half of 1869 which he spent in Galezow were the most idyllic months in his life. Away from the hated authorities, materially secure, treated as a member of the family and in the peaceful atmosphere of the country, he was able to pursue his studies. He continued to write his second manuscript on sociology, which he began in 1862. (The first manuscript had been lost during his attempt to escape from exile.) Among the books which he read at this time were Buckle's *History of Civilization in England*, vol. II, Mignet's *Histoire de la Révolution Française* (1824) and Condorcet's *Esquisse d'un tableau historique de progrès de l'ésprit humain* (1795). Condorcet confirmed his belief in the idea of progress as progress in knowledge and as basis for social change, and convinced him that the history of civilization was the history of enlightenment.[57]

During his formative years Limanowski had experienced these internal conflicts. His hostility toward Russia resulted in the conflict between his patriotism and his quest for social justice; and his attitude toward Catholicism as the bond between gentry and the non-Polish masses clashed with his agnosticism, fostered by Buckle. The first conflict was characteristic of a Pole living under the Russian regime: was social radicalism the result of Russian influence? Was socialist propaganda incompatible with patriotism and struggle for independence? The second conflict was rooted in the situation of the gentry in the north-eastern borderlands, where the peasants identified Catholicism with Polish nationality. These peasants were subject to religious oppression by the Russians but, on the other hand, only recently (that is, prior to the emancipation edict of 1861) they were serfs belonging to Polish landlords. Thus they had grievances against both Russians and Poles. During his brief visit to Podgorz in the summer of 1869, Limanowski saw his native region as the battleground between Poland and Russia.[58] Therefore, in order to "counteract the designs of Muscovy," he, virtually an agnostic, had to distribute religious gifts in the villages in the area, which he purchased for this purpose in Warsaw, in order to bring nearer his goal of restoring the old Polish-Lithuanian Commonwealth.[59]

Apparently now bored by the country life near Lublin, Limanowski returned to Warsaw in the early fall of 1869. Back again in Warsaw, Limanowski gave lessons, worked in a lawyer's office as well as in a library, and was employed on a part-time basis by the old Prince Stefan Lubomirski as reading companion. Among the works which he read to the Prince were those published in Galicia and forbidden by censorship in Russian Poland, which described Russian repressions against the Poles in the wake of the 1863 uprising. Except for the progressive *Weekly Review*, which provided a start for gifted writers like Henryk Sienkiewicz and Aleksander Swietochowski, the leader of Warsaw Positivism, Limanowski's access to Warsaw periodicals was closed, ostensibly because they had already a sufficient number of contributors prior to his return from exile. It was in the *Weekly Review*, therefore, that he began his career in journalism in 1869, attracting attention by his reviews of Polish literature which, on the model of Dobroliubov, were concerned with the social and psychological message rather than the aesthetic form, in the spirit of "social realism."[60]

However, during 1869 he also reviewed works on philosophy and politics, such as J. S. Mill's *Auguste Comte and Positivism* (1865), as well as the biography of George Washington by Cornelis de Witt.

Limanowski was interested in the problem of morality in politics; stressing Washington's patriotism, he presented him as a man of strong moral principles.[61]

He also concerned himself with economic problems. For instance, he reviewed *Des sociétés cooperatives de consommation et de crédit* (Paris, 1868), which was a speech by E. Villedieu, and *Realisme social* (Paris, 1868) by T. Juinier—to show that divergent schools of economic thought agree that the existing financial anarchy should be remedied. (These writers alleged that four-fifths of poverty in France was due to private credit. The remedy, as proposed by Villedieu, was provision of cheap credit by means of credit unions or, as proposed by Juinier, a nationalized central bank.) Thus, in this context like the revolutionary left in exile, Limanowski combined his agitation for cooperative associations—based on the concept of mutualism or collective possession—with agitation for collective ownership (nationalization), without stating his choice of allegiance. He was convinced by Saint-Simon that the "spirit of association" was gaining ground in Europe, as evidenced by the rise of the cooperative movement since 1848 in England, France and Germany, and by the various congresses, conventions, exhibitions, and associations that had been organized from that time.[62]

Advocating credit unions for Russian Poland in another context, Limanowski wished to combine them with the operation of nonprofit wholesale marketing boards created for the benefit of both the producer and retailer, in order to enhance the economic development of the country by means of a regulated internal trade. Limanowski condemned the concept of liberty as postulated by the protagonists of the doctrine of "laissez faire" as tantamount to a surrender to blind forces, similar to those which ruled the primitive societies. He rejected social Darwinism in favor of social justice. Influenced by Proudhon's mutualism, Limanowski was convinced that history bore witness to the gradual victory of reason and morality in both personal life and in public life (i.e., altruism and consensus), which in economics meant the victory of the principles of mutual assistance and organization.[63]

In the first of his two articles on labor legislation in industry (which he published in 1869 and 1890, respectively), Limanowski advocated a limitation of the working day for both adults and children, on utilitarian as well as moral grounds.[64] However, he was apparently more interested in the peasant problem than in industrial relations. In 1869 he wrote the first of his numerous articles on the agrarian situation in Poland and Europe. It is evident that, in spite of his theoretical commitment to socialism in exile, Limanowski was still in favor of individual landownership. Writing under the influence of Sismondi and Mill, he idealized yeoman farming—again on utilitarian as well as moral grounds. (Impressed by the Swiss economist and historian Sismondi, Mill saw peasant farming as a remedy for the distress in Ireland.) Limanowski agreed with Mill that in comparison to a latifundium employing hired labor, a peasant farm was more productive since it involved a more intensive exploitation of land and more efficient exploitation of labor. A peasant proprietor worked both better and longer than a hired farm hand and was interested in improvements. Thus Limanowski praised yeoman farming as a school of middle-class virtues: industry, thrift, foresight and perseverance. He also considered it an excellent school in citizenship.[65]

In 1869-1870 Limanowski was still in fact a liberal with a vaguely populist program; his social philosophy differed little from that of the nascent Positivist camp in Russian Poland. Apparently, he temporarily abandoned the idea of a social revolution as the means to national liberation. He now wished to draw closer to the peasants, in order to enlist them in the armed struggle for independence, and contemplated marriage to a peasant girl. His unfulfilled dream of such a marriage with a girl from the people was reflected in his unfinished novel entitled *Two Paths (Dwie Drogi)*, written in this period.[66] Yet the heroine of his novelette *The Girl of the New World (Dziewczyna nowego swiata)*, which the Russian censor refused to pass in 1869, was hardly an ordinary peasant girl. When portraying the girl, Helena, he had in mind a pretty and emancipated young woman he had met in exile in Voronezh. This sketch reflected Limanowski's longings for an ideal world of human relationships based on sincerity, brotherly love, equality and mutual trust. According to Limanowski's story, the gatherings at Helena's apartment were "like the gatherings of the early Christians as described by Gibbon."[67]

Limanowski longed to become a political activist and to have his name recorded in history. However, in 1870 the atmosphere in Warsaw was not conducive to either legal or illegal political activity. The insurrection forced the Russian government to emancipate the peasants on better terms than had happened elsewhere in Central and Eastern Europe. The measure had weakened the gentry economically while facilitating industrialization. The Polish intelligentsia now became hostile to the idea of insurrection and devoted its energies to economic development and to cultural tasks. "Let us get rich and acquire education" was the principal slogan of the period, the motto of "organic work." Meanwhile, Russian Poland was becoming just another Russian province, with Russified schools and administration. The new civil and military officials behaved in an arrogant and tactless manner. Limanowski felt helpless when confronted with the arrogance of the Russians; he became very depressed by the situation in Warsaw.

However, upon the outbreak of the Franco-Prussian War, there was an eager, but naive expectation that French troops would march into Poznania, as in 1807; yet September brought the news of Sedan. The subsequent announcement of a republic in France revived hopes in Poland of a German defeat. On hearing a rumor that a Polish legion was being created on French soil, Limanowski applied for a passport to go abroad. On October 2, 1870 he departed from Warsaw for

CHAPTER III

SOCIALIST AGITATOR: THE FIRST PERIOD IN GALICIA
(LVOV, 1870-1878)

On his arrival in Cracow in October 1870, Limanowski learned that the Germans had surrounded Paris; he now doubted whether a Polish legion would be formed in France. His first impressions of Galicia were unfavorable. He found it difficult to find employment in Cracow; and he was repelled by the atmosphere of conservatism prevailing in this city. Reluctant to return to the more oppressive regime of Russian Poland, however, he decided to stay in the capital of Galicia, Lwow (Lvov or Lviv), the atmosphere of which he found congenial.[1]

The period of eight years which he spent in Lvov was very important in Limanowski's life. He married in Lvov; here his first two sons were born; and here he was awarded a Ph.D. degree. It was in Lvov that a definite pattern was established in his career as a political writer. To earn his living, he turned to journalism; in addition, he began to devote much of his spare time to theoretical writings, and to the study of Polish history after the partitions. Also, from this period dated his championship of Galicia as the "Piedmont of Poland."

In the history of Galicia, two important events had taken place around the middle of the nineteenth century: in 1848 the abolition of serfdom by the partitioning power Austria, and by the end of the 1860s the achievement of an autonomous status within the constitutional Austro-Hungarian monarchy. In 1861 Galicia secured a local diet which met in Lvov. To elect the deputies to this diet and to the central parliament in Vienna, the population in Galicia was divided into curias on the basis of class; the voting was open and in the peasant curia indirect. The defeat of the Galician liberals, who aimed at a fully autonomous Galicia within a federal Austria, resulted in the reconciliation of Vienna with the conservative Polish nobility in Galicia, mistrustful of the authorities since the massacre of about 2,000 landowners by the peasants in 1846 (when local Austrian officials had encouraged the rebellious peasantry to deliver the apprehended insurgents to the authorities).[2]

Of the three regions comprising the old Commonwealth—Austrian, Prussian and Russian Poland—Galicia was the most backward province—economically, socially and culturally. The great landlords in Galicia were not usually interested in new methods of agriculture, since they possessed a plentiful supply of cheap labor. Moreover, peasant holdings granted under the provisions of the Emancipation Edict of 1848 often proved insufficient for bare subsistence, and this situation was aggravated in subsequent years by repeated equal divisions of the land among the proprietors' heirs, which was not subject to a law of primogeniture as, for instance, in England. In Galicia, the landless proletariat was numerous, and dwarf farms of less than ten morgs (i.e., about fifteen acres) constituted two-thirds of the peasant holdings. Impoverished, addicted to alcohol, and exploited by rural money-lenders, the peasant could not find employment in industry, due to the backwardness of the region. The dominant features of rural Galicia were primitive agriculture and peasant overpopulation. Hostile to the gentry, the Galician peasantry was passive and apathetic. This hostility of the Galician village was mainly due to its economic dependence on the landlord, and was aggravated by the latter's monopoly of brewing as well as the conflict over the access to forests and pastures retained by the gentry after peasant emancipation.

There were two other areas of conflict between peasant and his lord in Galicia, in the second half of the nineteenth century. First, mindful of the massacre of 1846 and fearful of stirring up the peasants, the nobility opposed any proposals for economic, political or cultural reform in the village. Secondly, a dual system continued for local administration in the rural districts, and a dual standard of taxation of peasant as opposed to gentry land. In each case it was the gentry who were favored and this gave rise to almost daily friction. Also, there was the cultural gulf. At the beginning of the constitutional era the Galician peasantry was almost completely illiterate. Even in later years, when a basis was already laid for primary education, Galician peasant children were only rarely able to gain access to secondary schools and universities.[3]

The middle class was not numerous in Galicia and did not threaten the political ascendancy of the nobility. Very backward in agriculture in comparison with Prussian Poland, Austrian Poland lagged behind Russian Poland in industrialization. With the advent of railroads, Galicia became a market for goods manufactured in the western provinces of the Monarchy; this resulted in severe competititon for the local handicrafts. Individual efforts to foster industrial development proved insufficient, and liberal opinion in Galicia demanded government aid. It was not until the last two decades of the century that the provincial government intervened by establishing a central bank, a statistical bureau as well as technical schools, and by granting subsidies to private enterprises.[4]

Notwithstanding its socio-economic backwardness, Galicia became the most favorable region for the preservation of Polish culture and nationality. In the 1860s the Polish language replaced the German in Galician administration, courts and schools. Under the federalist ministry of Hohenwart, which came into power early in 1871, steps were taken to establish in Cracow a Polish Academy of Learning *(Akademia Umiejetnosci)*; and universities and theaters became completely Polonized. Still German in appearance when Limanowski arrived in the fall of 1870, the capital of Galicia, Lvov, was completely transformed by the time he was leaving it for Switzerland eight years later.[5]

Shortly after his arrival in Lvov, in January 1871, Limanowski became a proofreader with the *Lvov Daily (Dziennik Lwowski)*. It was the organ of Franciszek Smolka's democrats, whose program, formulated in 1848-1849, contemplated full autonomy for Galicia, which Smolka hoped to achieve by a tactical alliance with the other Slavs of the Habsburg Monarchy. From 1863 Smolka envisioned the transformation of Galicia into the "Piedmont of Poland." In 1868 he founded a political party known as the Democratic Association.[6]

Smolka's democrats manifested some concern for the welfare of the masses. They formed cooperative associations, issued cheap editions of books and lectured in small towns on economics and history, as well as sponsoring gymnastic clubs and organizing voluntary fire brigades. Moreover, they distinguished themselves as agitators in the cause of compulsory and universal primary education. In all Galician towns they created cultural clubs for the artisan class, the so-called "Stars *(Gwiazdy)*."[7] However, while becoming concerned with education and welfare of the masses and representing the most radical political trend in Galicia in the early 1870s, they had no desire to foster an autonomous workers' movement and sometimes even opposed universal suffrage from fear of the uneducated populace.[8]

However, these Galician democrats considered themselves heirs to the tradition of Polish insurrections, arguing with the conservative "Cracow school" of Polish historiography, which considered the fall of Poland to have been due to the weakness of the central authority, to anarchical tendencies, and to lack of political realism on the part of the Poles themselves.[9]

Hoping to enhance national unity by strengthening the nation culturally and economically, Smolka's democrats—like the liberal intelligentsia in Russian Poland—envisioned "organic work" as a basis for independence in the future. Generally speaking, after 1863 the democrats in Poland no longer believed in the efficacy of insurrections. They no longer had a program of radical social reform, which would have enabled them to ally the Polish cause with the revolutionary movement in Europe, as in the past. During the period of his first residence in Galicia (1870-1878), Limanowski temporarily abandoned the idea of an armed struggle in favor of "organic work" and, therefore, on the national issue then agreed with Smolka's democrats.[10]

However, political literature printed abroad by the revolutionary left was readily available in Galicia. Thus shortly after his arrival in Lvov, Limanowski was able to familiarize himself with the ideology of Worcell, whose name was engraved in his memory since his university years in Moscow. It was from Tokarzewicz, writing in the *Commune*, that Limanowski learned then for the first time of the existence of the Polish *émigré* socialist organization, the Polish People. Impressed by the example of Worcell, Limanowski now felt that he had resolved the erstwhile conflict between his patriotism and his sense of social justice.[11]

Although in constitutional Galicia the situation was theoretically more congenial for dissemination of socialist propaganda than elsewhere in Poland, the socialist movement was virtually non-existent here at the time of Limanowski's arrival. It was not so much the bureaucracy as delayed industrialization which impeded the devel= opment of a labor movement in this province, at the beginning. The printers, who were the most intelligent and influenctial of the artisans in Galicia, had founded non-political mutual aid societies for themselves in Lvov and Cracow, in 1817 and 1850, respectively. When Limanowski arrived in Lvov, the Lvov "Star" was the most important workers' club in Austrian Poland. The "Star's" sponsors, Smolka's Democratic Association, had also established at the beginning of 1869 a fortnightly paper for workers, the *Craftsman (Rekodzielnik)*. Its editorial policy was determined by Dr. Tadeusz Skalkowski, a disciple of the German liberal economists Franz Hermann Schulze-Delitzsch and of the Polish philosopher Karol Libelt, who advocated establishment of producers' cooperatives and credit unions to protect the workers against exploitation.[12]

The printers, some of whom were insurgents in 1863, became Limanowski's first converts to socialism. Having met some of them early in 1871, in his capacity as a proofreader, Limanowski in the "Star" preached cooperation to the printers and opposed the slogan of "self-assistance" understood as a private enterprise (as he had done in his articles published in Warsaw periodicals in 1867 and 1869). While the controversy in Germany was between Lassalle's concept of "mutual assistance" (cooperatives with state credit) and Schulze-Delitzsch's "self-reliance" (cooperatives without state aid), in Poland the concept of "mutual assistance" denoted cooperation in either form, or simply workers' self-organization. The *Craftsman* espoused the liberal concept of Schulze-Delitzsch, while Limanowski advocated both forms of cooperation.[13]

Lecturing in the "Star" before a working class audience on March 12 and April 2, 1871, Limanowski idealized manual labor and preached cooperation on moral as well as economic grounds—as a fulfilment of Christian idealism and of the lofty moral principle of mutualism or brotherhood. Like Worcell and Mickiewicz, he envisaged socialism as Christianity applied to social relations. Speaking in concrete terms, Limanowski pointed to examples of successful cooperation: the efforts of Robert Owen; the Rochdale experiment of 1844; the development of mutual aid societies in Europe since the beginning of the century; and the St. Crispin's cooperative shoe factory in United States with a branch in Canada, too. Finally, Limanowski endorsed profit sharing by workers in capitalist enterprises as advocated by Cobden in Great Britain and Libelt in Poznania.[14]

In his lectures, Limanowski predicted the eventual rise of the principle of labor to dominance in social relations, resulting in a "reconciliation of intellect with manual work." This, in concrete terms, he defined as enhanced status for the worker, with each individual seeking to acquire at least one manual skill. He suggested a close collaboration of intellectuals and labor in the common cause of social progress, implying the need for workers' organization in Galicia. However, sympathetic to socialism in general, Limanowski provided no clear and definite program. Probably, his vagueness and reserve was partly due to the Galician situation (where socialism was still an alien social creed)—as suggested later by some of his future political friends in the Polish Socialist Party. The lectures published as a pamphlet proved suitable for agitation among the industrial workers, preparing the ground for the advent of socialism to Galicia. However, initially Limanowski made little impact on the working class audience, and the appearance of the pamphlet was almost ignored by the press in Galicia. Limanowski published it at his expense in 500 copies, a hundred of which he sent to Russian Poland.[15]

Occasionally, Limanowski commented on local events in Lvov *Craftsman*—the first workers' paper in Poland—to which he also contributed one article on cooperation in Galicia. Printing news about the foreign labor movements, by 1870 this paper began to champion the striking artisans in Lvov. The first to strike in Galicia were Lvov printers; they demanded a standardized rate of wages. Their success in having their demands met by their employers encouraged the bakers also to strike; the bakers' strike ended in complete victory for them, too. In 1871, upon Limanowski's suggestion put forward in the *Craftsman*, striking tailors founded a producers' cooperative, but it soon became bankrupt. Moreover, an unsuccessful strike of the cartwrights, which took place that same year, had serious consequences. Aware of the rise of the workers' movement in Lvov, the authorities credited it in part to Limanowski's propaganda, and in part to the example of the Paris Commune; the police in Lvov arrested some of the strikers and trade-unionism in Galicia suffered a temporary setback.[16]

Limanowski's pamphlet alerted the police in Lvov; it also attracted the attention of Polish students in Warsaw, Kiev and St. Petersburg. In the long run, both developments proved ominous for his future career in Galicia. Having read the pamphlet, the police commissioner Sobolak summoned Limanowski. In the course of being questioned, the latter concluded that he was suspected of having participated in the Paris Commune. Yet, for the time being, Limanowski escaped harassment by police.[17]

From 1870 to 1873 Limanowski acted as the Lvov correspondent for the *Weekly Review*; he also contributed to a women's magazine, the *Ivy (Bluszcz)*, published in Warsaw, too. Many of his contributions to these papers were reports of public meetings in Lvov, which he recorded in shorthand. His articles stressed the importance of the Galician democratic press as a champion of the underdog, criticizing acts of injustice and shortcomings on the part of public bodies and individuals. Also, Limanowski publicized such Galician associations as the "Stars" and the Pedagogical Society, the latter of which concerned itself with welfare of rural teachers and promoted elementary education among the masses.

His contributions to the *Weekly Review* were usually brief, consisting of his own opinions and impressions; those to the *Ivy* were more factual and comprehensive. He informed women in Russian Poland about Galician schools, agriculture and industry; he was at times

ridiculed for this by the Lvov correspondent to the Cracow conservative newspaper *Time (Czas).* For instance, in one of his columns he
deplored the situation of female workers in Galicia, where a capable
dressmaker earned five times less than the lowest paid male clerk. Limanowski stopped contributing to the *Weekly Review* when it came
under the influence of Swietochowski, whom he considered a Russophile. However, he defended the *Review* in the *National Gazette
(Gazeta Narodowa)* when the former was charged by the Lvov correspondent to the *Time* with allegedly pursuing an anti-national
policy.[18]

After Limanowski's arrival in Galicia from Warsaw—where he had
lived from day to day without financial security—he hoped to be
able to support himself by writing. However, here too his articles
were mutilated by censorship. It was difficult for him to find a publisher for his scholarly works. Those that did tolerate his opinions
were ones that sooner or later went bankrupt, because they lacked
sufficient financial backing; one such ephemeral sheet was the *Literary Daily (Dziennik Literacki),* where Limanowski published the
introduction to his study of Plato's *Republic.* Nevertheless, he was
able to support himself from proofreading during the first two
months of 1871. He was about to be promoted to become an editorial
assistant on the *Lvov Daily* when it ceased to appear, owing him a
substantial sum in outstanding wages.[19]

In May, 1871, he secured employment with the *Polish Daily
(Dziennik Polski).* This paper was the organ of a group of democrats
to the right of Smolka and led by Florian Ziemialkowski, who were
centralists and sought alliance with the Germans in the Monarchy.
Six months later, in connection with Polonization of educational institutions in Galicia, Limanowski became secretary, for one year, to
Rudolf Ginsberg, who distinguished himself by his fostering of economic development in Galicia and was a professor of applied chemistry
at the Polytechnic Institute in Lvov. During late 1871 and early
1872 Limanowski, in addition to his secretarial duties which he performed in the evenings, worked as an editorial assistant at the *Polish
Daily*; in his spare time he researched in libraries, attended public
meetings or lectures and reported the proceedings of the Galician
diet.[20]

In the spring of 1872 he applied for a vacant post in the City Archives; the successful candidate, however, was the future Austrian
statesman and leading historian of the "Cracow school" Michal
Bobrzynski. To compound Limanowski's disappointments, he was
dismissed from the *Polish Daily,* ostensibly on grounds of economy.

He suspected he was being penalized for his sympathy with the workers' movement in Lvov. However, he soon secured a clerical position in the local General Hospital. It was his second experience of office routine, which he disliked intensely. Yet he could not refuse to take the position for lack of an immediate alternative. While working in the hospital, he completed in his spare time a study of Plato's *Republic* and began writing his treatise on Thomas More and Thomas Campanella, which was given several favorable reviews, notably in Lavrov's *Forward (Vpered).*[21]

Shortly after securing employment in the hospital, he was invited to join the staff of another daily in Lvov, the *National Gazette*, whose editor-in-chief Jan Dobrzanski pursued an intermediate course in politics, remaining on good terms with both the gentry and the middle class, and leaned toward the federalism of Smolka. Limanowski reported in the *National Gazette* the proceedings of the City Council as well as all public meetings in Lvov; he also analyzed current events in Russia and Poland.[22]

Already in this period he began to endorse the national cultural aspirations of the eastern borderland peoples (e.g., use of native language in schools, courts and administration), as a democratic end in itself as well as the means of preventing their Russification. Notwithstanding his earlier loss of faith in the outbreak of a Russian revolution, by the mid-1870s Limanowski again anxiously awaited developments in Russia. Unable to express his opinions on this subject openly in the *National Gazette*, instead he commented on the Russian revolutionary *Zemlia i Volia* in "Corespondence from St. Petersburg," in Lavrov's *Forward* published in Zurich and London. Limanowski apparently confused at times nationalist aims—which might well be pursued by a reactionary as much as by a liberal government—with the personal rights of individuals; in one of his articles he branded as hypocritical the concern of Tsardom for the Balkan Slavs, prior to granting constitutional freedoms for its own people. Like many of his compatriots, Limanowski oscillated between sympathy for the southern Slavs and enmity toward Russia.[23]

While employed at the *National Gazette* , Limanowski received a salary which was double that which he had earned hitherto; he also secured more free time to pursue his studies. Meanwhile, the federalist Hohenwart Ministry had fallen in November 1871. The victorious centralists effected an amendment to the electoral law in March 1873, whereby the deputies to the federal Council of State (*Reichsrat*) were to be chosen directly, bypassing the provincial diets. The

Galician Jewish voters—hitherto federalist supporters of the Poles—
contributed to this change in the federal law by switching their support to the Ukrainian centralists, thereby arousing the indignation of
the staff at the *National Gazette*.

By their political stand, the Jews unwittingly drew the attention
of Galician journalists to the acquisition of peasant land by some of
the Jewish money-lenders. Opposed to anti-semitism, Limanowski
feared that an emotional attack on such Jews in the press might incite the populace to senseless violence; he searched for a means to
counteract the loss of peasant land without engaging in a hostile
press campaign against the Jews. In Limanowski's opinion, it was
necessary to raise the peasant's intellectual, economic and social
status, in order to make him capable of defending his holdings.

Therefore, encouraged by the writer and former member of the
insurgent National Government in 1863, Agaton Giller, who befriended him at the *National Gazette*, Limanowski wrote an article
in which he appealed to the upper classes to take part in a program
of social work in the village. According to Limanowski, there were
three aspects to the peasant problem: 1) It was necessary to make
the peasant aware of his economic and legal situation, to interest
him in education, and to make him chary of the evils of usury and
alcoholism. The educated classes should try to reach the peasant by
means of village assemblies, special periodicals, and agitation by persons whom the peasant could trust. 2) The intelligentsia should form
rural centers to educate the peasant for citizenship, to develop his
initiative, and to teach him new agricultural techniques. These centers would become the meeting ground for the peasantry and the
educated classes. 3) As a remedy against exploitation by money-
lenders, Limanowski advocated forming credit unions, to which end
he solicited government aid, in addition to appealing to private individuals. Unrealistically, he suggested that men of every political
orientation in Galicia should collaborate in improving life in the
village. However, he might well have had to follow the socially con-
ciliatory policy of the *National Gazette* on this issue.[24]

Limanowski's program was akin in spirit to that of the *émigré*
Democratic Society; by improving the lot of the Galician masses and
educating them, he hoped to awaken their patriotism. In a similar
attempt to rally the peasantry, in 1872 the Galician democrats had
conducted a fund-raising campaign to establish a system of universal primary education. The campaign took place on the centennial
anniversary of the first partition and was sponsored by Limanowski's

employer, the *National Gazette*, along with the *Polish Daily*. On this occasion, Limanowski wrote an article in which he suggested that the peasants themselves ought to contribute to the fund, too; to this end, he proposed selling them religious pictures, rosaries and prayer books. Evidently, he still considered that the peasants' devotion to the Roman Catholic religion could be utilized as an instrument of national politics. The sum collected fell far short of the objective. However, on November 5, 1872, the diet voted to establish a network of compulsory primary schools; and it provided funds for teachers' salaries and for the expansion of the existing facilities. Limanowski considered this law to have been the most important hitherto in the history of the Galician diet.[25]

Also, in 1872 the democrats had founded in Lvov a branch of the Association for Rendering Education Aid to the Duchy of Teschen, to provide a scholarship fund for worthy Silesian peasant boys attending Galician high schools and universities, and to strengthen patriotism in the Duchy by effecting a cultural *rapprochement* between the two provinces of the Austrian Monarchy, in which there was a very substantial Polish population. In January 1873, Limanowski too had joined this organization and on October 12, 1873, he lectured on its behalf. It was his contention that the compulsory primary system of education established in Teschen had indeed reawakened Polish patriotism among the masses; this in his view was one of the most significant achievements of the democratic trend derived from the cult of folk traditions, which was an important ingredient of Romanticism in Central Europe. He pointed out that in Austrian Silesia the upper classes had assimilated with the colonizing Germans. Hence, there was a pressing need to create in this region a native intelligentsia, who would give a conscious direction to the masses and defend them against Germanization. Echoing Lelewel, he charged that it was the Polish aristocracy who lost Teschen; it was now up to Polish democracy to regain it. However, he was not urging then insurrection in Austria; he merely wished to enhance the status of the Poles vis-à-vis the Germans in this region. In his lecture he concerned himself for the first time with the preservation of Polish nationality in the Polish western borderlands; later on he was to become a spokesman for the entire Polish *irredenta* in the west.[26]

Meanwhile, financially secure due to his employment at the *National Gazette*, and with free time at his disposal, Limanowski decided to obtain a Ph.D. degree and to qualify for a univeristy teaching position. In addition to majoring in philosophy, from October 1872 to July 1873 he studied mathematics, chemistry, physics and botany at the University of Lvov. In order to qualify for the doctoral degree in philosophy, he had to pass examinations in minor subjects, either chemistry and physics, or history. He chose the "abstract" sciences of chemistry and physics rather than a "concrete" discipline like history, in accordance with Comte's concept of sociology as "social physics." In 1876 Limanowski obtained the degree on the basis of his dissertation entitled *Sociology of Auguste Comte (Socjologia Augusta Comte'a)*.[27]

He was now certain of a respectable career as a lecturer in philosophy. In this capacity, he hoped to pioneer sociological studies in Poland, and thereby to prove that socialism was the logical outcome of progress. Already under the surveyance of police Limanowski, having antagonized the influential Galician conservatives, was surely naive to suppose that he would be allowed to preach socialism in the state university at Lvov.

Hoping eventually to obtain a university teaching post in Lvov and financially secure while employed at the *National Gazette*, Limanowski decided to marry. In October 1874, he married a Polish gentlewoman, Wincentyna Szarska, whom he had met in Pavlovsk. Wincentyna's first fiancé was killed in the course of the insurrection of 1863. She herself was exiled to Russia for patriotic activity during the uprising. Limanowski had not seen Wincentyna since his departure from Pavlovski in the fall of 1867; however, living many miles apart, they kept in touch by correspondence.[28]

Shortly after his wedding, Limanowski was again faced with financial problems. He contemplated returning home, but soon abandoned the idea as impractical and decided to apply for Austrian citizenship. However, early in 1877 he was dismissed from his post with the *National Gazette*, as the result of having inadvertently slighted its new publisher, Karol Groman, while reporting an electoral campaign in Lvov. (Limanowski was incapable of servility toward an employer.) Meanwhile, he had alienated some potential employers in Lvov. Fortunately, his mother aided him financially, and he derived some income from lecturing in mathematics at Professor Ginsberg's School of Distilling. Also, the Limanowskis supplemented their income by translations; Wincentyna translated works by prominent French and German novelists on her own, and jointly with her husband she rendered into Polish the *Geschichte des 18. and 19. Jahrhunderts* (1838-1848) by the German historian F. C. Schlosser.[29]

In the meantime, from August 1875, Limanowski became host to emissaries from radical student circles in both Russia and Russian

Poland. There was only one university in Russian Poland in the post-1863 period, the University of Warsaw, which had been Russified in the late 1860s. Therefore, a large number of Polish students attended universities in Russia, especially in St. Petersburg, Kiev and Odessa. These students maintained close contacts with Russian revolutionary circles, and with the Southern Union of Russian Workers. In 1875 they began transplanting their organizations to Russian Poland. Reacting to the Warsaw Positivists' acquiescence in the *status quo*, they formed the first socialist groups in Warsaw during the period from 1875 to 1878. This new revolutionary left in Russian Poland was hostile to the tradition of national conspiracies and insurrections. Also opposed to the romantic conception of Polish history, culture and social life, the Positivists unwittingly opened up a new intellectual vista to this first generation of Polish "Marxists," who thus considered themselves as the "children of organic work"—heirs to the materialist Positivist philosophy associated with the industrialization of Russian Poland, which took place during the 1870s and 1880s.[30]

The first visitor from Russian Poland was a young lawyer Kazimierz Hildt, whom Limanowski introduced to his socialist friends in Lvov. Subsequently, he was host to the medical student Jan Hlasko, on his way to London to meet the exiled leftist leader General Wroblewski. The Hlasko mission was sponsored by Polish socialist circles in St. Petersburg, Kiev and Moscow. On his return, Hlasko encouraged Limanowski to establish a learned socialist journal in Lvov, to stimulate development of the social sciences in Poland; to this end Hlasko pledged the support of his comrades in St. Petersburg and Warsaw.[31] However, the project was not implemented.

For some time now Limanowski wished to become the founder of the first Polish socialist party. He talked vaguely about socialism with both workers and intellectuals, and to acquaint them with a socialist ideology he popularized Schäffle's *Quintessence of Socialism* in the *Week* (*Tydzien*), which was the most radical but non-socialist periodical in Lvov around 1878. (Albert Schäffle, the Austrian liberal statesman and sociologist, was for a time an advocate of evolutionary socialistm. He envisaged nationalization of major industries; a certain equalization of income; guaranteed housing and employment; and granting of credit by the state to individuals in need. The most Utopian feature of his program was the provision for distribution of goods from public stores according to individual merit.) Limanowski secured converts to socialism among intellectuals like the poet Boleslaw Czerwienski who became the author of a famous socialist hymn, the "Red Banner." Generally speaking, however, the young people in Galicia were afflicted by a Hamlet-like state of indecision; they feared to break away from the gentry tradition of reconciling peasant and lord, and to boldly champion the people's cause.[32]

In contrast to prevarication of this kind, Limanowski—writing in the *Week*—now assigned the task of guiding the peasants exclusively to the democrats, despite the fact that the slogan of the paper for which he was writing was the aristocrat Krasinski's motto: "The Polish people ought to follow the Polish gentry *(Z szlachta polska—polski lud)*." Idealizing the common people, Limanowski wished to free them from tutelage by the educated classes; the editors of the paper tolerated him, while disassociating themselves from his views. In the *Week* he appealed to the reformist tradition of Staszic, Kollataj and Kosciuszko, and admonished the gentry not to alienate the people from the national cause. Limanowski pointed to the legacy bequeathed by Kosciuszko to Polish patriots—to work in the cause of social justice, thereby providing a strong foundation for independent nationhood. This was also the theme of the first of his numerous articles devoted to Kosciuszko, which was written for a children's magazine in several instalments from January through March 1877.[33]

Meanwhile, in November 1876, Limanowski was host to Edmund Brzezinski and Erazm Kobylanski, former students at the St. Petersburg Technological Institute. Brzezinski, who was en route to Vienna where he was to study medicine, handed to Limanowski a translation of Lassalle's *Workingman's Programme (Arbeiter-programm)* of 1863, to be revised by Limanowski for publication in Galicia. The other visitor, Kobylanski, an emissary of Polish and Russian revolutionary circles, was in transit to Switzerland to make contacts there with Polish, Russian and Ukrainian exiles, and to arrange the smuggling of censored books from the West to Russian Poland via Galicia.[34]

Soon after Kobylanski returned to Lvov on May 31, 1877, with a large number of Russian books as well as publications by Mykhailo Drahomanov (a former professor at Kiev University, who was a populist and Ukrainian patriot), the bellhop in his hotel reported him to the police. While searching Kobylanski's room, the police found an envelope addressed to Limanowski. Thereupon, they ransacked Limanowski's house as well, appropriating some of his papers and books. In addition to arresting both Limanowski and Kobylanski, the police apprehended a number of local Ukrainian radicals, among them an associate of Drahomanov, Mykhailo Pavlik, and the writer Ivan Franko. All the prisoners were charged with participating in an alleged "conspiracy" in Lvov.

In 1877, upon the outbreak of the Russo-Turkish war, the authorities in Lvov became alarmed by the rise of socialist and nationalist agitation among both Ukrainians and Poles as well as by the contacts of the Galician Poles with Russian Poland. Russian agents infiltrated the police in Lvov; moreover, at this time the Austrian and Russian authorities generally collaborated in the suppression of socialism. After spending two months in prison, Limanowski was released on bail supplied by his brother Lucjan.[35]

In the course of a preliminary hearing, the charge against Limanowski was dropped, due to insufficient evidence to convict him. Thereafter, he became once again optimistic about his chances of securing Austrian citizenship. He was deluding himself, for he was unaware that the police had already recommended to the Viceroy that he be banished from Lvov.[36] However, the trial of those who remained under arrest which took place in mid-January 1878, resulted in the acquittal of many of the defendants. This provided favorable publicity for socialism; and the trial led, too, to *rapprochement* between Polish and Ukrainian radicals who, contrary to the suppositions of the police, had not been acquainted with each other prior to their arrest.[37]

Limanowski's imprisonment enhanced his popularity among workers and students as well. After his release from prison and during 1878 Limanowski earned his living mainly by writing in the *Week*. When a certain professor of the Agricultural Academy at Dublany near Lvov, Juliusz Au, wrote a pamphlet in which he charged that socialism was a "social disease" and argued that the advent of capitalist economy had benefited the common people in the West, Limanowski criticized Au's view in the *Week*.[38] His reply to Au attracted the attention of a student at the University of Warsaw, Stanislaw Mendelson, upon whose suggestion Limanowski wrote a pamphlet in defense of socialism. Therein he defined it vaguely as the blueprint for a society which would honor labor, would be based "on the principles of justice and equality, would oppose all forms of exploitation, and provide the means for complete emancipation of the people."[39]

Like Comte, Limanowski discerned these four potential features of social development: 1) the social polarization of capitalists and laborers due to the economic concentration of capital; 2) society

based on the principle of industry or labor; 3) spiritual alliance between intellectuals and workers; and 4) the rise of altruism in human
relations.[40] In his treatise Limanowski combined Comte's evolutionist
and deterministic philosophy of history with the Marxian-Lassallian
concept of the political mission of the proletariat. While seeing scientific grounds for socialism in the science of "sociology," he based his
claim that the future belonged to socialism on the authority of such
individualists as John Stuart Mill and Herbert Spencer.[41]

Limanowski believed that economic development was determined
by custom and ideology.[42] Contradicting Marxism, in his treatise he
argued that socialism was not an ideology that reflected a concrete
system of economic relations, but was an absolute quest for abstract
justice, which had begun in very early times. Thus, Moses, Licurgus
and Plato were theorists of socialism![43] Limanowski's teleological
approach to socialism was contrary to the positivistic empiricism
which he professed.[44]

He admitted that a revolutionary upheaval was necessary under
despotic rule. On the other hand, echoing Lassalle, he was a protagonist, under a liberal regime, of the struggle to gain direct, secret
and universal suffrage as a basis for a peaceful parliamentarian transition to socialism.[45]

In his treatise Limanowski urged that the adoption of socialism
was in the Polish national interest, denying that it was a cosmopoli-
tan trend. He argued that loving mankind did not exclude patriotism,
just as patriotism did not exclude loving one's family. Professing to
seek an appropriate status for Poland within the international community, based on its "moral" qualities rather than on force of arms,
Limanowski justified armed struggle for independence as the prerequisite for socio-economic transformation in Poland. He believed
that every true Polish patriot was bound to support socialism in the
cause of national unity and that every true patriot was a potential
socialist, due to his love of the people as an end in itself. Conversely,
every true socialist, in Limanowski's opinion, was a potential patriot,
because patriotism in Poland was also a means of establishing a regime
of social justice.[46]

Arguing thus, Limanowski assumed voluntary sacrifice on the
part of the upper classes. Yet further on in his pamphlet he deplored
the class egoism of the Polish gentry in the past, as an obstacle to
progress and independence.[47] He believed that socialism was the fulfilment of the triple slogan of the French Revolution, i.e., *liberté,
égalité, fraternité*, adopted by the erstwhile Polish democratic movement. This slogan, in his opinion, was akin to the idealism of the
gentry democracy in the old Commonwealth. Thus socialism in
Poland—derived in his account from the ideology of the democratic

ment and, indirectly from the gentry idealism of the more remote past. Thus socialism was for Limanowski compatible with the Polish "national spirit."[48]

In this pamphlet, for the first time, he declared himself a convert to the idea of socializing agriculture, being influenced, it would seem, by the situation in Poznania and Galicia where Polish peasant land was passing to outsiders (Germans and Jews).[49] Appealing to the native tradition of collectivism, which he drew from Lelewel's writings, Limanowski was now convinced that the concept of private ownership had not yet taken root in the Polish village and, therefore, the transition to socialism would be easier to accomplish in Poland than in the West.[50] Assuming the good will of the gentry, he was prepared to coerce them into acceptance of agrarian socialism if nevertheless they opposed it, just as the Democratic Society in their Manifesto of 1836 had been prepared to accept the use of force should the gentry resist the implementation of peasant emancipation with land.[51]

Notwithstanding his attempt to ground socialism in the "positivistic" sciences (i.e., in history and sociology), Limanowski in his pamphlet adopted the metaphysical concept of a "Polish mission" (i.e., moral leadership) employed earlier by Polish Utopian socialists and philosophers, to justify socialism as a specific social program to be realized in Poland.[52]

Meanwhile, the first socialist trial in Poland (discussed above), which took place in Lvov early in 1878, gave rise to intensified socialist agitation in both Galicia and Russian Poland. Thus in February Hildt visited Lvov again, this time for the purpose of publishing his translations of Lassalle's *Capital und Arbeit* (1864) and *Die indirecte Steuer* (1863), as well as a life of the Polish radical leader General Dabrowski, to which Limanowski appended a biographical sketch of Wroblewski.[53] On arrival in Lvov from Warsaw in April, Mendelson and Hildt secured permission to pick up the books which the police had initially confiscated from the expelled Kobylanski. In this connection, Limanowski acted as an intermediary with the police. Thus, a police commissioner once referred to him fondly as "our Mr. Schäffle," voicing respect for "harmless" scholars like Albert Schäffle, the Austrian liberal statesman and sociologist, in opposition to the more formidable working class "Bebels" (a reference to August Bebel, one of the leaders of the German Social Democratic Party founded in 1875). Reassured of being granted Austrian citiczenship, Limanowski again applied for a position in the State Archives.[54]

However, after Mendelson and Hildt had been arrested on May 3, 1878, the police again searched the home of Limanowski. Upon the

expulsion of both on May 17, he became fearful of a similar fate
awaiting him in the near future.[55] Nevertheless, in June he met with
a number of political friends, including the two well-known Ukrainian patriots and populists Franko and Pavlik, the printers Mankowski
and Daniluk, the poet Czerwienski, and the pharmacist Adolf
Inlaender—to discuss future plans for the dissemination of socialist
ideas. In August, Limanowski formed a socialist committee comprising the above-mentioned individuals, shortly after the printers in
Lvov had established a new organ *Labor (Praca)* to replace their now
defunct *Type (Czcionka)*, founded in January 1872, just before the
demise of the *Craftsman*. In 1879 the committee was to become
known as the editorial board of *Labor* and a center for dissemination
of socialist propaganda in Galicia.[56]

Meanwhile, in September 1878, Limanowski met Ludwik Warynski, formerly a fellow student of Brzezinski and Hlasko at the St.
Petersburg Technological Institute. A conspirator by temperament,
Warynski came to Lvov in August to escape arrest in Warsaw, but he
found the attempt at "legality" by socialists in Lvov uncongenial
and he, therefore, departed for Cracow in December.

Around this time, Bismarck's Anti-Socialist Law of October 19
began to affect Galicia, too. Although similar legislation did not
exist in Austria-Hungary, the bureaucracy in Galicia now felt able to
limit the constitutionally granted freedom of association to existing
political parties as well as to non-political organizations. The Galician
police kept Russian subjects under close surveyance; the penalty for
illegal activity on their part was expulsion, or even being handed
over to the Russian authorities.[57]

In July, 1878, the authorities expelled Giller (who had come to
Lvov in 1871, when the liberal Hohenwart Cabinet welcomed Polish
exiles to Galicia from abroad). Already in 1874 the Russian authorities in the former Congress Kingdom attributed to Giller's initiative
the non-existent "National Committee" in Lvov, which allegedly co-
operated with Russian socialists in Zurich and Polish socialists in
London, in anticipation of a socialist revolution in Europe. In 1876
he did form a committee at the *National Gazette*, but this group
merely awaited developments that might issue from the Balkan crisis.
From March 1876, the followers of the writer Waclaw Koszczyc
(which was the pseudonym of Walery Wolodzko) contemplated an
uprising against Russia. Yet neither of these two groups had any
funds, contacts or influence. A supporter of the liberation of the
Balkan Slavs at this time, Limanowski opposed the aims of Koszczyc's

pro-Turkish "National Confederation." Moreover, he resented Koszczyc's insurrectionary agitation as potentially disruptive of the nascent labor movement, arguing that the national strength should be conserved for the imminent outbreak of a revolution in Russia.[58]

He himself was soon to be banished from Galicia, although he did not associate with either Koszczyc or Giller. A note found during a search of a German socialist, which stated that Limanowski was a socialist sympathizer and directed the bearer to him for the purpose of making new contacts, provided the Galician police with a pretext for the long-desired expulsion order. Summoned before the Viceroy, Alfred Potocki, Limanowski learned that it was his ill-advised contacts with Kobylanski and other "militant socialists" which incriminated him, rather than his writings. He was granted permission to delay his departure by one month, in order to prepare for exile.[59] Following tearful farewells by many acquaintances, Limanowski left Lvov on October 1, 1878, accompanied by his wife and their two little boys.[60]

The expulsion order forcing Limanowski to leave Lvov at short notice on October 1, 1878 was a great shock to him. Settled in Lvov for a number of years, he had to sell all his household effects quickly and at a loss and he departed with only a small sum of money into exile, though no longer alone as he had been on his way to north Russia, but with a wife and two little boys, of whom the younger—an infant not yet one year old—was ill.[1]. He would have liked to stay among compatriots in Prussian Poland, where the cost of living was cheaper than in the West, but he was barred from Poznania by Bismarck's Anti-Socialist Law which was about to come into effect.

Around this time, most political exiles from both Western and East-Central Europe rallied to Switzerland, due to its central location in Europe, its high standard of living, and its liberal regime. There was a large group of French ex-Communards; the leaders of the German Social Democratic Party with its organ *Vorwärts*; and a numerous colony of Russian *émigrés*. Some of Limanowski's acquaintances from the Russian parts of Poland had also settled in Switzerland, in Geneva. Among them were Kazimierz Hildt and Kazimierz Dluski. Both had met Limanowski in Lvov as emissaries of student circles in Russian Poland.

Limanowski, too, decided to settle in Geneva, being motivated by a number of considerations. First, he hoped to edit there a new socialist periodical. Secondly, he was counting on his former Russian friend from the University of Moscow, Anton Trussov, who owned a printing shop in Geneva, to provide him immediately with the means to earn a living. Trussov had just finished printing a Russian translation of a controversial book by a British doctor, G. R. Drysdale, *The Elements of Science or Physical, Sexual and Natural Religion* (London, 1854); and he was to arrange with Drysdale for Limanowski to render it into Polish. As Limanowski's English was inadequate for the purpose of translation, he was to base himself on this Russian edition. Thirdly, he wished to meet Colonel Zygmunt Milkowski and

General Walery Wroblewski, two prominent members of the post-1863 emigration, who resided in Geneva. Finally, he was attracted to this city, because he felt that it was located in the most liberal and progressive canton in Switzerland.[2]

It was Trussov, Dluski and Hildt who met Limanowski and his family at the Geneva railroad station early in Octoberber 1878. Until the arrival of Mendelson and Warynski in the spring of 1880, Dluski led the local Polish socialist colony. (Hildt died prematurely from consumption in 1879.)

Soon Limanowski began to attend all gatherings of foreign socialists in Geneva. Here he was befriended by the Swiss Marxist Johann Philip Becker. Formerly an active member of the Swiss section of the First International, Becker was now the editor of *Le Précurseur*, the organ of his "Partie du Peuple Travailleur," and a popular figure with the German and French *émigré* socialists in Switzerland. The Russian populist colony in Geneva was divided into two mutually hostile sections, a minority group led by the "Jacobin" Tkatchev and centered around his organ the *Tocsin (Nabat)*, as well as the majority which was then leaning toward anarchism. The majority included future members of the Marxist "Emancipation of Labor" group, and well-known revolutionary populists like Stepniak-Kravchinsky and Debogorii-Mokrievich. Limanowski attempted to be on friendly terms with all these Russians and was especially close to Zhukovsky, the leader of the Russian "anarchists."[3]

Nevertheless, he also became a friend of General Wroblewski who associated with Zhukovsky's rival, Tkatchev. Alled with the Russian "anarchists," Dluski and his Polish political friends, unlike Limanowski, were hostile to the General.[4] Limanowski defied these younger Polish socialists by collaborating with some of the non-socialist democrats among the Polish and Russian *émigrés* in Geneva. Shortly after his arrival, Limanowski paid a visit to the writer and democrat Milkowski, who was a member of the local Polish Society for Mutual Assistance. Founded in 1868, this Society comprised about ten members, all former insurgents in 1863. Limanowski joined it, in spite of the opposition on the part of Dluski's group, and became its librarian.

On November 30, 1879, the occasion of the forty-eighth anniversary of the uprising of 1830, he spoke before the Society. He told his audience that in the Russian Empire it was the socialist movement which would be an effective instrument in the struggle for Polish independence. He still believed in a revolution in Russia as the best means to achieve it. However, the non-socialist audience was hostile to his point of view.[5]

Limanowski was a regular customer of Elpidin, the Russian book-seller in Geneva. The son of a Russian Orthodox priest and member of the revolutionary *Zemlia i Volia* in the 1860s, Elpidin had taken part in the so-called "Kazan conspiracy," which was an abortive attempt on the part of a few students and army officers—both Poles and Russians—to start a rebellion among the peasantry along the Volga, in conjunction with the Polish uprising of 1863. In Geneva Elpidin edited the Russian liberal-democratic organ *Common Task (Obschchee Delo)*, which appeared from 1877 to 1890; one of his collaborators was the anarchist Zaitsev, a former co-editor with the Russian "nihilist" literary critic Pisarev of the *Russian Word (Russkoe Slovo)*, published in St. Petersburg from 1856 to 1866. It was Elpidin's contention that a people usually got the kind of government it deserved; Limanowski, however, argued that this was certainly not the case with the peoples under a foreign yoke.[6] Meanwhile, he was not in actual penury only because he was aided by his aunts in Belorussia. He doubted whether editing the proposed socialist periodical—which in fact appeared only toward the end of 1879—would yield him sufficient income to support himself and his family. Therefore, he insisted during 1879 on a final settlement of the family affairs. Due to his political activity, Limanowski could not return home to claim his share of the family estate. Therefore, he suggested that his brothers and his mother should divide the estate among themselves, and pay him in cash a sum equivalent to the value of his share of the property. Later in the same year he was to learn from the *Frankfurter Zeitung* that the Tsar himself had allegedly read his *Socialism (Socjalizm)*; now definitely convinced he would never secure a permit to return home, Limanowski began to press his brothers to reach a settlement, so that he might thereby receive from them a sum of money sufficient to establish him in some commercial enterprise abroad.[7]

In spite of the financial aid being rendered to him also by Becker and Wroblewski, Limanowski considered his first winter in Geneva a trying experience for himself as well as his family. In Switzerland he continued to contribute to the *Week*. Mindful of the household communes which had proved successful among Polish exiles in Russia, Limanowski in one of his articles in the *Week* deplored that the Poles in Geneva did not practice such communal living as a remedy for the high cost of living and domestic help.[8]

The income which he derived from his regular contributions to the *Week*, which was re-organized under a new name—the *Polish Week (Tydzien Polski)*—in February 1879, was very small; his wife earned more by her translations of contemporary French novels. Limanowski also attempted to collaborate with periodicals published in Warsaw, but was usually unsuccessful, in part due to the severe censorship in Russian Poland and in part to the opposition of editors themselves to his ideology. However, he began to write again for the Warsaw *Weekly Review*, this time in the form of "Correspondence from Switzerland" as well as "Letters on Contemporary French Literature" which included French Canadian literature, too. His contributions were made at the request of the publisher of the paper, Adam Wislicki. Limanowski was interested in the naturalist trend in French literature, reviewing the works of writers like Zola, Daudet and Loti, as well as those of the feminist Madame Leo (Leonia Champseix). To her Limanowski pointed as a desirable example for Polish women to follow.[9]

His "Correspondence from Switzerland"[10] was divided into two parts. First, he acquainted his readers with the general characteristics of a given Swiss canton or city, like scenery, climate, cultural or political and socio-economic affairs, or concentrated on a certain political or social problem—for instance, labor legislation—while in the second part he analyzed current events in Switzerland. He always researched his subject thoroughly, in order to persuade his readers of the validity of his own political viewpoint. Writing about Switzerland, Limanowski praised it as the only "welfare state" then existing in Europe, where the government was truly concerned with the well-being of the masses. He assessed Swiss political institutions as admirable and worthy of imitation. Inspired by the Swiss example, he now glorified political freedom as the most important factor in producing general progress as well as the welfare of the common people. He was convinced that the worker in Switzerland was not interested in any violent social upheaval, both because he enjoyed a high standard of living and benefited from the country's progressive labor legislation. Though a haven for international revolutionary organizations, Switzerland itself, in Limanowski's opinion, was nevertheless immune to social revolution.[11]

Again, Limanowski was impressed by the vitality of the Swiss federation. He concluded that common history and common institutions tended to produce in Switzerland a similarity of outlook and aspirations, shaping a single "national character," in spite of the ethnic and religious diversities of the constituent peoples of this federal union.[12]

In addition to writing for the *Polish Week* and the *Weekly Review*, Limanowski during 1879 and 1880 also reported on the developments in the contemporary Russian and Polish socialist movements for Eduard Bernstein's *Jahrbuch für Sozialwissenschaft und Sozialpolitik*.[13] In commenting on Polish affairs it was Limanowski's contention that the masses themselves must become an independent and class-conscious political force. At times Limanowski tended to idealize the masses, especially the French; though this was in contradiction to his realistic appraisal of their behavior in the course of a social upheaval. He felt that if only Poland had such articulate and intelligent workers as in France, the Poles would soon become able to achieve economic prosperity and a politically independent existence.[14]

He felt that the task of creating an autonomous political movement among the masses should be taken up by the Polish democrats—specifically Galician democrats, in this context—but he was not clear as to whether the democrats should adopt a socialist program. Perhaps he could not be explicit while writing for the non-socialist *Polish Week*. He did stress nevertheless that the cause of Poland demanded an alliance with the international socialist movement. According to Limanowski, the democrats in Poland had now lost their former ability to diagnose social ills; they were no longer advocates of a popular reform cause which would place them alongside the masses and distinguish them from the conservatives. It is not clear from this article precisely what political role Limanowski wished the Polish democrats to assume. Were they to become allies or leaders of the people? Like the *émigré* democrats prior to 1863, Limanowski oscillated between the notion of the people as the instrument and the notion of the people as the agent of progress.[15]

During 1879 and 1880 Limanowski derived his income from his translations of *The Elements*. On receiving a small inheritance in the spring of 1879, Limanowski purchased Polish type abroad, set up a printing enterprise in partnership with Trussov, and began to translate *The Elements*. Aided by his wife in the task of translation, Limanowski was generously rewarded for his effort by the author Drysdale who, having read Limanowski's treatise on More and Campanella, was particularly impressed by Limanowski's treatment of More's *Utopia*; when *The Elements* appeared in the Polish translation by Limanowski, Drysdale considered it *"admirable."* (He corresponded with Limanowski in French, but was able to read Polish with the aid of a dictionary.) With two other editions to its credit by this time—in a number of different languages—this book became very popular in Russian Poland.[16]

A major reason for its popularity lay in the fact that Drysdale, a disciple of Ricardo, J. S. Mill, and especially Malthus, advocated in it the controversial notion of birth control as the only means to eradicate poverty. Limanowski agreed with Drysdale that birth control was essential for improvement of the living standards of the masses; it was his contention that a socialist regime which ignored the problem of overpopulation would only make poverty universal. He assumed that at present the industrial worker was not interested in limiting his family, because he had no goods to pass on to his heirs; he argued that family planning—an aspect of overall socialist planning—would be possible only under a socialist regime, because only then would it be possible to reconcile individual interest with public interest by an intensive education of the populace as well as by a redistribution of income.[17]

Meanwhile, by setting up a Polish printing press Limanowski hastened, albeit indirectly, the appearance of the long-awaited socialist periodical which was named *Equality (Rownosc)*. However, because Trussov and Limanowski paid their workers on the basis of the high standard rates in Geneva, Dluski's group, who were sponsoring the project, decided that they could not afford the cost of printing in Trussov's and Limanowski's shop. Eventually, they set up a rival Polish printing enterprise to publish *Equality* as well as other political literature. However, Limanowski became a member of the editorial board of the *Equality*; and in the third issue in the fall of 1879 there appeared his article on the agrarian problem. The article, which was inspired by a recent prediction of famine in Galicia made by the conservative *Time* published in Cracow, is significant in revealing Limanowski's views on this issue as they had developed around 1880.[18]

In it Limanowski explained that the recurrence of famine in Galicia was a result of rural overpopulation and this, in turn, was due to the predominantly agrarian economy of the region, where 83.5% of the population derived its livelihood from agriculture. He attributed the low productivity of Galician agriculture to the prevailing socioeconomic conditions, rather than to the nature of the soil or to the character of the people. In Galicia 2% of the population owned 40% of the agricultural land, whereas 23% held the remaining 60%. Seventy-three per cent of the peasantry were landless. The predominance of dwarf farming ruled out improvements which could have produced sufficient food to satisfy at least local needs. Limanowski correctly noted that there was a much greater "land hunger" in Galicia than in Russian or Prussian Poland.[19]

However, he was opposed to any measures which would limit the excessive subdivision of peasant land in Galicia; he felt this would only increase the rural proletariat and thus magnify the problems of the overpopulation in the village. He was convinced that even the most equitable redistribution of the agricultural land in Galicia would only result in perpetuation of the unproductive dwarf farms of about five morgs each (i.e., less than ten acres). This would eventually produce the reassertion of inequality in the village. Limanowski alleged that of all the regions in the former Polish-Lithuanian Commonwealth Galicia was now the most "ripe" for agrarian socialism. His contention was that abolition of the rural proletariat and increased productivity would be possible only under a system of land socialization within the framework of the rural commune, where all income from the communal soil—apart from taxes paid to the state and the local government—would accrue to the peasantry. He envisaged the pooling of individual peasant holdings to form large cooperative farms; but he failed to state clearly what would become of gentry land under his scheme, some of which at least would have been needed to absorb the numerous rural proletariat in Galicia. At the same time, without the rural proletariat, gentry farming would have been almost impossible in Galicia. Yet Limanowski expressly opposed forced expropriation. Probably, he naively anticipated that the gentry would voluntarily part with at least some of their land, without compensation. He concluded his article with the following vague remark: "Let us remember that we are not contemplating a division of land but, rather, a socialist organization of labor."[20]

Despite his contributing to *Equality* and being a member of its editorial board, Limanowski clashed with Dluski's group due to their opposition to the slogan of independence. Shortly after his arrival in Geneva in October 1878, Dluski's group began transforming the so-called "Program of Polish Socialists" drafted earlier that year by their comrades in Warsaw. The new version was printed in Geneva in the first issue of *Equality*, which appeared in the fall of 1879. It was now named the "Brussels Program," in order to confuse the police in Russian Poland.[21]

The main body of this document consisted of eight points. Point 1 aimed at an all-round development of the individual in the socialist society, as preached by Marx. Point 2, under the influence of Proudhon and his followers, envisioned socialization of the means of production and of labor in industrial and rural associations and, in the spirit of anarcho-syndicalism, contemplated collective possession of the means of production by the workers, rather than by the nation—as in the partly Lassallian "Gotha Program" formulated by the German Social Democratic Party in 1875. Point 5 envisioned a universal revolution and Point 6 advocated, for tactical reasons, a loose federation with foreign socialists. Point 7 merely hinted at political action to implement socialism, assigning this task vaguely to "the people themselves, conscious of their rights and interests, under the moral leadership of a popular organization." Finally, Point 8 expressed the principle of "*moral* consonance of means and ends" (my italics). Thus, Points 2, 5, 6,7 and 8 show an unmistakable influence of anarchism. Although Dluski had been exposed to the "Gotha Program" while collaborating with the Southern Union of Russian Workers in Odessa in 1875, the influence of the Lassallian ideology on the "Brussels Program" was slight. It was confined perhaps to Point 4, which vaguely demanded "social equality" derived from the idea of "*égalité*," as expressed in the triple slogan of the French Revolution.

Moreover, the authors of the "Brussels Program" were more cautious than those of its German predecessor when defining the criterion for distribution of income; the "Gotha Program" had employed the vague formulas of "fair distribution" and "equal rights" and for this, among other points, it came under Marx's attack. According to Point 3 of the "Brussels Program," "each individual has the right to benefit from the product of associated work, which right would be defined in the future on the basis of science by the workers themselves." (This appeal to "science" as a remedy for social ills was characteristic of Warsaw Positivism.) Finally, the "Brussels Program" differed from the "Gotha Program" in that it appealed to peasants as well as to city workers, considering both to make up the "people."[22]

Limanowski had taken part in the discussions which preceded the publication of the "Brussels Program" and included the Russians Lavrov and Zhukovsky; there is no evidence to suggest that he disapproved of this document as such. However, he opposed the elimination by the *Equality* group of several patriotic causes, which appeared in the original. The *Equality* group, like Limanowski when he was in Galicia, objected to insurrectionary conspiracies. They wished to conserve the national strength for a social revolution which would abolish social as well as national oppression; such revolution they anticipated in the very near future. Yet, unlike Limanowski, they failed to reconcile patriotism and socialism in their ideology. Theirs was an apolitical program, without any immediate political demands and without reference to independence. Therefore, the "Brussels Program" did not fully satisfy Limanowski. However, as a pragmatist, he felt that it was futile to argue over words and shades of meaning, and hoped that in time Dluski's group would abandon their anti-patriotic standpoint.[23]

Nevertheless, in November 1879, shortly after the appearance of the revised program, Dluski prepared an editorial for *Equality*, en-titled "Patriotism and Socialism," in which he attacked the Polish democratic tradition. In it he charged that hitherto the upper classes in Poland had been utilizing the people as their tool in the struggle for independence. Having given credit to the true patriots in the past for their idealism, Dluski rejected the idea of insurrection as now re-actionary and, as such, incompatible with socialism.[24] Thereupon, Limanowski threatened to resign from the editorial board, should the article appear as scheduled. After a lengthy dispute, Limanowski agreed to remain on the condition that the article would be signed by Dluski as an expression of his personal opinion, rather than that of the editorial board as a whole. However, Limanowski's collabora-tion with the *Equality* group became henceforth only nominal.[25]

At this time, Warynski and thirty-four of his comrades—including Mendelson and Kobylanski—awaited trial in Cracow; they had been arrested there in February 1879. The prisoners were held by the Ga-lician authorities on the charge of "high treason," eventually reduc-ed to "disrupting public peace." In the course of the trial, which lasted from February 16 to April 15, 1880, and influenced consider-ably the development of socialism in Galicia, the defendants were subjected to unconstitutional pressure by the police, who attempted to extort confessions from them in order that the jury might return a verdict of "guilty." Limanowski publicized these irregularities in Prince Petr Kropotkin's *Le Révolté* and in Becker's *Le Précurseur*, and sent his articles to the Minister of Justice in Vienna, as well as to the Viceroy in Lvov, in an effort to help the accused. The indict-ment as printed in the official *Lvov Gazette (Gazeta Lwowska)* in-cluded numerous references to Limanowski and the previous social-ist trial which had taken place in Lvov in January 1878. During a public session the prosecutor read passages from Limanowski's *Socialism*, stressing that some of the defendants, for example Waryn-ski and Mendelson, were critical of Limanowski's pamphlet and ob-jected to his patriotic socialism. The intention of the authorities since the arrests and in the course of the trial was to make public opinion hostile to the accused on account of their lack of patriotism. Hence Limanowski's reluctance to make his dispute with the *Equality* group public by an open break over patriotism; he feared that this would play into the hands of the authorities and harm the case of the defendants. However, the jury brough in a verdict of "not guilty," though two defendants—Mendelson and Kobylanski—were subjected to brief imprisonment for illegal residence in Cracow, after having been expelled previously from Galicia. Those among the acquitted Polish socialists who were Russian subjects had to leave Galicia immediately. Eventually, Warynski, Mendelson and Kobylanski joined the *Equality* group in Geneva.[26] Of his young socialist comrades, Limanowski was to remain on good terms longest with Kobylanski, due rather to their similar political views than to personal affinity.[27]

The final split between the *Equality* group and the handful who supported Limanowski's orientation took place immediately after the fiftieth anniversary of the November (1830-1831) insurrection. In November 1880, two celebrations of this event were slated in Geneva; one organized by the *Equality* group and the other by the Polish Society for Mutual Assistance with Milkowski as chairman and main speaker. Limanowski abstained from attending either of these meetings, so as to remain impartial.

The meeting sponsored by the *Equality* group was the largest public assembly organized by émigré Polish socialists in the West during the 1880s. In their formal invitations which they issued to foreign guests, the *Equality* group stressed that the old democratic slogan "Long live Poland!" had been superseded by the new socialist slogan "Proletarians of all nations, unite!" Most of the foreign guests were anarchists or sympathizers of anarchism. One of the speakers at the meeting was the famous Prince Kropotkin. The other two Russians who spoke were Vera Zasulich and Nikolai Zhukovsky. In principle, they all agreed that the struggle for the liberation of Poland must be subordinated to the international struggle for socioeconomic liberation of all mankind, the principal goal of both socialism and anarchism. But Marx and Engels, and a few other former members of the General Council of the now defunct First International, glorified the Polish democratic tradition in a letter which they sent to Geneva from London especially for the occasion and which ended with the old democratic slogan "Long live Poland!"[28]

The most prominent speech from the Poles present at the meeting was that of Dluski, who admitted the influence of patriotism among the Polish workers, but this patriotism—he charged—the upper classes had exploited to safeguard themselves against social revolution. He ended his speech with the cry, "Down with patriotism and reaction!"; "Long live the International and Social Revolution!"[29]

Hitherto, Limanowski attempted to reconcile his friendship with the post-1863 patriotic socialist *émigré* Wroblewski and his regard for democrats like Milkowski and Giller, with loyalty to the *Equality* group. Now he felt compelled to dissociate himself from the latter. The Geneva meeting itself was a watershed in the history of Polish socialism; it resulted in the polarization of the two opposing tendencies in the Polish socialist movement—the internationalist left and the nationalistic right.[30]

From 1878 to 1883 the Polish revolutionary group in Geneva were to act as the "general staff" for the labor movement in the home country. (In 1882 Warynski of the Geneva group organized in Warsaw the first Polish socially revolutionary party, the Proletariat, which he initially led until his arrest in Septmber 1883, and which was finally suppressed by the Russian authorities in 1886.) The activity of the *émigré* left was generally confined to printing socialist literature, to be smuggled into Prussian and Russian Poland, where socialist activity was perforce conspiratorial. However, in Galicia there was even a legal paper, *Labor*. Therefore, the conspiratorially oriented Geneva group, led by Mendelson after the departure of Warynski, assumed that their principal task in Galicia was to combat the democratic socialism of the legally operating *Labor*, with which Limanowski sympathized.[31]

Immediately after the departure of Warynski from Lvov in December 1878, the editorial board of *Labor* (Mankowski, Czerwienski, Daniluk, Ludwik Inlaender, Franko and Pavlik) attempted to organize here a legal socialist party, but their efforts to this end were frustrated at that time by the authorities and the employers of Galicia.[32]

In January 1881, Czerwienski, Inlaender and Franko prepared a pamphlet entitled "Program of the Polish and Ukrainian Socialists in Eastern Galicia," and sent it to Limanowski for printing. In the course of printing, the word "Ukrainian" was omitted from the title; but Limanowski claimed that the error was unintentional and was due to an oversight on the part of his partner Trussov. However, Trussov denied this allegation. Limanowski altered the title to read, again incorrectly, the "Program of the Galician Socialists." Therefore, due to this misunderstanding with Limanowski, the editorial board of *Labor* sent their next program (of May 1881) for printing to his rivals—Mendelson, Dluski and Warynski.[33]

During the next ten years the Galician socialists attempted to reconcile the social radicalism of Warynski with Limanowski's nationalistic ideology aimed at resurrecting a federal Poland within the boundaries of 1772.[34]

Limanowski clashed on the issue of boundaries with the prominent Ukrainian exile Drahomanov, who published in Geneva a Ukrainian periodical, the *Community (Hromada)*, which appeared from 1878 to 1882. In the first issue of *Community* Drahomanov, while surveying the struggle against Polish rule in the Ukraine, condemned the post-1832 Polish democrats, as well as contemporary patriotic socialists like Limanowski, for wishing to resurrect the old Poland within the eastern boundary of 1772. Drahomanov protested, among other things, Limanowski's reference in his *Socjalizm* to Galicia as a "Polish land." In response to this criticism, Limanowski wrote a letter on November 28, 1878, explaining his position, and Drahomanov eventually printed it in the *Community* in 1882.

In his letter Limanowski declared himself in favor of holding a plebiscite in the eastern borderlands, after the liberation of Poland from foreign yoke, to determine whether the borderland nationalities wished to maintain their connection with Poland. He admitted their right to secede, if they so desired. He told Drahomanov that as a social democrat he was against any form of national oppression. However, as a Pole born in the eastern borderlands among a Latvian population, he hoped for a revival of the Union of Lublin; also, as he explained, the Swiss federation had convinced him that several nationalities could happily coexist within a modern state.

Moreover, he assured Drahomanov that statesmen in the future reborn Poland were bound to pursue an enlightened policy on the nationality issue. He was convinced that, as he was himself, they too would be aware of the rise of national consciousness throughout Eastern Europe, and would realize that disaffected ethnic groups—in a country like Poland existing between two powerful unfriendly neighbors, Russia and Germany—might well threaten the internal order and external security of the state by collaborating with such enemies. Limanowski told Drahomanov that, due to these considerations, Polish statesmen were bound to recognize the need for granting broad autonomy to all nationalities in the new Commonwealth. Anyhow, it was Limanowski's belief, that in the new Poland it would be the bond of a common ideology—socialism—which would unify the constituent nationalities.[35]

But Drahomanov was not satisfied with Limanowski's reply; his concept of a Slav federation which was based on Russia was not at

all the same as Limanowski's idealized Union of Lublin. Drahomanov contended that Polish socialists must discard their harmful myths
based on an idealized historical tradition, in order to find a common
ground with the Ukrainian socialists in the struggle for social and political liberation.[36]

When Limanowski began to give lectures in Geneva around 1880,
with the purpose of convincing his friends from Russia that the 1863
insurrection was a socially progressive struggle, it was Drahomanov
the Ukrainian who was his most vehement critic and who doubted
whether the insurgents were actually in favor of social justice and
nationality rights for the Ukrainians. The Great Russians, especially
Kravchinsky and Zhukovsky, were more sympathetically inclined
toward Limanowski's lectures, as well as toward Polish aspirations
in 1863, than was Drahomanov.[37]

Drahomanov indeed felt that the enmity which the Ukrainian
peasants had manifested in 1863 toward the Poles was justified; they
had seized the insurgents and handed them over to the Russian authorities. He was convinced that the insurgents in Ukraine were drawn
almost exclusively from the Polish gentry and, as such, were enemies
of the Ukrainian people. Limanowski, on the other hand, contended
that they were usually city artisans and *déclassé* nobility, rather than
members of the landed gentry class. Therefore, in his opinion, the
Ukrainian people should have believed in the promises of the National Government in 1863, for it was, in the words of Limanowski,
"morally responsible for the outcome of the insurrection." We may
add that since the overwhelming majority of the Poles in the Right-
Bank Ukraine were gentry, it was unrealistic of Limanowski to dismiss their attitude as being of no significance for implementing the
insurgents' program—peasant emancipation with land and cultural
autonomy for the Ukrainians. Limanowski believed that, had the
Ukrainian peasants supported the Polish insurgents and they had
won, the gentry in the Right-Bank Ukraine would have been paid
compensation for peasant emancipation with land from the national
fund; and thus the burden of compensation would not have been
borne exclusively by the peasants themselves, as actually happened.
In Limanowski's opinion, it was in the interest of the Ukrainians,
too, to struggle for freedom against the Tsarist autocracy alongside
the Poles. He assumed the sincerity of the left "Reds"—who formed
the most socially radical faction among the insurgents—believing in
their commitment to the idea of social justice and national rights for
the Ukrainians. He expected the left "Reds" to have been in a position

to implement their program, after a joint Polish-Ukrainian victory. In the existing conditions, his speculations were entirely unrealistic.[38]

On the basis of these bi-monthly lectures, Limanowski in 1882 published a history of the 1863 uprising.[39] Always interested in the events of 1861-1864, he read avidly on the subject and listened to eyewitness accounts as well. Therefore, when he was asked to write a monograph on the uprising by a Lvov bookseller and publisher, Adam Bartoszewicz, Limanowski readily accepted Bartoszewicz's terms. Due to his reputation in Galicia—where the authorities considered him a "militant socialist"—Bartoszewicz published the first edition of Limanowski's book anonymously.[40]

This work represents the first systematic history of the 1863 insurrection in Polish historiography. An earlier attempt had been made by Agaton Giller who, however, produced an account that was fragmentary and incomplete, and his interpretation had not satisfied either the "Reds" or the "Whites."[41]

In his book, Limanowski attacks the prevailing current of "loyalism" and "organic work." Thereby he was pioneering a radically new historiographical trend in Poland, stressing the interconnection between the movement for independence and agrarian reform. His interpretation—relating the past to the present politically and combining the romantic tradition of struggle for independence with a radical social orientation—was to remain the most popular one in the Polish democratic camp until the outbreak of the First World War. The new generation which matured in the 1870s and early 1880s in Poland had little knowledge of Polish history since the partitions. The epoch preceding 1863 was often misunderstood and misrepresented. Limanowski attempted to remedy this. Although his pioneering effort was ignored in "official" historiographical circles in Poland, the reaction to his book on the part of the Polish young people was most enthusiastic. The monograph, as well as his later writings on Polish struggles for independence, assured him more popularity than anything else he had ever done.[42]

Limanowski wrote on the basis of recorded documents, oral testimony and from his own recollections; his bibliography was not extensive. In the nature of an outline only, this monograph—as well as his later writings on the 1863 insurrection—contains some minor errors of fact. However, his principal merit was to utilize for the first time the *émigré* literature of the left-wing. Like Lelewel, Limanowski believed that historical truth was the product of a scientific method combined with commitment to the right cause. Writing as an "epigone" of the left "Red" political camp, Limanowski alleged that the insurrection could have succeeded with mass support; that peasant participation was considerable and could have been even greater but for the counter-revolution of the "Whites"; and that the agrarian reform of 1864 in Russian Poland—effected on better terms to the peasants than elsewhere in East-Central Europe—was forced upon the Russian Government by the socially radical Polish insurgents.[43]

In contrast to the democrat Giller, who was a spokesman for the right "Red" orientation, Limanowski spoke for the left "Red" political camp. He objected to Giller's patronizing and suspicious attitude toward the peasants, and to his half-hearted commitment to the idea of a *levée en masse* in 1863. It was Limanowski's contention that in the cause of national unity the gentry should have supported the social program of the left, should have radicalized the peasant, and identified themselves with the people. He felt that only a mass "people's revolution" would have succeeded in the struggle against the partitioning powers. He dismissed the possibility of a social revolution directed by peasants against the gentry, the thought of which worried the conservatives and even democrats like Giller. Because he criticized the "Reds"—as well as the "Whites"—for their conduct of the insurrection, Limanowski's friendship with Giller became somewhat strained after the appearance of his monograph in 1882. However, in spite of his opposition to Limanowski's socialism, Giller valued him much above the Polish internationalist socialist group in Geneva; he appreciated Limanowski's patriotism and was always ready to acknowledge his good intentions.[44]

In 1882 we find Giller writing to Limanowski: "I am very grateful that you have recognized the need to remind our young people of their duty to be patriotic, by means of your "Social-Democratic Library."[45] The "Social-Democratic Library" was a series of pamphlets which Limanowski began issuing in October 1880, both in order to augment his income and to develop his own social ideology as an alternative to that professed by the *Equality* group. Valuing the old Polish democratic tradition, Limanowski had come to feel that the democratic movement which inspired the uprisings of 1846, 1848 and 1863 must adapt to the new socio-economic conditions by making democratic socialism its own creed; otherwise, it would perish. In his pamphlets he defended the old democratic dogmas, which aimed at enlisting the people in the struggle for independence, as well as advocating the socialist ideals of Lassalle.[46]

From 1881 to 1883 Limanowski published in all five pamphlets in this series. He provided forewords respectively to a Polish translation of Wilhelm Liebknecht's speech of October 22, 1871 at Crimmitschau, Saxony, defending democratic socialism, and to a Polish translation of Becker's *Manifesto to Farmers (Manifest do ludnosci rolniczej)*, issued under the auspices of the Swiss branch of the First International and presenting a Marxist interpretation of the agrarian problem in Europe. (I shall deal with this problem in my next chapter.)[47] Also, in this series Limanowski published three original pamphlets by himself: a brief biographical sketch of Lassalle and, more significant, two theoretical works—*Patriotism and Socialism (Patriotyzm i Socjalizm)* and *Political versus Social Revolutions (Polityczna a spoleczna rewolucja)*. Of the series, the last two pamphlets were most widely read.[48]

In *Patriotyzm i Socjalizm* Limanowski, influenced here by Spinoza—perhaps also by Hobbes—argued that it was the instinct of self-preservation, egoism, that was at the root of man's sociability (which he equated with patriotism). Limanowski identified man's social instinct and patriotism; and he distinguished between an unselfish patriotism and chauvinism. Patriotism represented for him a basic social bond, without which neither the individual nor a nationality could survive. Like many nineteenth century thinkers from Mazzini and Mickiewicz onward, Limanowski argued that patriotism was always consistent with the highest form of internationalism, just as egoism was compatible with patriotism. Limanowski did not realize that, like Hobbes, he was also stating here a theory of clear-cut individualism, incompatible with his notion of society as a single organism.[49]

Limanowski, however, recognized that the more equitable the distribution of income the happier the people, and the more "organic" is the social bond within a nation. For him socialism represented a means of enhancing national unity by removal of economic inequality and oppression. He argued that, at the basis of all oppression was economic oppression: "Private property is at the root of economic oppression. . . . [It is] a terrible weapon, by means of which man exploits man physically, mentally and morally. . . . To change this terrible weapon . . . into a means to . . . freedom and brotherhood is the real task of socialism."[50]

Yet he still laid stress here on the primacy of national liberation. He was convinced that the common people in Poland suffered more under foreign rule—Russian and German—than did the educated

classes. By forcing the people to attend schools where the language
of instruction was alien to them, a foreign regime lowered educational standards and impeded intellectual development of the masses;
by conducting its business in an alien language it oppressed the
people; and it armed brother against brother. Restating here the old
democratic argument directed against the "reactionary" powers of
the Holy Alliance, Limanowski justified reconstruction of Poland
not only as a prerequisite for the welfare of the masses in that country, but also as essential to general progress in Europe as a whole.[51]

When he wrote his *Socialism* in Lvov in 1878, Limanowski had
been not merely defending socialism, but writing an apologia for
himself as a socialist; conversely, in his *Patriotism and Socialism* in
Geneva in 1881 his primary object was both to defend patriotism
and to write an apologia for himself as a patriot. In this last work he
contends, as in Lvov, that patriotism and socialism do not contra-
dict but, rather, reinforce each other. True patriots, who love the
people as the mainstay of the nation, perforce become socialists. On
the other hand, true socialists, because of their love for the people
must necessarily become patriots. But Limanowski no longer stresses
here, as he had done in his *Socialism*, that the abolition of the foreign yoke was necessary for victory of socialism in Poland; there is
no longer any organic interconnection between his two ideals of socialism and independence.[52]

This transition in Limanowski's view implied that—again disillusioned with the Russian revolutionary movement, due to suppression of the *Narodnaia Volia*—he was now about to revert to his earlier
belief that insurrection was the only means of securing independence.[53] Upon the suggestion of Kobylanski and Zygmunt Balicki—
both future activists in the rightist National Democratic Party in
Russian Poland—Limanowski decided to create a new socialist center
in Geneva, in order to rally Polish socialists who, like himself, envisioned insurrection as the immediate object of Polish socialism. To
this end, he drafted a program of his own in August 1881, giving
primacy to the goal of national liberation within the boundaries of
1772. He did not wish to imitate Western socialists by confining his
program to the industrial proletariat, being aware that the majority
of the people in Poland were peasants, and he followed the democratic tradition by appealing to all the working people in Poland. In
fact, he failed to distinguish clearly between the workers and petite
bourgeoisie.

It was Balicki who suggested the name for the new organization—the Association of Polish People (i.e., in effect the fourth *émigré* Polish People). Limanowski at first opposed this name, because he contemplated the revival of the old Polish-Lithuanian Commonwealth, comprising peoples other than Polish. He felt that it would be necessary to collaborate with them to shake off the foreign yoke, and he continued to hope that they would willingly federate with Poland. Led by Balicki, the majority of his group, however, favored revival of an ethnic Poland. After a lengthy debate, a compromise formula on the issue of boundaries was drafted, "Self-determination within boundaries of voluntary gravitation."[54] This formula came under attack, for its vagueness, from the editorial board of *L'Aurore (Przedswit)*, the successor to the organ of Limanowski's socialist rivals, *Equality*.[55]

In his program Limanowski fused the Lassallian concept of the mission of the "Fourth Estate" with the idea of Poland's mission, hoping for a Poland that would "march" in the vanguard of progress. In the preamble he viewed recent Polish history in terms of a struggle for independence rather than in terms of class struggle; he condemned the Polish bourgeoisie for its "loyalism" and "organic work" in addition to blaming it for exploitation of the workers. In the main part of the program Limanowski advocated nationalization of the means of production, the establishment of cooperatives, and the abolition of all privilege and inequality, as well as postulating political decentralization on the basis of self-governing communes. He now considered national liberation as important as socioeconomic liberation (unlike some of his supporters in Galicia, for whom independence was merely a means of struggle for socialism).

This program was an emotional document, with anarchistic undertones; it made no provision for immediate political demands. Thus Limanowski—unrealistically—intended to produce an umbrella socialist program for Poland that would rally all Polish socialists under his nationalistic banner. (He had started his Association with a membership of around five!) Hence, also his provision that the slogans to be used for socialist manifestations should be determined on the basis of local needs. His tactics were, however, the usual socialist ones, i.e., economic and political agitation (strikes, protests, demonstrations), the formation of both legal and illegal associations, and oral as well as written propaganda. But a program formulated in this vague manner had nothing to offer to the socialists in Galicia, who already had their "Program of the Galician Workers' Party" of May 1881, which was more logical and practicable than Limanowski's.[56]

This document was the successor to the "Program of the Polish and Ukrainian Socialists in Eastern Galicia" of January 1881, which—as discussed above—was sent to Limanowski for printing. The latter program was in turn based on a number of "minimum" clauses drafted by Mankowski, Czerwienski, Inlaender and Daniluk of the editorial board of *Labor* in December 1879, consisting of two parts: 1) pertaining to labor legislation and labor relations and 2) dealing with political issues like universal suffrage and reform of the press law. The program of January 1881, spelled out the aims and principles of socialism in general and specified the tasks to be accomplished by the socialists in Galicia. It envisioned nationalization of the key industries under the existing regime, and a peaceful transition to socialism. The demand for producers' cooperatives in this program reflected the influence of Lassalle. Its authors, i.e., Czerwienski, Inlaender and Franko, tended toward Limanowski's nationalistic socialism, but they differed from him in their appeal to the city worker and in their opposition to the revival of Poland within the eastern boundary of 1772.[57]

Due to its length and predominantly theoretical content, the program of January 1881, proved impractical for the purpose of mass agitation. Therefore, it was rewritten by the same authors in May 1881, and renamed the "Program of the Galician Workers' Party," mentioned above. It concentrated on issues in Galicia and stressed immediate demands based on the model of the "minimum" program of the Austrian Social Democratic Party. There were the usual provisions for freedom of speech, assembly and association, as well as for universal suffrage. However, Point 7 envisioned a broad autonomy of both urban and rural communes, thus tending toward anarchism—under the influence of Warynski rather than Limanowski.[58]

Due to Limanowski's misunderstandings with socialists in Lvov in the past, they were reluctant now to grant him a mandate to represent them at the international congress held in October 1881, at Chur (Coire), Switzerland. This meeting was an attempt on the part of Belgian socialists to revive the now defunct First International; it was presided over by Becker with an attendance of approximately sixteen delegates and several guests and was finally adjourned without accomplishing its purpose. Socialists in Galicia had wished to send their own delegates to Chur. Unable to do so, those in Cracow designated Warynski, who also represented Prussian Poland. (No delegates from Russian Poland were present, where the socialist

movement was temporarily suppressed.) At Limanowski's own request, the editorial board of *Labor* granted their mandate to him.[59]

En route to the congress, Limanowski visited Zurich, where he
met Bernstein, Becker, the French socialist Benôit Malon, and Count
Schneeberger. The latter, a Freemason synpathetic to socialism, was
a friend of Becker and resided in Vienna. They all accompanied Limanowski to Chur. Just before their arrival there, a landslide occurred
in a nearby village, causing considerable property damage. Thereupon, Limanowski suggested to his companions a token contribution of money toward the relief fund. Malon objected that their resources should not be expended for non-socialist purposes. Limanowski did not insist, but he resented then—and he was to resent
also on future occasions—the distinction which was often made, as in
this instance by his colleagues, between socialist aims and mere phil-
anthropy.[60]

As the representative of *Labor* as well as of his own organization,
the Polish People, Limanowski at the Chur congress (where he was
chosen secretary) presented both Galician programs of January and
May 1881, and his own program of August 1881. As a spokesman
for the Polish People, Limanowski clashed with Warynski and with
Dluski, the latter of whom attended merely as a guest and acted,
because of his fluent French, as the principal spokesman for the
L'Aurore group. Although they valued the Polish democratic movement in the past, Warynski and Dluski now considered the idea of
insurrection reactionary as well as Utopian—because they were convinced that neither the Polish middle class nor the peasants in Poland
desired independence—the bourgeoisie was contented with the status
quo for the sake of access to the Russian market, while the peasants
feared a return to serfdom, which they associated with the idea of an
independent Poland. At the congress Warynski criticized Limanowski's program, because it was formulated so as to apply for the whole
of Poland; instead, he advocated alignment of Polish socialists with
the socialists of the partitioning powers in each of the three regions.
Warynski objected to Limanowski's omission of a "minimum" program, pointing out that political decentralization, as proposed by
Limanowski, was not, as such, a guarantee of individual rights. Also,
Warynski's opinion was that Limanowski tended to confuse political
oppression with national oppression. He pointed out that Limanowski was now harking back to the slogans of the Utopian socialist
wing of the earlier Polish democratic movement; his program's appeal
to idealistic youth among the Polish intelligentsia and to the "populist

democrats'' was tantamount to a betrayal of socialism, because now democrats were necessarily either in the socialist or in the anti-socialist camp. Limanowski contended that insurrection in Russian Poland was inevitable, due to the intensified process of Russification which began in the 1860s; Warynski answered that Limanowski's stress on class solidarity in Poland, necessary for the struggle for independence, was incompatible with a social revolution. He considered national oppression "a detail to be solved in the course of a general solution of the social question." However, convinced that the Polish proletariat was patriotic, Limanowski felt that Polish socialism must appeal to the patriotism of the masses to flourish.[61]

Limanowski came under attack from Warynski, too, for his "unscientific" approach to socialism; Warynski opposed the parallelism of Limanowski's goals of independence and socialism as being contrary to Marxist historical materialism. He charged that Limanowski subordinated economics to politics, thereby subordinating socialism to the struggle for independence. On the other hand, Limanowski attributed the internationalist orientation of the *L'Aurore* group to the influence of Bakunin rather than Marx. Although valuing highly Marx and Engels as theoreticians of socialism, Limanowski approached their ideas in a critical spirit. He dismissed Marx's "surplus value" theory as metaphysics; and he interpreted the ambiguous Marxian legacy so as to admit patriotism and a peaceful transition to socialism. Although Limanowski accepted class struggle in the past as an historical fact, he rejected social revolution in reference to contemporary Poland, because he considered it disruptive of the national unity essential for a successful struggle for independence. He felt that Marxist economic determinism was conducive to fatalism and that this was bound to impede armed struggle for Polish independence. Limanowski's objection at this time to assigning priority to the economic factor in history by Marxists did not stem from a theoetical consideration on his part; rather, he thought exclusively in terms of tactics. While he believed that Marxists rigidly adhered to their theory, he felt himself that theory should be pragmatically used as a tool of analysis, and for the purpose of reflection, rather than solely as a guide to action.[62]

At the congress Dluski and Warynski made an attempt to persuade the other delegates that Limanowski was merely a patriot and not a socialist. Dluski alleged that Polish socialists had nothing in common with patriots; they were indifferent to the idea of restoring Polish independence. Malon then asked him, "Alors, vous rénoncez à votre nationalité?" Embarrassed, Dluski explained that he was not denying his nationality, but was merely opposing Limanowski's insurrectionist tactic aimed at reconstruction of the old Polish-Lithuanian Commonwealth. It was Dluski's contention at this time that socialists should not concern themselves with politics. The congress, however, resolved that the issue between Limanowski and Dluski (and Warynski) lay outside its sphere of competence. Yet it is worthy of note that Kautsky supported Dluski and Warynski, while Engels agreed with Limanowski. And Marx himself, too, understood Polish patriotic socialists like Limanowski.[63]

Unlike Warynski, Limanowski combined a pragmatic approach to socialism as a theory with a dogmatic commitment to the idea of armed struggle for independence. A writer rather than an organizer, Limanowski was to remain somewhat distant from the modern industrial workers; he was never to acquire practical experience of their economic needs. On the other hand, Warynski, as leader of the party Proletariat in Russian Poland, would strive for the immediate betterment of working conditions; unlike Limanowski, he proved more concerned with the day-to-day problems of the workers than with the anarchistic concept of "liberalization of the human spirit." But it must be stressed that notwithstanding some anarchistic tendencies in Limanowski's theory, he was always hostile to anarchism as a movement, and as an "apolitical" ideology. He dismissed the anarchistic concept of complete decentralization as impractical; he opposed anarchists like Bakunin for their hostility to the idea of reviving the historic Poland within the boundaries of 1772; and he came to resent the anarchist stress on universal revolution.[64]

Limanowski no longer awaited social upheaval in Europe—and Russia—as a means of reconstructing Poland; he was convinced that a peaceful transition to socialism was more likely in the independent countries of Europe than a violent social revolution. Moreover, the example of Switzerland had persuaded him that political democracy was a desirable end in itself, too. In 1882 he was confirmed in his evolutionary approach to socialism by further acquaintance with the organicist ideology of Schäffle. Limanowski, who at times deplored the egoism of the upper classes, nevertheless became convinced that society was organic; he rejected the Marxist concept of society inevitably split by class antagonism. How did he reconcile his organicism with socialism? He explained that the social organism as a whole was bound to become concerned with the injustice being done to its parts.[65]

He did recognize some of the obstacles to socialism, however. Limanowski now felt that it was difficult to legislate social transformation by fiat, because social change had as its premise long years of "arduous toil" aimed at overcoming the resistance of vested interests, superstition and custom. In his last "Social-Democratic" pamphlet entitled *Political versus Social Revolutions (Polityczna a spoleczna rewolucja)*, 1883, Limanowski defined "social revolution" as an evolutionary movement toward an egalitarian and just society to be based on collective ownership of the means of production. On the other hand, he interpreted the term "political revolution" to mean struggle for national liberation or, sometimes, a struggle for civil and political liberties.

Limanowski noted that most nations in Europe were sovereign whereas, apart from piece-meal reforms, none of them had embarked as yet on a really thorough program of social transformation. Mindful of the development of political liberalism in the West, he concluded that it was more feasible to accomplish a political change than a social reform;[66] and that the struggle to shake off a foreign yoke represented the "easiest" and most "popular" of all "revolutions."[67] But the distinction between "political" and "social" revolution that he made in his pamphlet was, in his opinion, merely a theoretical one. Each successful political revolution—or insurrection—brought social changes too; and the greater these were the more powerful and effective was the revolution.[68] In Poland, reiterated Limanowski, insurrection was the prerequisite for political democracy and for transition to socialism. Should a non-socialist party be formed in Poland with the goal of an armed struggle for independence, then Polish socialists ought to collaborate with it in the national cause, while at the same time attempting to influence its social program.[69]

He argued that a constitutional regime, as such, provided no safeguards against national oppression. For instance, in Prussian Poland schools and courts served the ends of Germanization; he noted also that the Irish, who enjoyed similar political freedoms to those of the British, strove nevertheless for national independence.[70]

Limanowski always felt that independence was not just a means of struggle for socialism, but a moral imperative in its own right. Now he assumed explicitly that independence was the primary goal for socialism in Poland. Henceforth, in practical politics he would subordinate socialism to the struggle for independence. His patriotism would find expression at times totally unrelated to his socialist goals; rather than stress that political freedom and independence were necessary for the emancipation of the proletariat in Poland, he would merely assume that in order to become "respectable" and popular socialists ought to be patriots and that, further, independence was as important to the Polish proletariat as to the nation as a whole. Hence his optimism that the people would understand this and follow the educated classes in the struggle for independence.

Limanowski's writings proved fairly popular among socialists and democrats in Poland, particularly in Galicia, even though in Galicia—and elsewhere in Poland—the influence of the Polish people had to compete with the socially revolutionary internationalism of Warynski.[71] According to one official account of the Polish socialist movement operating in the territories administered by Austria, fourteen months after the split had taken place in Geneva within Polish *émigré* socialism, the situation in Vienna and in other Austrian or Galician towns reflected the situation within the Polish socialist movement abroad. The official report notes with satisfaction that the mutual hostility between the followers of Limanowski and those of Warynski had now everywhere rendered Polish socialism ineffective.[72]

There is a consensus in Polish historiography that the Polish People played a very insignificant role within the Polish socialist movement, though it represented a political trend which in the 1890s was incorporated into the ideology of the Polish Socialist Party (PPS). However, in the early 1880s we do find some flickers of activity on the part of the Polish People. In 1881 they asked some of their compatriots in Paris to intervene with the French authorities, so as to prevent the extradition of the Russian Lev Hartmann who, having plotted against the life of the Tsar, had found asylum in France. The group also issued a proclamation to the French press, written by Limanowski and dated March 6, 1882, which protested the execution of several Russians by Tsarist authorities and expressed solidarity with the Russian revolutionaries in their struggle against the autocracy.[73]

Alienated from the extreme Polish left, Limanowski nevertheless still felt closer to it than to his fellow countrymen on his right. A collection of essays had been published in 1882, on the occasion of the twenty-fifth anniversary of Milkowski's writing career. This book contained two articles directed against Polish socialism; one by a Warsaw economist W. Wscieklica and the other by Swietochowski, the leader of Warsaw Positivism. In an attempt to defeat the Polish Marxists (i.e., the *L'Aurore* group and the Proletariat) by their own weapon, Marxism, Wscieklica accused them of disregarding the lessons of their master; he alleged that the situation in Poland was not yet conducive to a revolution, where the Polish working class was small and fairly well-off. Also, he attributed to them lack of patriotism—on the basis of the article by Dluski, discussed above. Thereupon, in solidarity with the entire socialist camp in Poland, Limanowski in 1883 wrote an open letter to Wscieklica, in which he compared Dluski's patriotism with that of Swietochowski. This comparison of course favored Dluski. Limanowski argued that although—like Swietochowski—Polish Marxist socialists now opposed an armed struggle for independence, they were nevertheless patriots both in virtue of their concern to fight the economic exploitation of the masses, who formed the mainstay of the nation, and because they had not ceased to value Polish national culture and traditions.[74]

Meanwhile, the *L'Aurore* group shifted their headquarters to Paris, where in May 1884, they started to publish a new theoretical organ, *Class Struggle (Walka Klas)*. Even before their departure from Geneva, Limanowski had ceased to associate with his former comrades. Nevertheless, in the summer of 1884 he was approached by Maria Jankowska and Stanislaw Kunicki (a member of the Executive Committee of the *Narodnaia Volia* and the leader of the "Proletariat" after Warynski's arrest in September 1883), in their capacities respectively as delegate of the Polish socialist group in Paris and of the home organization based in Warsaw. They told him of their recently concluded alliance with the now declining *Narodnaia Volia*. But Limanowski considered this tantamount to subordinating the Polish Proletariat to the Russian organization. In spite of this overture, no *modus vivendi* between Limanowski's followers and the *L'Aurore* group proved feasible henceforward, because he differed from them fundamentally in his approach to Polish politics. During the balance of his residence in Geneva in 1884 and 1885, Limanowski—preoccupied with his domestic affairs—was quite inactive politically.[75]

CHAPTER V

BOLESLAW LIMANOWSKI AND THE BIRTH OF THE POLISH LEAGUE
(THUN AND ZURICH, 1885-1889)

Soon after his arrival in Geneva late in 1878, Limanowski resigned himself to being permanently barred from returning home by the
Russian authorities, but still hoped to return to Galicia eventually.
At times he experienced a longing to abandon city life in Geneva, in
order to devote himself to farming. However, he was determined to
ensure a good education to his boys, and he felt dutybound to complete his history of recent social movements in Poland and in the
West, which he had begun in Lvov, in 1877. Therefore, he dismissed
his longing to settle in the country as an impractical idea.[1]

In the spring of 1881 Limanowski obtained 2,000 rubles from an
aunt; this sum compensated him partly for his share in the family
estate, easing his difficult financial situation. His brothers Jozef and
Lucjan had now agreed to divide the estate among themselves, having assessed his share at 10,000 rubles. He received half of this
amount in 1883—i.e., 5,000 rubles—which was equivalent to the
then considerable sum of 20,000 Swiss francs.

In 1885 he decided to apply this capital toward the purchase of a
photographic studio which was offered for sale in Thun, near Bern.
To this end, he formed a partnership with a certain Romuald Kom-
powski, a Pole whom he had met while in exile in north Russia, and
with a photographer from Lvov named Breiter.[2] Limanowski arrived
in Thun in June 1885, a month after his partners opened the studio
for business. The delay in his arrival was due to his beloved wife's
grave illness; since 1883 she had been suffering from cancer. Wincentyna died on November 15, 1885, at the age of forty-seven. Just
a few days earlier she had begged her medical specialist to prolong
her life for a few more years, for the sake of her children and "pour
courir et travailler." These words became engraved in Limanowski's
memory, because they summed up her personality so well. He realized that he had lost also an intellectual companion and mentor, for

Wincentyna Limanowska was among the best educated Polish women of her time and her husband's writings owed much to her criticism and insight. Now Limanowski was left alone with three little boys, the youngest of whom was four years old. His precarious financial situation complicated his life, too; this was due to the lack of a sound business sense of his partners, to the heavy financial obligation which he assumed when purchasing the studio, to the severe competition which he encountered in Thun and, finally, to the expenses incurred by him because of the lengthy illness of his wife.[3]

To dull the pain of his personal loss, Limanowski occupied himself with his history of social movements in Poland and in the West. He wrote during the early morning hours, i.e., between 4 a.m. and 8 a.m., when his mind was at peace. Later on, he was tormented by business problems which impeded him from concentrating; nevertheless, he attempted also to write some articles and reviews, to augment his income. His financial situation continued to be difficult; at times he even lacked a postage stamp for correspondence with Lucjan. It was this faithful brother who sustained Limanowski during his difficult months in Thun, by sending him small sums of money in regular installments.

In writing his history of social movements in Poland and in the West, Limanowski used sources which he obtained from his Russian friends in Geneva, Trussov and Elpidin. He also imported books from Lvov; his distant cousin Boleslaw Wyslouch, who had resided in Lvov since 1884, sent him works by Marx.[4]

Late in 1885, Wyslouch invited Limanowski's contributions to his projected organ, the *Social Review (Przeglad Spoleczny)*, the first issue of which appeared in Lvov, in January 1886. Twenty years younger than Limanowski, Wyslouch was born in Polesie (i.e., western Belorussia) of Polish medium gentry stock. As a student in St. Petersburg he had come in contact with Russian populism and embraced a peasantist creed rather than Marxism.[5]

Wyslouch developed his ideology in the *Social Review* in a series of articles entitled "Programmatic Sketches." He defined his program as being "populist." He aimed first of all at the welfare of the masses, but he wished to enlist them in the cause of Poland and progress, too, attempting to reconcile thus Polish national aspirations with a democratic social creed. So far he agreed with Limanowski. However, Wyslouch opposed Limanowski's idea of resurrecting the old Poland within the boundaries of 1772. He was critical of these boundaries—both in the east and in the west; he deplored the fact that the eastern border had partitioned the Ukraine, while the western border had excluded Silesia and Prussian Mazuria. He pointed out that, unlike Ukraine and Belorussia, these lands were inhabited by a Polish peasantry; Silesia was even then experiencing a national revival.

Wyslouch argued that the union of Lithuania and Poland, cemented in Lublin in 1569, proved beneficial only to the Polish and Lithuanian gentry—whether materially or culturally. By sanctioning the incorporation of Volhynia, western Ukraine and Podolia into Poland—the western Russian provinces conquered by Lithuania in the fourteenth century—the Union of Lublin had actually alienated the majority of the population in these provinces, i.e., the Ukrainian peasantry. He felt that the Ukrainian masses resented alien institutions being forced on them by the Poles. Wyslouch was not convinced that the alleged cultural superiority of the Poles justified their rule over the Ukrainian and Belorussian peoples; he pointed out that the Germans had claimed cultural superiority to justify their rule over the Poles, too. He believed that potential democratization, and a consequent rise in national consciousness among the borderland masses, were bound to accentuate there the gulf separating the peasantry from the Polish gentry. Moreover, he felt that it would be unfair to deny the right to national self-determination to the eastern borderland peoples. Like Limanowski, Wyslouch hoped for a federation of Poland-Ukraine-Lithuania; however, he stressed the right to self-determination, rather than the legacy of the Union of Lublin, as a means thereto.[6]

Limanowski argued against Wyslouch that the concept of nationality was not equivalent to the concept of "race"; it was the common struggles for social justice and independence which had united the Swiss and attached the Alsatians to France. Hence, shared history often forged a stronger bond than a linguistic or religious affinity. Limanowski did not realize that the situation in the borderlands differed considerably from that in Switzerland or France. Idealizing the Polish rule over the borderland peoples in the past, he argued that a cultural type of regional autonomy should now satisfy their needs. When discussing the nationality problem in Poland, Limanowski put his trust in due process of law as a means of assuring justice to the national minorities, after Poland regained independence. On the other hand, in regard to Polish-German or Russo-Polish relations he insisted that complete Polish independence, even at the expense of an armed struggle, was the only solution.

Limanowski objected to Wyslouch's "ethnographical" point of view; he alleged a certain chauvinism on Wyslouch's part. For instance,

Limanowski pointed to Wyslouch's proposed solution of the Jewish problem in Poland, i.e., the assimilation of a minority of them and emigration of the rest of the Jewish population. Limanowski was in favor of a truly voluntary process of assimilation of Polish Jews; and he opposed any means to expel them—on moral grounds, on grounds of Polish national prestige, and on grounds of practicability, i.e., cost. It is interesting to note that he considered liberalization of Russia as a means of solving the Jewish problem in Poland; he believed that, if permitted, Polish Jews would emigrate to Russia *en masse*. He recollected that, while he was an exile in south Russia nearly twenty years earlier, Jewish artisans had been granted the right to settle in Russian towns outside of the Pale; he noted then that within one year a street in Pavlovsk became peopled exclusively by Jews. Limanowski hoped for a constitutional regime in Russia, which would safeguard the right of all national minorities in the Russian Empire; therefore, he considered the situation in Russia "not merely a Russian problem, but also our problem and a problem of the entire Europe."[7]

Nevertheless, he was ambivalent in his attitude toward the internal situation in Russia; at times he hoped for a constitution and reforms from above, and on other occasions he anticipated a revolution in Russia from below. For instance, while commenting in the *Social Review* on the then unpublished memoirs of the *narodnik* Debogorii-Mokrievich, a fragment of which appeared in that journal, Limanowski disagreed with the former's pessimism as to the possibility of a revolutionary upheaval in Russia in the near future.[8]

Limanowski agreed with Debogorii-Mokrievich that the Russian revolutionary movement reached its culminating point about the beginning of 1881; at this time the latter noted unusual revolutionary "enthusiasm" among students throughout Russia, about half of whom had come to believe in the need to overthrow the Tsarist regime. This revolutionary "enthusiasm" Debogorii-Mokrievich dismissed as straw fire. However, Limanowski considered mass enthusiasm, rather than the actual strength of the *Narodnaia Volia*, the key to a successful uprising in Russia during the 1880s. He was convinced that immediately after the assassination by the *Narodnaia Volia* of the Tsar on March 13, 1881, its prestige was enhanced in Russian society at large; in Limanowski's opinion, during the early 1880s there was a basis for a successful uprising in Russia. He felt that "one thousand armed and determined revolutionists," aided by a wave of popular enthusiasm, might have incited then a mass peasant rebellion along the Volga and the Don, under the slogan of "land and freedom *(zemlia i volia)*."

Due to the arrests in the army, which took place in Russia in the wake of the assassination, Limanowski concluded that the Tsarist army was unreliable and that part of it at least was likely to support the would-be insurgents in the Volga-Don region. He knew that the leaders of the *Narodnaia Volia* were obsessed with the idea that only such an uprising as they might start in the capital could result in the seizure of power by them in St. Petersburg. Moreover, he realized that, in fact, they considered an uprising based on the masses in St. Petersburg a totally impractical project and, instead, hoped for an army revolt at the seat of power. No army revolt occurred in the capital or elsewhere. Thus the *Narodnaia Volia* made no attempt at this time to rise either in St. Petersburg or in the Volga-Don area.

Limanowski went on to argue that the south-eastern part of European Russia represented the best location to start a revolution; in his opinion, only in this region would a popular movement have a chance of unhampered development. He contended that such a popular movement—if organized in the early 1880s—would have enlisted the Ukrainians in the struggle against Tsardom; and it would have incited Poland, the Caucasus and Siberia to rise as well.

In 1887, he continued hoping for a rebellion along the Volga. We might note that both the earlier and the subsequent history of Russia proved these speculations of Limanowski to be wrong. Unlike the leaders of the *Narodnaia Volia*, he forgot that the four major peasant uprisings along the Volga or Don organized by the Cossacks in the seventeenth and eighteenth centuries had been futile; and he did not realize that in a unitary centralized state like Russia in order to be successful a revolution must both start and win at the seat of power. On the other hand, his approval at this time, at least in principle, of a revolutionary Russo-Polish alliance against Tsardom (as originated by the "Kazan Conspiracy") displeased those Polish patriots who, like the writer Wladyslaw Studnicki-Gizbert were greater Russophobes than Limanowski.[9]

In addition to his polemics with Wyslouch regarding the future borders of Poland, and his comment on the situation in Russia, Limanowski concerned himself in the *Social Review* with more general political issues like suffrage. He wrote in an article on this problem in the *Review*, in comment on the views of the French parliamentarian, industrialist and disciple of Fourier, J. B. André Godin (1817-1888), who had approached him in the matter.[10]

Limanowski assumed that democratization of the suffrage was a primary goal for every legal socialist party in Europe, including the *Labor* group in Lvov. He argued that enfranchisement of the masses represented the first step toward the implementation of a socialist program; in Galicia it was bound to result in a strong patriotic socialist party, which would exert a positive influence on Polish politics in the other parts of Poland. He agreed fully with Godin that universal suffrage would safeguard the rights of and ensure education for the common people, would result in legal as well as political equality, and would facilitate socialization of the means of production. Limanowski felt that to limit suffrage to the upper classes was tantamount to legalized exploitation of the people; he noted that socio-economic inequality was more pronounced in those countries where political inequality was sanctioned by law. Like the British philosophical radicals Limanowski believed that a legislature which was elected on the basis of a very broad suffrage and controlled the executive, was bound to reflect a "general will."[11]

Aware that universal suffrage was an ideal which was then not yet realized anywhere, Limanowski stressed that the right to vote should be extended to all—including women. Moreover, he felt that in their quest for a truly representative parliament, which would reflect a "general will," reformers must face the basic issue whether a deputy should represent his locality, or whether he ought to be a representative of a more general "national" interest. He realized that local affairs would be closer to a deputy than any "national" interest, and yet he was ready to admit that—ultimately—the "national" interest must prevail, should local and "national" interests conflict. At the same time, Limanowski was convinced that provision for political decentralization, which would guarantee autonomy to minority nationalities, was just as vital as ensuring equitable representation in parliament. He pointed as an example to the ineffectiveness of the Polish deputies in the German Reichstag to prevent Germanization of Prussian Poland, arguing that the only means of safeguarding regional autonomy in a multi-national state was a federal parliament on the Swiss model, with a lower house representative of the nation as a whole and a Senate elected on behalf of the constituent provinces.[12]

Around the time of his collaboration with the *Social Review* (1886-1887) Limanowski—like many Polish patriots at home and abroad at this time—began to hope for the outbreak of a European war, which would result in the liberation of Poland; it was the Bulgarian Crisis of 1885-1887 which now re-awakened such hopes for liberation among patriotic Poles. The appearance of the *Social Review* was in itself symptomatic of intensified political activity among the Poles, due to this new crisis in the Balkans.

Late in 1886 Limanowski, influenced by the international situation, wrote a number of letters to Dr. Julian Lukaszewski, formerly a representative in Prussian Poland of the National Government in 1863, now residing in Rumania. In this correspondence, Limanowski urged the need at this time to unite all Poles living abroad; he proposed an official *émigré* Union to represent Poland diplomatically and to act as an intermediary between the large Polish colony in the United States and the home country. Apparently, he now contemplated a kind of national "embassy" abroad on the model of Prince Adam Czartoryski's Hotel Lambert in Paris, which had ceased its activity after the Franco-Prussian War. But Lukaszewski considered Limanowski's project an impractical idea in view of the political fragmentation among the Poles.

Limanowski also suggested to Lukaszewski that the proposed organization might administer a "National Treasury," i.e., a fund made up of voluntary contributions to finance the next uprising; the idea of such a fund had originated with Giller, who died in 1887. In fact, a "National Treasury" did come into existence in 1887. In that year Milkowski published a pamphlet entitled *The Case for Active Defense and a National Treasury (Rzecz o obronie czynnej i Skarb Narodowy)*, in which he urged the Poles to prepare for a struggle for independence by just such measures as this, and he was critical of "organic work" and Warsaw Positivism for their ineffectiveness. Limanowski eagerly endorsed this central idea contained in Milkowski's pamphlet, and it became the program of a new *émigré* organization, the Polish League *(Liga Polska)*, founded by Milkowski in Geneva, in August 1887.

The new *émigré* Polish League aimed at the restoration of Polish independence within the boundaries of 1772; it harked back to the ideology of the original Democratic Society, but was very vague on social issues. The leadership of the League stressed class solidarity in the cause of national liberation, hoping for the outbreak of an Austro-Russian war and favoring alliance between the Poles and the Habsburg Monarchy.[13]

Convinced that Polish socialists ought to participate in non-socialist organizations aiming at independence, Limanowski joined the League as an ordinary member. Meanwhile, his former associate in the now defunct Polish People, Kobylanski, who by now had ceased to be a socialist and instead had become an activist in the League, was anxious to place Limanowski on the executive of this new Polish *émigré* group; however, Kobylanski's efforts to this end proved unsuccessful because, as a leading socialist, Limanowski was *persona non grata* to the persons running the League, who were anxious to attract the older generation of *émigré* democrats hostile to socialism.[14]

What prevented the League from becoming just another ephemeral *émigré* group was its popularity among students in the home country. In 1886 Milkowski sent Balicki to Warsaw to agitate among students; Balicki soon created in Russian Poland the Union of Polish
Youth (known as "Zet"), which copied the conspiratorial model of
Freemasonry and became affiliated with the League.[15]

In addition, a group of patriotic intelligentsia, associated with a
new populist organ the *Voice (Glos)*, which began to appear in Warsaw in 1886, was sympathetic to the League. Around this time, in
Russian Poland (as in Galicia) peasants conflicted with landlords
over access to forests and pastures. However, fearful of the stringent
censorship in Russian Poland, the *Voice* made no attempt to champion the local peasantry. Rather, it was interested in the Polish western "irredenta," i.e., Upper Silesia, Pomerania (*Westpreussen*), Ermeland and Mazuria, where peasants were predominantly Polish-speaking.
Ideologically, the *Voice* resembled the contemporary *Social Review*
in Lvov; it espoused a kind of peasantist patriotism. Until about
1890 its contributors included a number of progressive writers; however, from the beginning the *Voice* was edited by Jan Poplawski, a
leader of the future rightist National Democratic Party.[16]

At Poplawski's request Limanowski contributed several articles to
the *Voice*.[17] These articles, as well as his treatment of the agrarian
situation in the *Weekly Review*, and later on in *La Diane (Pobudka)*,
show his preoccupation with the "Great Depression" (1880-1900) in
Western European agriculture, which was chiefly due to the immense
increase in imports of cheap North American grain, made possible by
application of an up-to-date technology to large-scale farming on the
newly opened virgin lands in the United States—and eventually in
Canada, too—as well as to revolutionary improvements in transportation.

In Switzerland Limanowski believed that the plight of the farmers
was aggravated by the tax burden imposed by this "welfare state."
Influenced by Becker's *Manifesto to Farmers*, Limanowski became
convinced that peasant farming—in its present form—was doomed
everywhere, due to the general trend toward capitalist concentration
of production, which had followed in the wake of industrialization,
as well as advances of agricultural methods and improvement in
transport, and was especially apparent in North America. Like the
Marxist Becker, Limanowski now advocated establishment of producers' cooperatives as a temporary remedy for this otherwise imminent process of expropriation of European peasant farms by capitalist

agrarian enterprises. He suggested two forms of agricultural cooperation—first pooling by the peasants of resources and sharing profits
(he had witnessed it in Switzerland), and secondly cooperation between small producers and large producers. He believed that "where
the agrarian crisis was complicated by national oppression," it was
possible for large landowners and peasant proprietors to collaborate, in order to resist jointly the challenge of expropriation by alien
capital, a proposal which had been advocated by the philosopher
Libelt for Prussian Poland in his *Coalition of Capital and Labor.* [18]

Limanowski stressed that the trend to agrarian cooperative socialism in reaction against capitalist expropriation, which he thought
he discerned throughout Europe, was not a "metaphysical" process.
He considered agrarian socialism a utilitarian ideology which found
its sanction in the historical "myth" of the ancient European peasant commune, as preserved in Russia, and he disagreed with the German economist Baron Haxthausen that the collective principle as
embodied in the Russian *mir* was a metaphysical "inexorable force." [19]

Like Marx himself and the Russian Marxist Plekhanov, Limanowski now believed that the commune in Russia was doomed by the
imminent rise in all countries of large-scale capitalism, in agriculture
as well as in industry. He felt that this process would result in an
ever-growing rural proletariat throughout Europe, with no vested
interest in the status quo, and that it would eventually lead to a
modern type of agrarian collectivism, with the workers becoming
shareholders of the proceeds from the publicly owned estates. [20] Obviously, Limanowski now considered cooperation by small farmers,
as discussed above, a mere palliative; rather, in his opinion, the remedy for the agrarian crisis in the long run was large-scale collectivism.

Limanowski was in favor of socializing most of the agricultural
land—including gentry estates as well—on economic, social and political grounds. He now assumed that—notwithstanding the imports
from North America—the contemporary agricultural production in
Europe fell short both quantitatively and qualitatively of the standard
necessary to feed the European peoples adequately. The reason for
this, in Limanowski's opinion, was that small proprietors, who still
dominated agricultural production in Europe, could not afford the
large investment of capital necessary to rationalize and mechanize
their farms. He believed that only large-scale socialist farming, combined with equitable distribution of grain and other foodstuffs, and
involving a large investment of capital as well as rational planning,
would adequately feed all of Europe. He felt that such intensive cultivation would be economically superior to the yeoman peasant type

of farming, because it would replace much human labor by mechanization; it would increase soil productivity by widespread use of artificial fertilizer; and it would facilitate drainage, irrigation and combination of farming with industrial enterprises like brewing. He concluded that small individual land ownership was contrary to the interest of most of the farmers themselves. It resulted in numerous landless proletariat and, on the other hand, it demanded exhausting toil on the part of the peasant farmer and his family, which was harmful to both his mental and physical health.[21]

Limanowski admitted that under socialist organization of labor the peasant would be less motivated to work diligently and to make improvements than an independent farmer, but he was convinced that the worker in a socialized agricultural enterprise would be more conscientious than a hired hand or an insecure tenant farmer. Furthermore, Limanowski felt that agrarian socialism would provide the means of granting access to land to landless peasants who might otherwise migrate to the cities; it was a remedy for the proletarianization of the peasant, a way to transform him into a patriot by identifying him with land and thus to give him a sense of belonging, and a way, too, to train him for full-fledged citizenship through participation in local communal self-government. In Limanowski's view, communal landownership was bound to prevent economic inequality and the danger of social strife resulting therefrom in the village.[22]

A conservative might well have agreed with Limanowski as to these merits of agrarian socialism. However, while Limanowski contemplated socialization of both gentry and peasant land, Polish (or Russian) conservatives, who found the idea of agrarian socialism appealing as a means of preventing social unrest, wished to confine socialization to peasant holdings only. Nevertheless, like some conservative writers in Poland, Limanowski still admired the principle of the ancient Slav commune. When he advocated the modern type of agrarian collectivism, at times he did not distinguish clearly between the principle of modern agrarian collectivism, i.e., cooperative farming which is characterized by collective possession of the common land.[23]

While residing in Thun Limanowski frequently travelled on business to nearby Bern. Here he became acquainted with a group of Polish students, and it was due to their efforts that he was invited to come to Zurich early in 1887, to celebrate the twenty-fourth anniversary of the 1863 uprising; his hosts were local Polish students organized in the Society of Polish Youth. They asked him to preside over their meeting; his speech met with their approval, except for his reference to the alliance of the insurgent Poles with Russian revolutionaries in 1863. Out of about one hundred Poles studying in Zurich, the majority were democrats; there were also two small socialist circles of a nationalist and an international orientation, respectively. This last group was led by a brilliant recent arrival from Galicia, Feliks Daszynski (brother of the more famous Ignacy), who was soon to die prematurely from consumption.

The student democrats now persuaded Limanowski to move to Zurich. They were ready to guarantee him a regular income from lecturing on modern Polish history in the period after the partitions. Having sacrificed his entire capital to get rid of his photographic studio and to settle the debts which he had incurred in Thun, Limanski was forced to support himself and his boys by writing and lecturing; he continued, however, to receive additional financial aid from Lucjan as compensation for his share in the family estate.

After taking up residence in Zurich in April 1887, Limanowski began to take an active interest in local Polish student affairs, and on May 20, 1887, he was chosen honorary chairman by the student democrats. Apparently, Limanowski moved for a time (ca. 1887-1889) somewhat toward the right; he associated exclusively with the non-socialist democratic majority of students and was a member of the local non-socialist Polish Society for Mutual Assistance, which he joined shortly after his arrival in Zurich.[24]

In December 1887, Polish students from Zurich, Geneva and Bern met in Zurich to found an umbrella *émigré* student organization, which they called The Union of Polish Student Societies Abroad. It continued to exist until 1914 and came gradually to include Polish students in France and Germany as well. At this founding congress of the new organization Limanowski lectured on Galicia, stressing the important role which this province now played in Polish national life. Noting the backwardness of this region—which was larger than Switzerland or Belgium—he nevertheless praised Galicia as a basis for free development of Polish culture and scholarship. He asserted the need for nationalization of the means of production in Galicia; he felt this would accelerate both material and cultural progress in this region, which he continued to consider the "Piedmont of Poland."[25]

After Limanowski had finished his lecture, a report on the situation in Russia was presented to the congress; this led Feliks Daszynski to urge an alliance with the Russian revolutionary movement (then in fact virtually nonexistent). The idea of a Russo-Polish alliance was challenged by Balicki; and a conflict ensued between the "socialists" and "patriots." In an attempt to mediate the quarrel, Limanowski declared himself in favor of a Russo-Polish revolutionary alliance, in which the Poles would assume "moral leadership." But his somewhat condescending attitude toward the Russians was resented by Daszynski and his political friends, who accused Limanowski of favoring an alliance with conservative Austria.

This antagonism toward Limanowski on the part of Daszynski stemmed too from his awareness of Wyslouch's recent talks with the Austrian General Neipperg in Lvov, in which Limanowski was also indirectly involved. Earlier that year, in the course of the Austro-Russian crisis over Bulgaria, Wyslouch had demanded of Neipperg a permit for Limanowski to return to Galicia, as one of the conditions for offering Polish collaboration with the Austrian intelligence against Russia. These negotiations proved abortive; nevertheless, a rumor persisted among some of the Polish students abroad that Limanowski would have been willing to become an Austrian agent. After the congress, Daszynski composed a satirical pamphlet ridiculing Limanowski, along with Balicki and Milkowski; it appeared in 1888 and was widely publicized both at home and abroad.[26]

In reply, Limanowski published his lecture on Galicia early in 1889 in *La Diane (Pobudka)*, which was the organ of a new Polish *émigré* socialist organization—the National-Socialist Commune—founded in Paris in 1888 by a new arrival from Russian Poland, Stanislaw Baranski (1859-1891).[27] Like Limanowski's program of August 1881, that of *La Diane* combined the goals of national liberation and socialism. However, it was a more precise and realistic document, as well as less emotional than that of the fourth Polish People, and it laid less stress on social solidarity than Limanowski had done. Also, unlike its predecessor, *La Diane*'s program stipulated freedom of action for the Polish socialists even when allied with democratic parties. Yet it stressed "external" association with alien socialist parties and, generally, independence from foreign resources and influences. And both programs defined future Polish boundaries by means of a vague formula. *La Diane*'s program sought to conceal opposition to an ethnic Poland among some of the members by stating that "the borders of the future Commonwealth would be defined by the revolution itself."[28]

There is no evidence to suggest that Limanowski directly influenced drafting *La Diane*'s program. Its authors, like Baranski, appeared to be more socially radical than he was; they attempted to combine advocacy of an insurrection with their goal of a social revolution. True,

Limanowski envisioned an insurrection as an agent of social reform; however, like the leadership of the League, Limanowski now subordinated his idea of a social change to the demands of the struggle for independence. Yet it is not quite clear what was meant by "social revolution" in the program of *La Diane*. We may conclude that Limanowski's ideology differed little from that of the *La Diane* group. By 1889, the year in which *La Diane* began to appear, the revolutionary *Narodnaia Volia* had almost ceased its activity in Russia; most of the *émigré* Poles, including the *La Diane* group, no longer viewed Russia as the source of immediate social revolution. Their hopes, like Limanowski's, now centered on Austria, because of its growing antagonism toward Russia. In the home country the ideology of *La Diane* appealed mainly to the students organized by Balicki in "Zet."[29]

Balicki, having effected a union of Polish students in Russian Poland, now concentrated his efforts on transforming the League into an umbrella *émigré* organization. It was due to his intervention that the Commune joined the Polish League as a single body in 1889. According to the terms of this union, the Commune retained its internal self-government, but the agreement in effect subordinated it to the Polish League which, in turn, became obligated to aid the Commune, especially by circulating its organ in Poland. We may note that the union did not actually affect Limanowski's affiliation, since he was already enrolled as a member of the League.[30]

Although Limanowski influenced the ideology of *La Diane* indirectly, he differed from the younger contributors to this paper in that he always wrote in a tactful and dignified manner; the younger members of the Commune, however, attacked respectable personages in the democratic camp like the Galician Tadeusz Romanowicz, and they repelled many liberal Poles by their fervid anti-religious propaganda as well.[31]

Of the articles which Limanowski contributed to *La Diane*, the most noteworthy are his three theoretical articles—on the nationalization of land, of industry, and of the means of distribution. The first forms his most comprehensive discussion of the agrarian problem before 1918.[32] In it Limanowski combined a Marxist approach to agriculture with the ideology of John Stuart Mill and the "single-tax" school of Henry George. Influenced by the second volume of Mills' *Principles of Political Economy* and by George's *Progress and Poverty*, Limanowski in this article came out in favor of taxing land rather than industry; he endorsed the idea of taxing land rent, i.e., appropriating the "unearned increment" of land by the state. Limanowski argued here that the entire state revenue—or at least the major share of the tax burden—should derive from land so as to eliminate indirect taxes and to ensure prosperity for industrialists, especially the small entrepreneur. He was convinced that this would result in more employment for the workers in industry and would benefit society as a whole, also forcing speculators to return idle land to cultivation. This would lower the price of agricultural land, providing additional employment on the land; and it would indirectly benefit the industrial proletariat, which would become less numerous and thereby more expensive to hire.

However, in the final analysis, Limanowski came to the conclusion that such imposition by the state of a land rent tax would restrict, but would not do away with the monopoly of the great landlords who would remain in possession of the means of production and would continue to exploit their workers. In these circumstances, the small farmers would be forced to employ the most up-to-date agricultural methods, or perish. By adopting the "single tax," the legislator would intensify the competition and the process of capitalist concentration of land; he would stimulate the trend to expropriation of the weak by the strong and would thus aggravate the inequality of opportunity in agriculture. Limanowski decided that in the long run only socialism could provide a sure remedy for the impending agrarian economic crisis which would intensify social injustice in the village.[33]

He believed that it would be possible to safeguard individual rights under socialism by democratic rule, by providing an equal start and the continuance of equal opportunities for all, as well as by political and economic decentralization. He was convinced that a decentralized system whereby public land would be leased to local associations and communes by a Ministry of Agriculture would serve local needs well, and would avoid depersonalization of human relationships characteristic of centralized bureaucracies. It is not clear how he reconciled economic decentralization with overall state planning.[34]

Limanowski wished to accomplish the socialization of land in Poland by peaceful means. Yet it is clear that he did contemplate a general expropriation of gentry land with compensation by the state. However, he was prepared to confiscate only the land held by traitors to the Polish cause—in the event of an insurrection—or the land purchased by individuals from one or the other of the partititioning powers, which had originally been public land owned by the Polish state.[35] In the course of an insurrection, he envisioned nationalization by decree of private estates where the land was tilled by hired labor, and transformation of workers into shareholders in these estates. He stressed that he was opposed to a compulsory appropriation of peasant-cultivated land by the state. He wished to leave small private ownership intact where local conditions of cultivation permitted doing so. Evidently he contemplated a dual system of land holding, at least for the time being—one in which there were both peasant proprietors and huge cooperative farms.[36]

Limanowski was convinced that socialization of land, rather than discouraging the Polish peasant, might well become the slogan which would arouse him to struggle for Polish independence. He refused to believe that the common people were motivated exclusively by a "vulgar material interest"; in his opinion, it was the sheer idealism of the masses in Europe which, for instance, had produced and upheld the Crusades during the Middle Ages. He noted that contemporary democrats in Poland expected the people to rise for patriotic reasons and without material reward; at least socialism offered them the means of improving their existence. He argued that in the past the collectivism of the first Polish People in exile, of the radical priest Sciegienny in the Congress Kingdom, and of the bookseller Stefanski in Poznania had all had appeal for the peasant masses.[37]

In his article on "Nationalization of Industry" Limanowski expressed a belief in the efficacy of labor legislation under capitalism, considering it to be the first step toward nationalization. He envisioned a nationalized economy run by a Ministry of Industry at the center, with appropriate departments on the lower levels, but in the spirit of anarcho-syndicalism he demanded complete autonomy as to organization of work and salary scale for each plant, which was to be leased to the workers. Each plant was to pass on its final surplus to the National Treasury to provide for such social services as nurseries, schools, hospitals and old peoples' homes. He made no provision here for a military budget and did not distinguish between light and heavy industry; indeed, he considered that national industry should give priority to consumer goods, and that exports should be restricted to those articles which would pay for imports necessary to satisfy the people's needs. He did not exclude a violent revolution as a means of nationalization of industry for countries other than Poland.[38]

He believed that in the course of the next Polish insurrection there would be agitation for nationalization of land, industry, distribution, and of all cultural institutions (e.g., theaters). He did not doubt the success of this agitation, nor did he speculate on the possibility of a civil war resulting therefrom, in spite of his mistrust of the

upper classes in Poland. He hoped for a just, classless commonwealth
to be realized in Poland without bloodshed, even though he was
ready to admit that this was a possibility elsewhere.[39]

Limanowski considered nationalization of the means of distribution a necessary complement to nationalization of industry and land.
He believed that private merchants were motivated solely by greed,
and he observed that they at times made profit by exporting their
goods while the local population starved; on some occasions, they
were even willing to waste their merchandise rather than lower their
price. Moreover, he noted a disparity in distribution between regions
in one country as well as between individuals. He felt that the remedy
for this was distribution of all consumer goods by the state. For instance, he pointed out that in Switzerland the cantons of Zurich and
Geneva exercised salt monopoly and provided inexpensive and un-
adulterated salt; and the Polish National Bank in the former Congress Kingdom owned steam flour mills which produced cheap and
good bread. He was convinced that distribution of food by the state
would be both practicable and beneficial to all, and that state planning of distribution would eliminate waste of resources and services.
He envisioned a central Ministry of Distribution with federal and
local organs, which would distribute food according to local demand;
any shortages were to be made up by those communes, counties or
provinces which had a surplus. The local communes—both rural and
urban—were to provide retail outlets and a home delivery service.[40]
Also, there would be, according to Limanowski, self-sufficient cooperative communities on the model of Godin's Fourierist "familister"
in Guise, and huge common kitchens serving suitable food for everyone.

Limanowski proposed that distribution of industrial goods should
proceed by the same channels as distribution of food. The Ministry
of Industry was to supply each commune, through the Ministry of
Distribution and its organs, according to local demand. He somewhat
naively believed that in addition to providing quality goods in sufficient quantity, such a system of distribution would be equitable and
would result in eventual elimination of money.[41]

These proposals could circulate freely only among the Poles in
exile. At home Limanowski's writings were banned, even in Galicia.
By 1890 the police in Cracow felt that Limanowski was the most important Polish *émigré* socialist leader; they considered his patriotism—
as well as the patriotism of other Polish nationalistic socialists—a dis-
guise to make socialism palatable to compatriots. Despite these hin-
drances, toward the end of the 1880s Limanowski's popularity grew

rapidly among democratic socialists in Galicia as well as among Polish students at home and abroad.[42]

His special appeal to the young people was due, first of all, to his works on Polish history, which acquainted them with Polish democratic traditions at a time when the dominant historiography was that of the conservative Cracow school and the teaching of Polish history in the high schools ended with the partitions. Secondly, he owed his popularity among students both at home and abroad to his histories of modern social movements and theories in Poland and in the West, *A History of Social Movements in the Second Half of the Eighteenth Century* (Lvov, 1888) and *A History of Social Movements in the Nineteenth Century* (Lvov, 1890).[43]

It was Limanowski's belief in socialism as a means to Polish independence which provided the motivation for his writing these two volumes. In order to ground socialism in the science of "sociology," he first studied the history of ideas, commencing with Plato. He considered this study began during the early 1870s, which eventually produced the two above mentioned works, merely a preliminary task to composing a treatise in "sociology." (He proceeded to prepare this treatise immediately upon completing these two books in 1888.) In both works Limanowski attempted to justify his faith in socialism, in progress, and in the eventual independence of Poland. Beginning with the mid-eighteenth century, he examined the main political and social trends in Poland and in the West, especially during the French Revolution and also during the later social and political struggles in France and Poland, in order to determine the influence of the Revolution on the "collective mind" of the modern era.[44]

In the first of these books, Limanowski—dating the modern era from the second half of the eighteenth century—assumed a continuous progress in Europe since that time, resulting from the struggle between the principle of "revolution" and the principle of "reaction" as embodied in the "old regime." He believed that Poland became a victim of this conflict when it was partitioned by the three "reactionary" powers. Limanowski felt that it was of great importance for the Poles to be acquainted with the events of the second half of the eighteenth century from a "progressive" standpoint, because he considered an alignment with the "forces of progress" by the Poles the only means of regaining independence. It was a period in Polish historiography, which he considered either neglected or written from a "one-sided" point of view. He assumed that his history was at once a pioneering contribution to Polish historiography, a political guide, and a prerequisite for his projected study in sociology.[45]

The first of the two books met with an enthusiastic reception in student circles in Warsaw, which pleased him greatly. A student correspondent described the book as timely and satisfying a real need. "Our youth having passed through a phase of so-called cosmopolitanism and having regained interest in the history of their country, feels that Poland must also have had men who were devoid of clergy and gentry class egoism; wishes to become acquainted with [Polish] slogans and ideas [in the past] which perhaps even now are still progressive; wishes to learn the causes and conditions under the influence of which these ideas were transformed and [yet] failed to produced the desired effect. We have had no books to date to give all this information."[46]

In his second book Limanowski envisioned the nineteenth century in terms of a gradual victory of progress. The two most important factors in producing progress were, in his opinion, industrialization as well as the rise to prominence of the natural sciences and positivistic liberal thinking.[47] This book compares favorably with contemporary, as well as with some of the later, works on socialism written in the West. It was well received not only in Poland, but also abroad, notably in Russia and in student circles in Prague as well. Malon's *Revue Socialiste* referred to it as "oeuvre de haute pensée et de serieuse érudition . . . . Remarkable ouvrage écrit par le plus éminent des écrivains socialistes polonais de ce temps."[48]

In writing the two books Limanowski usually adopted the method of a reporter rather than of critic, especially when dealing with non-Polish themes. Almost all the reviewers praised Limanowski's objectivity and erudition; they pointed out that the volumes contained a great deal of factual information—largely based on primary sources. Because they gave a comparative presentation of ideologies in Poland, Europe and America, these works were considered by most critics as being extremely valuable and indeed representing a pioneer contribution to historiography, especially the chapters dealing with Poland. Yet, composed of self-contained chapters, they lacked cohesion and synthetic treatment, i.e., they were collections rather than treatises on well defined single themes. Several reviewers felt, too, that Limanowski was at times uncritical of his sources; his treatment of the various ideologies was not given against the background of their environment; and he was imprecise in using the term "socialist," applying it to non-socialist democrats and even early nineteenth-century liberals like Staszic.[49]

Limanowski was now becoming a respected figure among the Polish *émigrés* scattered throughout Europe, as well as among the older generation of Poles at home, in addition of course to his wide popularity already alluded to—among Polish students. When the Polish Society for Mutual Assistance in Bucharest was invited to attend a commemorative meeting on the occasion of an anniversary of the 1794 uprising, which was to be held in Geneva on April 18, 1889, they chose Limanowski as their representative at that meeting. He was now considered somewhat of an authority on problems confronting Poles abroad; just to give one instance, he was consulted by a lawyer from Lvov—who was directed to Limanowski by a mutual acquaintance, Adolf Inlaender—as to whether it would be possible to set up large Polish colonies in South America.[50]

Meanwhile, Limanowski's financial situation was becoming almost desperate. In 1887 he married for the second time—albeit reluctantly; the proposal came from the lady! His new wife was his former pupil, Maria Goniewska, to whom he had acted as tutor on a landed estate in Galezowa Wola near Lublin in 1868-69. He had now again a wife to support as well as four boys, the youngest of whom was born in 1888. Limanowski attempted to secure either a university or a high school teaching position in Bulgaria. The delegate sent by the Polish colony in Bulgaria to the congress of the League, held in Geneva in 1889, tried to find such a post for Limanowski—but without success. It was then very difficult in Western Europe for a scholar to earn a living by writing. Limanowski felt that in Poland the situation in this respect was better, especially in Russian Poland; yet the topics in which he was interested were forbidden by Russian censorship. Russian censors in Poland rejected his entire articles; his signature was sufficient to arouse their suspicions and even his initials now attracted their attention. Therefore, he began to write anonymously. At this time, the conservative papers in Poland paid well, but Limanowski rejected their social creed and was thus barred from contributing to them. On the other hand, progressive periodicals often could not offer any remuneration, or at best paid nominal sums. Limanowski contributed, in particular, to a progressive scholarly journal in Warsaw, the *Ateneum*, which recompensed its writers fairly well, but was subject to a very stringent censorship. He was so discouraged by the frustrations of his life in exile that he sometimes even contemplated suicide. It was the writing of a treatise on sociology which succeeded in alleviating his dispair. In March 1889, Limanowski confided to his brother Lucjan: "Previously my socialism was based solely on conviction; now I am confirmed in my conviction by experiencing myself the iniquity of individual egoism institutionalized as a social system."[51]

However, a change in his dreary existence soon came to bolster his flagging spirits; he was chosen by the National-Socialist Commune to represent them at an international socialist congress to be held in Paris; and the Commune agreed to provide for all his expenses. At the international labor congress which met in London in 1888 it was decided to hold a further meeting in Paris the following year. However, the delegates of the British Trade Union Congress insisted that this gathering be non-political; most of the Continental socialists objected, with the exception of the French Possibilists (Communalists), and made arrangements to convene a congress of their own. In July 1889, Paris was to celebrate the one hundredth anniversary of the storming of the Bastille. A huge exhibition was arranged for this occasion and the two international socialist congresses met simultaneously, as planned; the one sponsored by Brousse's Possibilists gathered at Rue Lancry and the other was convened by Guesde's Marxists at the Salle Petrelle. Both congresses designated May 1 as the Labor Day, and agreed to agitate for a maximum working day of eight hours, but it was the Marxist gathering which resulted in the foundation of the Second or Socialist International. The Possibilists' International was stillborn.

Limanowski attended the Possibilists' congress, being instructed to do so by the Commune which feared cancellation of their mandate to the Marxist meeting by their rivals, the internationalist Polish socialist group. Actually, he would have preferred to go to the Marxist congress, and he looked forward to seeing there his acquaintances Bernstein and Liebknecht. The Possibilists were principally concerned with practical economic problems concerning the workers, which were unfamiliar and of little interest to Limanowski.[52]

At the Possibilists' congress Limanowski told his audience that for the Poles, especially for the majority of them who lived in Russian Poland, the resolutions of the congress were but a "beautiful dream." It was his contention that socialists of subjugated nations must assign priority to national liberation, for as long as there were countries in Europe, subject to an oppressive foreign rule which deprived them of the usual political rights as well as of the rights of their nationality, and handicapped them economically (as in the case of Prussian and Russian Poland), there could be no universally applied labor legislation. He proposed the following resolution:

> Whereas nationalities which are deprived of independence and political rights do not suffer alone, but also obstruct the normal progress of other nations;
>
> Whereas, as long as despotism flourishes, exploitation of the workers continues;

> The congress declares it the duty of every country to assure each
> nation in Europe national independence and political freedoms.

Due to an especially friendly response from the English delegation,
which proposed his candidacy as chairman the next day, Limanowski's resolution passed by acclamation.[53]

When the deliberations ended, the municipal authorities in Paris
provided a reception in the City Hall for the delegates to both congresses; it was the first official indication that the socialist movement was now considered politically respectable in France. Limanowski attended this reception and was also invited to a banquet
sponsored jointly by the Parisian Possibilists and Guesdists; here he
was invited to sit at the table of honor and, asked to speak, he addressed the gathering in French. Commenting on a Possibilists'
proposal to divide the executive of the contemplated International
into two sections (i.e., an "economic department" devoted solely to
working class affairs and a "political department" concerned with
general politics), Limanowski upheld this proposal; he believed it
would prevent misunderstandings like those which had resulted in
the demise of the First International. He was convinced that the
"economic department" would be appropriate for the working class
in the West and the "political department" would be useful in
Eastern Europe, where the basic need was political liberation, especially liberation from a foreign yoke. Limanowski admitted that as
a Pole he subordinated socialism, i.e., strictly working class affairs,
to the goal of securing Polish independence. However, he ended his
speech by exclaiming: "The idealism of the French Revolution will
be fulfilled everywhere; it will destroy despotism and all kinds of
privilege and will transform Europe into a free federation of Social
Democratic States!"[54] Limanowski also spoke at the Père Lachaise
cemetery where many of the Communards were buried. In addition,
he took part in a congress of the Union of Polish Student Societies
Abroad also held in Paris around this time. His entire stay in the
French capital represented a modest personal triumph for him;
everywhere he was cheered, honored and entertained; and he discovered that there was considerable demand for his histories of the
uprising of 1863 among the local Polish community.[55]

Due to his precarious financial position, Limanowski contemplated a move to Rumania in the fall, being disappointed in his
efforts to secure employment in Bulgaria. However, in October
1889, he learned from Dluski, now a member of the Commune, that
he had found employment for Limanowski in Paris as a bookkeeper.

The new employer, Stanislaw Krakow, formerly a companion in exile of Limanowski's brothers in Vologda in northern Russia, was a member of the Commune as well.

Prior to his departure from Zurich for Paris at the beginning of December 1889, Limanowski attended a farewell party given in his honor by a group of Polish students organized in the local "Zet." They presented him with a written euology, praising his efforts in the cause of both freedom and progress. Limanowski could not detect the names of any socialists among the signatories of the eulogy; he was not surprised, however, being already aware that the young socialist intelligentsia resented his popularity among Polish students at large.[56]

CHAPTER VI

BOLESLAW LIMANOWSKI AND
THE ORIGINS OF THE POLISH SOCIALIST PARTY
(PARIS, 1890-1893)

Soon after his arrival in Paris early in December 1889, Limanowski became the secretary of the local "Union of Polish Exiles," which comprised Poles of divergent political views. On June 28, 1890, he spoke on behalf of this association, and in the name of the National Socialist Commune as well, on the occasion of the solemn transfer of Mickiewicz's remains from the Montmorency cemetery in Paris to Cracow's Wawel Castle vaults, the burial place of distinguished Polish statesmen and authors. In his speech, given at the cemetery, Limanowski honored the poet not only as a great patriot, but also as a forerunner of the nationalistic socialism espoused by the Commune.[1] As Mickiewicz had done, Limanowski stressed the urgent need to ensure social justice to all in the new Poland, which they both expected to be reborn of a forthcoming defeat of despotism, militarism and "reaction" in Europe.[2]

In Paris Limanowski had assumed the role of a mediator between Polish socialists and the non-socialist members of the League and the "Zet." Upon his arrival, the Polish student democrats in Zurich (and those in Geneva as well), who were all members of the "Zet," attempted to place Limanowski on the executive of the National Treasury which was based in Paris. This fund-raising organization had come into being after the formation of the Polish League in 1887 (as mentioned above). However, Limanowski as a leading socialist still remained a *persona non grata* to the individuals running the League, who also controlled the "National Treasury," and they rejected his candidacy for this post.

Shortly afterwards—in March 1890—Milkowski attacked the younger generation of Polish socialists in Paris in two articles entitled respectively "Youth and Old Age" and "Our Socialist Youth," both of which he published in his *Free Polish Word (Wolne Polskie Slowo)*. Thereupon, a student correspondent from Zurich, Stanislaw Badzynski, confided in Limanowski that he feared these young socialists would retaliate in *La Diane*. Though Polish student democrats in

Geneva and Zurich were also hurt by Milkowski's articles, nevetheless they were anxious to avert an open conflict between Milkowski's Polish League and the more socially radical student "Zet," which was synpathetic to socialism at this time. According to Limanowski's correspondent, Polish student democrats in Zurich and Geneva were apprehensive of alienating Milkowski, because the latter was very popular in the home country and was therefore considered indispensable both to the League and to the "Zet." The student correspondent begged Limanowski to restrain the young Polish socialists in Paris from attacking Milkowski in *La Diane*. He was aware that Limanowski was now highly respected by the entire younger generation of *émigré* Poles in Paris; this would facilitate his task of mediation. In turn, Badzynski promised Limanowski that he would attempt to influence the other members on the executive of the League to prevail upon Milkowski to refrain from further polemics with Polish socialists. However, the coexistence between the socialist and the non-socialist elements in the League and in the "Zet" was precarious and, as will be shown below, was to end late in 1892.[3]

Meanwhile, toward the end of 1890. Limanowski—in his capacity as an honorary member of the Zurich student Society of Polish Youth (affiliated with the Polish Society in Zurich)—was invited to the forthcoming celebration of the twenty-fifth anniversary of this organization, which took place in Geneva. The students provided for all Limanowski's expenses. They received him enthusiastically; and he attended their banquet followed by an amateur theatrical play. Subsequently, the Union of Polish Student Societies Abroad decided to publish a series of pamphlets in honor of Limanowski, including a translation of Lassalle's *On the Nature of a Constitution (Uber Verfassungwesen)*, originally published by Lassalle in 1862.[4]

To mark the one-hundredth anniversary of the famous Polish Constitution of 1791, an important *émigré* congress took place in Switzerland in the spring of 1891. On this occasion, Limanowski participated in the two consecutive meetings convened consecutively in Zurich and in Rapperswil; he attended the congress upon invitation from the "Zet" and the Polish Society in Zurich. Limanowski spoke in Zurich on behalf of the "Zet." In Rapperswil, he defined the Constitution of 1791 as a progressive measure, which the majority of his audience approved.[5]

It was the intent of the organizers of this congress to take advantage of the occasion in order to form an umbrella Polish National

Union in Europe. To this end, Limanowski prior to the congress was charged with preparing a list of all Polish organizations in France and of their delegates, and to ensure that the latter had written credentials. He performed this task in his capacity as secretary of the Paris Union of Polish Exiles. The congress of 1891 indeed united all Polish *émigré* organizations, albeit only formally. In Paris it actually resulted in embittering an old feud reaching back to June 28, 1890, the time of the transfer of Mickiewicz's coffin to Poland, between a group led by Limanowski's employer, the industrialist Krakow, and the followers of a well-respected post-1863 democrat, Dr. Henryk Gierszynski, who was allied with the editorial board of *La Diane*.

The renewed antagonism between the two groups was due in part to a controversy as to the significance of the Constitution of 1791: Krakow's faction considered it sufficiently progressive for its times, whereas Gierszynski and his friends vehemently disagreed. Limanowski reluctantly identified himself with the minority following Krakow (who died in April 1892).[6] It was then discovered that Krakow's enterprise was on the verge of bankruptcy, and Limanowski lost his regular job. His financial situation in Paris, even while he was still employed, was quite difficult, because here it proved impossible for him to live on credit for an extensive period of time as he had been accustomed to do in Switzerland. However, he now had fewer dependents to support. Two of his sons, Zygmunt and Mieczyslaw, were studying in secondary schools in Galicia: the fifteen-year-old Zygmunt was in Cracow, at no cost to Limanowski, a companion to one of the sons of Dr. Gierszynski, whereas the oldest, the sixteen-year-old Mieczyslaw, was boarding in Lvov, under the supervision of Wyslouch's wife, Maria, and was supported mainly by the sums sent to Maria by his uncle Lucjan. But to compound his father's difficulties, in May 1892, Mieczyslaw was expelled from school for "an act of anti-religious propaganda," i.e., he was overheard to doubt the doctrine of transsubstantiation! Eventually, he was re-established in another gymnasium in Cracow.

Meanwhile, Lucjan kept sending money to his exiled brother; the total was to exceed the latter's share in the family estate. Limanowski's articles were constantly being mutilated by censorship in Russian Poland; and he continued to earn very little from this source. His new efforts to find employment in Rumania or Bulgaria proved again unsuccessful. Now anxious to secure a permit to join his sons in Galicia, Limanowski at the general congress of Polish exiles in Zurich in 1891 met a certain Polish journalist Smolski, who promised

him to ask the liberal and patriotic aristocrat Prince Jerzy Czartor-
ski, the nephew of the famous Prince Adam and a member of the
House of Lords in Vienna, to intervene on his behalf with the authorities in Galicia.[7]

During the summer of 1891 Limanowski had attempted to become librarian at the Polish Museum in Rapperswil; this appoint-
ment depended on the leadership of the League. He was unsuccessful, because of his political beliefs. Some months later, in 1892, he
was to apply for a similar position, this time available at the Polish
Library in Paris, administered by the conservative Polish Academy
of Learning in Cracow. It is not surprising that his candidacy was rejected, again due to his socialism.

However, he was now becoming a respectable figure among the
Polish exiles in Europe. His negotiations regarding the formation of
the Polish National Union in 1891, at which time he dealt with a
number of important personages, convinced him of this. Perhaps
some of the Polish conservatives now realized that he was not as
radical in Polish national affairs as he had appeared to them previously. Or perhaps the general change in political atmosphere in
Europe, where moderate socialism was becoming a more respectable
social creed than hitherto, affected Polish *émigré* politics as well.
Limanowski was now fearful of compromising himself with Polish
socialists by being on friendly terms with some of the Polish conservatives. He complained to Lucjan as follows: "Personalities do
play an important role in politics; and yet I would prefer to remain
always on the impersonal ground, the ground of sheer principle...."[8]

Meanwhile, at the request of the owner of the *Weekly Review*,
Adam Wislicki, Limanowski was preparing a book on Galicia, which
comprehensively described this region of Poland. Mutilated by censors, it appeared early in 1892. Though the remuneration which Limanowski obtained for it was insignificant, it eased somewhat his
difficult financial situation in Paris at this time.[9]

Perhaps the most important part of this book, for our purpose,
was the chapter on industry in Galicia. Limanowski concerned himself here with the general prosperity of this region which he continued to view as a future Polish "Piedmont" and, consequently, worthy
of receiving publicity in the other parts of Poland. He continued his
advocacy of both producers' cooperatives and credit unions for Galicia, joined in a single umbrella union to effectively resist foreign
competition and to secure cheap credit for its members. He was in
favor of consumers' cooperatives for the working classes in Galicia,

on the model of those existing then in Belgium. Limanowski went on to deal with and deplore the bad working conditions in Galicia, i.e., long hours, low pay, insanitary and unsafe buildings, and lack of social insurance. He felt that in addition to strengthening handicrafts there was urgent need in the province for development of modern large-scale industry, for which favorable conditions did exist, i.e., extensive natural resources and plentiful supply of labor. He had been impressed by the famous book written by the romantic liberal Polish industrialist Stanislaw Szczepanowski—*The Statistics of Poverty in Galicia (Nedza Galicji w cyfrach)*, published in 1888. In the words of Szczepanowski, he argued that "since the introduction of oil-drilling by the Canadian system [in Galicia] . . . it had become possible to compare the performance of the best of the Canadian workers with work done by our own skilled labor. We have a number of workers here who are by no means inferior to the Canadians. Moreover, it was in the oil industry that a technical intelligentsia made its first appearance in Galicia. This development is of momentous significance for the future [of this part of Poland]."[10]

Notwithstanding this plea for the industrialization of Galicia, Limanowski maintained his preoccupation with the problems of the Galician village. He believed that the prerequisite for industrialization in that province was an expansion of the internal market, i.e., improvement in the standard of living of the peasantry. He reiterated that the only solution to the agrarian crisis in Galicia was socialization of land. To this end, he advocated that the local government grant credit to the peasant communes, formed for the purpose of buying out the land which was offered for sale by large landowners, noting several actual instances of such collective land purchase by peasant communes in Galicia.[11]

Around the time of appearance of Limanowski's book, i.e., early in 1892, an important event took place in Galicia. At last, the socialists in this region were successful in creating a political party of their own, which they named the Social Democratic Party of Galicia. By combining the editorial board of *Labor* with the editorial board of a new organ, the *Worker (Robotnik)*, founded by the printer Mankowski in March 1890, they formed the executive of this new party. The Galician socialists had rejected the overture by the *La Diane* group in Paris to merge all Polish socialists in Poland and abroad into a single Polish Socialist Party; instead they affiliated with the Austrian Social Democracy and adopted its program. They held their first congress in Lvov from January 31 to February 4, 1892. At the congress, the delegate Hudec reminisced about Limanowski's erstwhile lectures on the labor problem, given in 1871.[12]

Almost concurrently to the congress, Limanowski planned with an emissary of the "Zet" in Warsaw, Roman Dmowski, a new *émigré* periodical which was to appear in Paris. Dmowski agreed to collaborate with Dluski and Limanowski in writing the introductory article to this periodical which was named the *Socialist Review (Przeglad Socjalistyczny)*.

On the other hand, under the direction of Jan Lorentowicz—who replaced the deceased Baranski in 1891—*La Diane* continued its hostile policy toward the internationally oriented Polish socialist groups; its attitude toward the Russian revolutionary movement was condescending; and it viewed the beginning of industrialization in Russia as a threat to prosperity of Russian Poland (industrialized after 1863 on the basis of access to the Russian market). Convinced that the connection with Russia was harmful, Lorentowicz stressed the need for an immediate separation of Poland from Russia, and advocated a Russo-Polish war fought in alliance with Austria. To prepare for this allegedly imminent war for Polish independence, Lorentowicz—as mentioned above—urged the unification of all socialist Polish groups at home and abroad. Eventually, his motivation in urging this unity became also a means of remedying the decline in popularity experienced by *La Diane*, due to its rivalry with the *Socialist Review* group.

The ideology of this new periodical—which only appeared in the early fall of 1892 and ceased publication after three issues—differed little from that of *La Diane*; yet, unlike *La Diane* (this was due to the influence of Limanowski, who for a time identified himself with the *Socialist Review* group), it stressed the significance of the peasant—both for socialism and for Polish independence. Moreover, in contrast to *La Diane*, whose supporters were sincerely socialist, some of the contributors to the *Review* aimed at exploiting the recent rise of a spontaneous mass movement among Polish factory labor as a means of furthering the socially conservative aims of the Polish League. The members of the League, in turn, were to reinforce temporarily the socialist ranks. The overt admission that this was to be merely a tactical alliance was due to the influence of Dmowski and Balicki, the future leaders of the rightist National Democratic Party, and to the financial backing of this paper by the League. Yet, in his prospectus for the *Review*, Limanowski, who was only nominally a member of its editorial board, evidently then took its socialism at its face value.[13]

Limanowski contributed two articles to the *Review*, a biographical sketch of Worcell and a chapter from his "Introduction to Sociology." This latter article was entitled "National Consciousness," and is of some significance in presenting his ideology as it developed to date. Viewing the national society in terms of a "spiritual organism," Limanowski attributed "a higher form of moral development" to those national societies which strove for social justice and were guided by an enlightened patriotism. Like Lassalle, Limanowski considered the proletariat the most "moral" class in society; he felt that it embodied the ideals of social justice and of the "general will." Yet, according to Limanowski, it was the intelligentsia which transformed the "subconscious" manifestations of a national life, i.e., folk customs and traditions, into a "conscious" patriotism, and it should articulate the "general will." Thus, Limanowski turned to the intelligentsia rather than to the workers for leadership. It is noteworthy that in his prospectus prepared for the *Socialist Review*, he wished to appeal mainly to "youth," i.e., the young Polish socialist intelligentsia. Again, there was an ambivalence in his ideology as to whether he considered the people an agent or an instrument of socialism and progress.[14]

Limanowski continued to be popular with the idealistic student youth in Switzerland organized in the "Zet," some of whom were sympathetic to socialism. However, the precarious co-existence of "socialists and patriots" within the one organizational framework of the League and of the "Zet" was about to end. In the spring of 1892 Limanowski learned from Balicki, who resided in Geneva, that due to the presence of Milkowski there, an especially acute ideological crisis had developed within the Geneva branch of the politically mixed Polish National Union; Balicki wished to remedy this situation by removal of this branch to Paris, away from Milkowski. Limanowski advised him that it would not be practicable to do so. And only six months later the birth of the Polish Socialist Party (PPS) was to initiate a process of ideological polarization—both in *émigré* and home politics—which would transform the League into a right-wing organization essentially loyal toward the Russian autocracy and bitterly hostile toward the PPS.[15]

Limanowski did not anticipate any of these developments. He was very surprised indeed to receive an invitation to attend what became the founding congress of the PPS, extended to him by two socialist emissaries from Russian Poland in November 1892. At the time he was cut off from the Polish socialist movement both at home and abroad. For many years he had had no contacts with the

Polish internationalist socialists; now he was also alienated from both the *La Diane* and the *Socialist Review* groups, and was not identified with any Polish socialist organization. Convinced that the contemporary socialists in the home country were hostile toward him, and fearful of becoming the divisive force in the new Polish socialist movement, Limanowski only reluctantly accepted the invitation to this congress, which was held in Paris from November 17 to November 23, 1892. Again he was most surprised to be chosen chairman at its first session and to learn that the *émigré* leader of the Proletariat II (the successor to Warynski's organization) Stanislaw Mendelson, who was Limanowski's erstwhile opponent, now chaired a commission which was to prepare a new socialist program assigning priority to independence. Therefore, as Limanowski was to reminisce later on, "the barrier separating . . . [him] from the socialists [now] collapsed."[16]

Henceforth Limanowski was to preside over every meeting of the congress. He seldom voiced his personal views, keeping himself in the background, acting as a mediator and generating enthusiasm for the task of unification at hand. For instance, at the congress he mediated a dispute concerning the tactic of political terror and the publicity to be given to terrorist acts committed by members of the new party; this dispute became the chief bone of contention and even threatened to disrupt the congress completely. But Limanowski succeeded in persuading the delegates that such theoretical disputes were senseless. During the final stage of the controversy, he pointed out that there was already a consensus to apply terror only in exceptional cases and not to use it as the sole means of political struggle. Limanowski argued that it ought to be determined pragmatically whether a terrorist act should be publicized or not; he insisted that each case be judged on its merit. Again and again he stressed that Polish socialists ought to set an example to the nation at large in their willingness to make sacrifices for the common good. Finally, a general consensus on all issues on the agenda was achieved at the congress.[17]

This congress of November 1892 represented a watershed in the history of the Polish socialist movement; it resulted eventually in the formation of the first modern mass labor party in Poland, the PPS. It is essential to examine now the background to the congress, in order to evaluate Limanowski's role in this important event in the history of Polish socialism. It is worthy of note that during the 1880s the socialist movement differed in each of the three regions of Poland. In

Prussian Poland delayed industrialization, national oppression and persecution of socialists under the Anti-Socialist Law—all hindered the formation of a local Polish socialist party. In the economically backward Galicia industrialization was retarded as well, impeding socialist activity (as already had been noted above). Of the three regions in partitioned Poland, by the early 1890s only Russian Poland possessed numerous factory proletariat, due to the industrialization which took place there during the 1870s and 1880s; however, the local socialist movement was split into three warring factions. There was the Proletariat II, founded in 1888 as the successor to Warynski's party (as mentioned above), which aimed at a constitution in Russia, along with autonomy for Poland, as a means of social revolution; it relied on political terror as its main tactic. The least numerous group was the Workers' Unity which broke away from the Proletariat II on the issue of terror in 1891, whereas the strongest force was the "economist" Union of Polish Workers founded in 1889 and based in the main Polish industrial center, Lodz. A spontaneous mass labor movement in Lodz, which had culminated in the bloody strike of May 1892, and caused as well numerous arrests, impressed the leaders of these three groups with the necessity for joint action. By 1892 the failure of "economism," i.e., giving the priority to purely economic demands, had removed in principle the antagonism toward the Proletariat II on the part of the previously "economist" Union of Polish Workers. In addition, the Workers' Unity had by this date come under the influence of the patriotism of Limanowski as well as of the "Zet" and of the League, which were both at that time interested in the socialist movement, as evidenced by their sponsorship of the *Socialist Review*. The new patriotism of the Unity brought it closer ideologically to the nationalistic socialists abroad like Limanowski himself. Meanwhile, as mentioned above, arrests had decimated the two larger organizations; this development as well as the intermediate position ideologically of the Unity in relation to the two larger socialist groups enhanced its significance, both in terms of numbers and in terms of ideology, in the socialist movement in Russian Poland. Moreover, as a result of the Franco-Russian *rapprochement*, hope was born among Polish socialists at home that the European war, which now appeared imminent, might well result in the speedy liberation of Poland. An indication of the increasing trend toward unity among Polish socialists had been the fact that for the first time, at the international socialist congress held at Brussels in 1891, the Polish delegation represented all three parts of Poland.

Notwithstanding a general tendency toward *rapprochement* within the home Polish socialist movement, due to the still strong animosities in Russian Poland the initiative for unification had to come from outside that area. The overtures to this end originated with Mendelson, the editor of *L'Aurore*—then the organ of the *émigré* Proletariat II based in London; the leader of the Social Democratic Party of Galicia, Ignacy Daszynski (the younger brother of the deceased Feliks); the self-styled foreign representative of the Union of Polish Workers, Stanislaw Grabski; and finally, with the editorial board of *La Diane* in Paris. However, it proved impossible to organize a congress which would be representative of all Polish socialist groups, as originally planned, due to the complicated situation in Russian Poland and abroad.

The congress which had met in Paris in November 1892 had theoretically reunited the socialist organizations in Russian Poland; in fact, it had merely merged *La Diane* and *L'Aurore* groups abroad (and it had resulted of course, too, in Limanowski's rejoining the Polish socialist movement). It is noteworthy that of the three main issues which had divided hitherto the Polish socialist movement at home, only the less significant tactical issue was settled at the congress (by Limanowski, as shown above). On the more fundamental issues, i.e., whether to use political tactics and whether to aim at independence, a consensus was reached among the delegates just prior to the congress, when they agreed to adopt a common political platform demanding Polish independence from "backward and reactionary" Russia, which they hoped to achieve by means of an alliance with Austria in a potential Austro-Russian war. The consensus arrived in Paris was facilitated by the fact that apart from the delegates of the Workers' Unity, who were patriots and proved to be quite influential at the congress (numerically, they dominated the commission which was to draft the new program), the other participants were representative of patriotic *émigré* groups only and possessed no definite mandates from socialist organizations in Russian Poland.

Thus, it is not surprising that a handful of die-hard internationalists recruited by Rosa Luxemburg from the remnants of the Proletariat II and of the Union of Polish Workers soon challenged the validity of the resolutions passed in Paris. In June 1893, this group—referred to later as the "old PPS"—formed the Social Democracy of the Kingdom of Poland (SDKP), which continued the internationalist socialist trend in Poland and favored a close alliance with the revolutionary movement in Russia. The PPS proper came into being in the fall of 1893 in two centers: in Wilno and in Warsaw. The Wilno organization was recruited mainly from radical erstwhile members of the "Zet" and of the League, while the Warsaw organization was based on former members of the Workers' Unity.[18]

The new program of the PPS, embodying the resolutions agreed upon in Paris, was published in *L'Aurore* in 1893. Its preamble was drafted by Mendelson. He rejected here his naive internationalism of the early 1880s and now—like Marx and Engels (as well as Limanowski of course, too)—he demanded both Polish independence and abolition of Tsardom as a means to victory of socialism in Europe. Attempting a Marxist survey of the situation in partitioned Poland and an assessment of the socio-economic transformation which took place specifically in Russian Poland after the uprising of 1863, he noted the appearance here of new social classes, i.e., proletariat, capital and petite bourgeoisie. However, Mendelson now felt that class antagonisms in Poland were complicated by the foreign yoke; like Limanowski, he derived Polish socialism from the erstwhile democratic movement for Polish independence and peasant emancipation. He believed that the PPS would succeed where the democrats had failed; the PPS would create an autonomous working class movement, thereby ensuring the liberation of Poland. According to Mendelson, in the new Poland the proletariat was to assume political leadership for the benefit of all. This was in accordance with Limanowski's Lassallian ideas that the workers' cause was also the "general cause" and that a labor party was the vanguard of progress.

The main program of the new PPS was divided into two parts: Part I spelled out the immediate "political" and "economic" demands, while Part II envisioned "gradual socialization of land, of the means of production, and of transportation." Although Limanowski did not himself collaborate in drafting this program, modelled by Mendelson—its main author—on the Erfurt Program of the German Social Democrats of 1891, it nevertheless agreed in general with his ideology (especially Part II and also the "political" section of Part I).[19]

The formation of the PPS was unlike the process by which the other socialist parties in Europe came into being. The PPS was formed abroad; most of its original founders were *émigrés* whose exile predisposed them to idealize the Polish past and to "worship" the memory of the Polish uprisings. It was this cult of Polish struggles for independence which distinguished the PPS from its counterparts in the West. And especially representative of the insurrectionary trend in the PPS was Limanowski, the chief advocate hitherto of Polish nationalistic socialism. An "epigone" of the left "Reds," (i.e., of the most radical orientation among the insurgents in 1863), struggling to make a meager living during an "unheroic" era, Limanowski by his writings had inspired thousands of Polish young people with the idea of struggle for independence, helping to maintain the insurrectionary tradition among the younger generation of Poles, especially abroad.[20]

Limanowski was not directly involved in forming the PPS. Nevertheless, his writings were instrumental in creating an intellectual atmosphere favorable to the new Polish nationalistic program among both the *émigré* and home socialists. It is noteworthy that in the home country Limanowski's ideology had reacted not only on the Workers' Unity which was also influenced by the "Zet," and whose leader Edward Abramowski was perhaps the first modern patriotic socialist in Russian Poland, but it also affected the leaders of the Proletariat II and of the Union of Polish Workers.

Limanowski, the *émigré* Commune (i.e., the *La Diane* group), and the Workers' Unity in Russian Poland—all were connected with the "Zet" and the League. In the case of Limanowski and the Commune these ties were also organizational, albeit they were somewhat tenuous. It was the patriotic ideology of the League which actually united the members of the early PPS; the PPS then was not a homogenous party. The majority in it had been closely associated with the League during its early, democratic phase; for them socialism usually represented a means to independence, rather than an ideal in its own right. Among these former members of the League and of the "Zet," some eventually came to accept socialism as their primary goal. Others—like Pilsudski—were to abandon the PPS just prior to the restoration of Polish independence, or shortly thereafter. (Limanowski himself was never to relinquish his membership in the PPS.) Meanwhile, the "Zet" and the League continued to exist. In 1893, having lost its radical members to the PPS, the League under the leadership of Dmowski evolved toward the right; at first known as the National League (*Liga Narodowa*), in 1897 it was renamed the National Democratic Party (*Stronnictwo Narodowo-Demokratyczne*), now appealing to affluent peasants, the gentry, and the bourgeoisie.[21]

Limanowski's "spiritual paternity" of the PPS, never questioned by any Polish historian, met with particular approval on the part of the National Democrats. Their leading historian of the interwar era, Waclaw Sobieski, recognized that the origins of both the PPS and the National Democratic Party were similar. Stressing the "insurrectionary epigonism" of the "spiritual fathers" of both these organizations,

Milkowski and Limanowski, he wrote: "Just as T. T. Jez [Zygmunt Milkowski] incorporated the insurrectionary legacy of 1863 into the ideology of [his League which was the precursor of] the National Democratic Party, so Boleslaw Limanowski combined that same ideological legacy with socialism, thereby becoming the 'foster father' of the PPS."[22]

CHAPTER VII

SPOKESMAN FOR THE RIGHT WING POLISH SOCIALIST MOVEMENT
PART I
(PARIS, ZURICH, PARIS, 1893-1899)

In December 1892, the *émigré* Polish patriotic socialists formed an organization affiliated with the PPS, the Union of Polish Socialists Abroad (ZZSP), the executive of which—known as the Centralization *(Centralizacja)*—was to be based in Paris. Elected to the Centralization, Limanowski immediately declined, explaining that he was already the secretary of the Union of Polish Exiles in Paris. Soon, due to the intervention of the Russian Embassy with the French authorities, the members of the Centralization were expelled from Paris, whence they went to London.[1]

Toward the end of 1893, Limanowski himself was to leave Paris for Lvov in quest of a permit to reside permanently in Galicia, being hopeful that the situation there had changed in his favor due to the recent legalization of the Galician socialist movement. Before his departure, Limanowski resigned from his executive post in the Union of Polish Exiles, thereby upsetting the *émigré* democrats in Paris; Dr. Gierszynski considered him the only Pole whom they could trust to become secretary of the Union.[2]

Limanowski left for Lvov in October 1893, accompanied by his third son Witold (who was to remain in Galicia along with his two elder brothers, away from his unsympathetic step-mother). In Lvov Limanowski for the first time met his distant relatives, the Wyslouchs. Very favorably impressed by Maria, he initially found his cousin too taciturn and reserved. Limanowski had been closely identified with the editorial policy of *La Diane* in 1889, at which time this paper disapproved of the ideology of Wyslouch's new peasantist and non-socialist organ, the *Friend of the People (Przyjaciel Ludu)*. Despite their political disagreements, Limanowski soon established a very cordial relationship with the Wyslouchs; he became a frequent guest in their home while awaiting the decision of the authorities in Lvov. He was also taken care of by an old-school democrat Bronislaw Szwarce, who had been a prominent member of the insurgent

Government in 1863. Exiled to Siberia, Szwarce had eventually succeeded in escaping; he now occupied a minor post in the Galician
civil service.[3]

Two weeks after Limanowski's audience with the Viceroy, Kazimierz Badeni, Limanowski was ordered to depart within three days;
this deadline was extended until the end of November, in accordance
with Limanowski's request. Badeni, it seems, was most anxious to
get rid of Limanowski; before his departure, the latter was ordered
to report to the Cracow police, immediately after arrival there on his
way back to the West.[4]

However, in petitioning for a permit to stay in Galicia, Limanowski was now strongly supported by some of the most respected personages in the Galician democratic camp like Smolka and Romanowicz. A committee composed of a number of prominent citizens of
Lvov intervened with the Viceroy, insisting that Limanowski be allowed to remain in Galicia for an indefinite period. He received an
enthusiastic reception from the students in Lvov, too. Yet Badeni
refused to change his mind; he explained that in excluding Limanowski from Galicia again, he had been guided solely by *raison d'état*.
Just prior to his departure, Limanowski was summoned to appear
before the Chief of Police in Lvov, who showed him his articles in
La Diane, with the pertinent passages—considered offensive by the
police—underlined in red. The Chief of Police in Lvov was quite ignorant of the actual situation abroad; he assumed that organizations like the Polish Society in Zurich, the umbrella Polish National
Union, or the National Treasury, were all sympathetic to socialism.
He offered Limanowski his support, should the latter decide to
petition the authorities again in the spring of 1894. Limanowski
interpreted this promise as a ruse by the police to get him out of
Galicia quietly, without any fuss. Apparently, the police feared
Limanowski's stubbornness and they were reluctant to use force to
expel him, so as not to create a scandal. He himself recognized that
Badeni still considered him a real threat to the cause of conservatism
in Galicia. Probably Badeni feared Limanowski's presence there during the incipient struggle for universal suffrage, at a time when the
recent Franco-Russian *rapprochement* had reawakened hopes for an
armed struggle for independence and social transformation in Poland.[5]

Before Limanowski's departure from Lvov, a prominent liberal
deputy to the Galician diet, Karol Lewakowski, promised to help
him to secure the position of librarian which was still vacant at the

Polish Museum at Rapperswil, now controlled by the National Democrats. In addition, due to the initiative of Szwarce and Romanowicz, the citizens' committee in Lvov, mentioned above, agreed to sponsor a short modern history of Poland by Limanowski, and they guaranteed him remuneration for this book.[6]

Limanowski decided to travel to Zurich to await there a reply from Rapperswil. In Zurich he commenced to write his brief history, which he entitled *One Hundred Years of Struggle by the Polish People for Independence*. Based on his lectures given in Zurich in 1887-1889, this book came out in Lvov in April 1894. It sold briskly, perhaps due both to its very low price and its appearance on the one hundredth anniversary of the 1794 uprising. It was twice reissued prior to World War I and became very popular reading among members of secret student societies in Poland during the intervening twenty years. However, financially the book proved a disappointment to Limanowski. Although it was published anonymously, it nevertheless immediately attracted the attention of the Lvov police, who soon guessed the identity of its author.[7]

Back in Zurich, Limanowski again interested himself in Polish student affairs. On December 24-26, 1893, he attended the VII Congress of the Union of Polish Student Societies Abroad, which was held in this city and made him optimistic as to the future of Poland. He now believed that Polish students abroad had become the ideological heirs to the former Polish emigrations: they related Polish past to the present; and they bolstered sagging spirits of Poles everywhere.[8]

Early in 1894, Limanowski was asked by Balicki to chair the important Zurich branch of the umbrella Polish National Union, which was based in Geneva. (Limanowski seems to have declined the offer of this honorary post.) Despite his continued association with non-socialist exiles, he was also in demand as a speaker before the Zurich branch of the ZZSP.[9] For his two lectures given before this socialist group early in 1894 Limanowski chose a chapter from his "Introduction to Sociology" which he was then preparing, entitled "The Influence of the Applied Sciences on Social Change."[10] His avowed aim here was to justify historically the belief, shared with nineteenth-century thinkers like Saint-Simon and Spencer, that every "military-bureaucratic" state, devoted to the pursuit of war, was bound to be transformed eventually into a new type of state, the industrial state, which aimed at the advancement of technology and was based on the principle of peaceful labor. Limanowski began by describing the primitive community of man. In his lectures he attempted to show how the initial division of labor, i.e., the separation of the hunters from the shepherds, would eventually, in the course of millenia, evolve into a peace-loving social democracy based on the collective ownership of the means of production. Such was Limanowski's ideal state.

It is interesting that in spite of Limanowski's nineteenth-century illusion that mankind was inevitably moving from war to peace, and from isolation to association, he was realistic enough in his lectures to foresee tremendous increases in the state budgets of the great powers necessitated on the one hand by the ever-growing military needs, and on the other hand by new expenses for social welfare.[11]

He pointed out that the growing expenses for both social welfare and military needs—as well as public criticism of existing methods of taxation—all led politicians and independent economists to explore new sources of revenue. Hence governments in Europe were departing from the unpopular indirect tax on necessities in favor of taxing income and inheritance. This development Limanowski even considered a form of "state socialism."[12]

Limanowski was not blind to the "extreme militarism" of contemporary Europe. But he argued rather ingeniously that this militarism was bound to result in socialism. He admitted that never in human history had military expenditures been as great as were those of the 1890s. This seemingly contradicted his optimistic belief in a forthcoming and lasting peace. But Limanowski expected that the burden of taxation, necessitated by huge military budgets, would become unbearable and eventually would result in revolt by the masses. He pointed out that both the new military technology and arming of almost the entire adult male population made the conduct of a modern war immensely complicated. A new war would involve organizing a very elaborate logistical support and, on the other hand, there was the potential threat to European governments by the armed masses. Limanowski went on to argue that the new universal short-term conscript service on the Continent was bound to destroy capitalism there in the long run, just as the appearance of the professional standing armies in the late Middle Ages was a harbinger of the demise of feudalism in Western Europe.[13]

These lectures were enthusiastically received by Limanowski's socialist audience. However, while awaiting a decision as to his employment from Rapperswil, Limanowski in addition to political activity had to concern himself with earning a living. In Zurich he derived income mainly from contributing to the Warsaw *Weekly Review* and to Wyslouch's *Week (Tydzien)*, which began to appear in 1893 as a supplement to his progressive daily, the *Lvov Messenger (Kurier Lwowski)* purchased by Wyslouch in 1887.[14]

After receiving a negative reply from Rapperswil, Limanowski had no choice but to return to Paris. During the subsequent thirteen years which he was to spend there (1894-1907), Limanowski—beset with financial difficulties and domestic worries as well as oppressed by the dreary routine of office work—led a frustrated and hopeless existence. Following his return to Paris, Limanowski and his family would have starved but for Lucjan's financial aid. Finally, during the summer of 1894 Limanowski secured a clerical position with the New York Life Insurance Company's branch office in Paris, which provided him with a small but regular income of 150 francs monthly. He was nearly to double this salary eventually. Limanowski's political beliefs were of no concern to his employer, while his immediate supervisor, who was an American born of an Italian mother, Adolph Davidson, actually sympathized with his Polish patriotism.[15]

The office job, however, interfered with Limanowski's political activity and his writing. He could no longer find time to contribute regularly to periodicals. Eventually he had to refuse invitations to write for the organ of the PPS in Berlin, the *Workers' Gazette (Gazeta Robotnicza)*, as well as for the *Bulletin Officiel*, which had the task of describing the activities of the local Paris branch of the ZZSP to the French public and was published by its "Literary Commission." This branch was to come into being only late in 1894. Then Limanowski was chosen a member of the "Literary Commission"—the activity of which will be discussed below—but he found no time to attend its meetings and could contribute his advice only. On weekdays, Limanowski, having risen at 5 a.m., usually wrote his books between 6 a.m. and 8 a.m.; at first it was his *Sociology* which he was preparing for printing—its first chapter, "La classification des sciences et la sociologie," appeared in René Worms's *Revue Internationale de Sociologie* in 1894—and following the completion of his *Sociology* in 1897 he began work on the final draft of his *History of Polish Democracy in the Period after the Partitions*, published in 1901. He had now almost no social contact apart from seeing his comrades in the ZZSP; he felt that he was living "like a hermit" and had little idea of what the Polish émigrés at large were doing. He was to confide in his brother Lucjan in August 1896: "I am truly living only in the morning when I am doing my own work."[16]

Shortly after his return to Paris, Limanowski began to correspond with the secretary of the Centralization, B. A. Jedrzejowski. Their friendly exchange of letters continued for many years. Limanowski not only attempted to answer the questions put forth to him by Jedrzejowski, but he also volunteered advice to the Centralization and became its confidential representative in Paris.[17]

Initially, their correspondence revolved around the problem of money. Prior to securing his position in the New York Life office, Limanowski literally lacked sufficient funds to buy even a postage stamp. The Centralization was desperately short of money too, and Jedrzejowski was to approach Limanowski for a loan, hoping that the latter might be able to arrange it with one of his Paris acquaintances. In his reply, Limanowski stressed that he was not in a position to obtain a loan for himself, let alone to borrow money on someone else's behalf. He complained that he had difficulty in collecting the sums owed him for his books—namely his brief *History of Lithuania (Dzieje Litwy)* which was published in both Paris and Chicago in 1895, and the *One Hundred Years of Struggle by the Polish People for Independence* mentioned above. Limanowski's "Jacobinism" became the bone of contention between him and the publishers of the latter work; and he was disgusted with wrangling over the changes proposed in the text. Finally, Limanowski won his point, having merely scored a partial victory. It now seemed unlikely that he would collect full remuneration for this book.[18]

In an attempt to recover the outstanding amount, Limanowski exchanged several letters with the leader of the now rightist National League, Roman Dmowski, in 1895 and 1897. Several hundred copies of his book were smuggled from Galicia for circulation in Warsaw, probably for the benefit of the League. In his letter of July 25, 1895, Dmowski had promised Limanowski to help him to recover the outstanding sum, but failed to live up to this promise.

In one of his letters to Dmowski, Limanowski evinced interest in Dmowski's new *All-Polish Review (Przeglad Wszechpolski)* which the latter began to edit in Lvov in 1895. In turn, Dmowski solicited Limanowski's advice regarding this paper. Yet, he rejected Limanowski's offer to write an article which would be sympathetic to the PPS. He did ask Limanowski to contribute a column in his paper— confined, however, to reviewing new books published in the West. After this correspondence with Dmowski, Limanowski lost all contact with him.[19]

On the other hand, Limanowski continued to be alienated from the Polish socialist left now led by Rosa Luxemburg, who was to become his chief antagonist on the issue of Polish independence. This issue was to become the bone of contention between the SDKP (and its successor the Social Democracy of the Kingdom of Poland and Lithuania—SDKPiL) and the PPS for a period of twenty-five years (1893-1918), i.e., from the foundation of the PPS to the transformation of the SDKPiL into the Polish Communist Party. In the gradually Russified former Congress Kingdom of Poland, which after the uprising of 1863 was renamed "Vistula Land," a Russian Governor General ruled supreme in both civil and military affairs. Back in 1851, abolition of the customs barrier between Russia and Russian Poland had opened the Russian market to Polish manufactured goods, resulting in industrialization of the former Congress Kingdom after 1863. Following the defeat of the insurrection of 1863-1864, the Polish bourgeoisie, deprived of any prospect of independence, concentrated instead on the exploitation of the Russian market as a means of economic advancement of Russian Poland as well as of individual enrichment. Rosa Luxemburg was to analyze this situation in her Ph.D. dissertation entitled *The Industrial Development of Poland* (1898). According to her thesis known as the "theory of organic incorporation," the industrialization of Russian Poland depended on access to the Russian market, resulting in an indissolluble political bond between Russia and Russian Poland.[20]

Three years before her Ph.D. dissertation was completed, in 1895 Luxemburg produced her first cohesive statement on the national question in her pamphlet entitled *Polish Independence and the Workers' Cause (Niepodleglosc polska a sprawa robotnicza)*, which she published under the pseudonym Maciej Rozga. Between 1895 and 1897 she restated her arguments on the national question, as contained in her pamphlet, in a series of articles published in *Die Neue Zeit* and *Critica Sociale*, chief theoretical organs of the German and Italian Social-Democratic parties, respectively.

Her anti-patriotism, as expressed in the pamphlet and articles, was based on two assumptions: 1) that patriotism and socialism were incompatible in general; and 2) that the concept of national self-determination did not apply in the context of the Russian Empire.[21] She felt that in Poland patriotism was bound to divert the workers from true socialism. Luxemburg considered patriotism an ideology appropriate to the middle classes. Yet, she pointed out, in Poland the bourgeoisie was not a revolutionary force; here patriotic socialists, in virtue of their ideology, had chained themselves to a bourgeosie which was itself impotent politically. She was convinced that, if threatened with social revolution, the bourgeoisie in Poland would abandon its allies in the Polish cause, the patriotic socialists, to collaborate with the counter-revolutionary Tsarist autocracy. Luxemburg

apparently believed that the Polish proletariat, as a class, was powerful enough to overthrow capitalism eventually, but by itself it would
be incapable of gaining independence for Poland. This argument was
illogical and unconvincing. Also, she implied that independence was
not a respectable aim for the workers; rather, she considered it an
aspiration of the Polish petite bourgeoisie.[22]

Rosa Luxemburg's chief Marxist antagonist in the PPS was Kazimierz Kelles-Krauz (1872-1905). Krauz, scion of a Baltic German
baronial family, was to acquire fame as a distinguished Marxist sociologist. Like Marx and Engels he argued that abolition of Tsardom
and restoration of Poland were necessary for the victory of socialism
in Europe; conversely he felt that its victory in Poland depended on
a favorable international situation. He rejected the concept of "permanent revolution," Assuming that the next social upheaval would
still be a "bourgeois revolution," Krauz insisted that it was wrong to
delay the realization of any one of its slogans, namely the slogan of
independence, until after the revolution had occurred, as Rosa Luxemburg wished to do. He objected to her "theory of organic incorporation" on economic grounds as well. Contrary to her thesis, he
believed that Poland's economic development was hampered by its
political incorporation into the more backward Russia. Krauz combined Marxism with a conscious idealization of the ancient Slav
commune. According to his *Sociological Law of Retrospection (Socjologiczne Prawo Retrospekcji)* published in 1898: "All movements
for social reform model their ideology on social practice in a more or
less distant past. . . . It is not a question of a faithful recreation of
the past, but only of an idealization which is inspired by a conviction. . . that principles of the past, if brought to life, must blend
with the achievements of the modern age. . . for the sake of a truer
synthesis and a better balance of conflicting interests."[23] This
theory was close to that of Limanowski; yet it was more realistic
than his, and it was less conducive to idealizing the Polish past.
Krauz was eventually to become Limanowski's associate and chief
antagonist in the Paris Branch of the ZZSP.

Meanwhile, the initial task of Limanowski as a confidential representative of the Centralization in Paris was to form a branch there,
and to report immediately to London any local activity on the part
of Luxemburg's agents, which might be hostile to the PPS or to its
affiliate the ZZSP. Rosa Luxemburg was to employ Millerand and
Jaurès ' organ, *La petite République*, which they had made available
around 1894 to all socialists in France irrespective of their ideology,
as one of her mouthpieces in Paris.[24] On September 28, 1894 this

paper published an article written by a member of Luxemburg's group. It alleged, among other things, that the SDKP was the only socialist party now in Poland. Limanowski immediately forwarded this article to the Centralization in London. In the accompanying letter he stressed that Luxemburg's "intrigues" must not be ignored; he also informed London of the local anti-PPS propaganda disseminated by Luxemburg's own organ, the *Workers' Cause (Sprawa robotnicza)*, published in Paris during the years 1893-1896. Thereupon, Jedrzejowski sent to Limanowski a reply to the offensive article, for publication in *La petite République*, and entrusted him with delivery, in person, of a letter drafted in London and addressed to Millerand himself. The necessary note of rectification appeared in the October 22, 1894 issue of this paper. In it the French were told that a Russo-Polish revolutionary alliance had been concluded by the *Narodnaia Volia* with the Proletariat in 1884, and that the PPS were continuing this alliance. Also, the note stressed that the independence of Poland was absolutely necessary to the cause of freedom in Russia as well.[25]

Soon Limanowski was to preside over the first meeting of the new Paris branch of the ZZSP. Since April 1894, the Centralization had attempted to enlist the so far recalcitrant *Socialist Review* group in Paris. It finally decided to join the ZZSP on October 5, 1894; until this date Limanowski was the sole member of the ZZSP in Paris. For personal reasons which he was not to disclose, he at first did not wish to become an intermediary between the *Socialist Review* group and the ZZSP; but now that the former had agreed to become the Paris Branch, he offered to act as the liaison between the *Socialist Review* group and London. (It is not clear what became of the former members of the *La Diane* organization which—unlike the *Socialist Review* group—had been represented at the founding congress of the PPS in November 1892.)[26]

Thus the Paris Branch of the ZZSP officially came into being on November 11, 1894. Its executive was to consist of two secretaries and a cashier, elected annually. Krauz became one of the secretaries. Two commissions were formed at the time: a "Financial Commission" and "Literary Commission," the latter of which was mentioned earlier. Its terms of reference were as follows. It was to concern itself with all problems pertaining to the publication by the Branch and by the ZZSP in general, as well as with ideological developments at home and abroad, especially in socialist parties. General meetings were to be held on the first Sunday of each month.[27]

As mentioned above, Limanowski could not spare the time to take an active part in the business of the "Literary Commission" and seldom contributed to periodicals during his employment with the New York Life Insurance Company. Previously, he had published a number of articles in Wyslouch's *Week*. Early in 1894 in these articles he discussed *inter alia* the two issues which were of greatest concern to him at the time, i.e., the situation in Galicia and developments in Russia.

Concerning the first issue, he criticized the current opposition by both the Galician conservatives and democrats to a broadening of the suffrage, demanded by the Ukrainians and the Polish socialists as well as by Wyslouch's populists. Limanowski considered this an ill omen for the future of Galicia as the "Piedmont of Poland," for he believed that this country in its quest for independence must rely first of all on her own resources and that, as earlier patriots like Kollataj and Kamienski had urged, it was imperative to enlist the masses in the Polish cause by means of a "mild" social revolution. Limanowski adopted here Kamienski's somewhat vague definition of a "social revolution" in Poland; he viewed it as "an organic function," to be accomplished with full social consensus if this were at all possible. Limanowski now became fearful that the current opposition by the upper and middle classes to universal manhood suffrage in Galicia might eventually even lead to a violent upheaval which would be detrimental to the Polish cause.[28]

Regarding the second issue, that is, developments in Russia, Limanowski of all Russian radicals contemporary to him was ideologically closest to Lavrov. (Lavrov resided in Paris around the time of Limanowski's sojourn there.) In May 1893, the Russian Marxist Plekhanov had attacked the Russian *narodniki*, including Lavrov, in an Open Letter in pamphlet form entitled "On Social Democracy in Russia (*O sotsialnoi demokratii v Rossii*)," which he addressed to the Polish publishers of Alphons Thun's *History of the Russian Revolutionary Movements (Geschichte der revolutionären Bewegungen in Russland)*, originally published in Leipzig in 1883. According to Plekhanov, the *narodniki* considered the struggle for political freedom a harmful "bourgeois" idea (though they themselves were forced by events to engage in such a struggle); "politics," he went on, was anathema to every "orthodox" Bakunist and *narodnik*. Plekhanov pointed out that this apolitical approach, which he considered to be a legacy of the Utopian socialists' opposition of "socialism" to "politics," combined with a native Russian brand of Slavophilism, was both impractical and incongruous in theory; every class struggle was perforce a political one. In a reply contributed to the *Week*, Limanowski contended that it was not necessary for Plekhanov to labor, in an unctuous manner, the obvious—that is, that a political liberalization was the prime need for Russia and a means to the country's social transformation. He objected to Plekhanov's ridiculing Lavrov as an eclectic "Utopian socialist" who was ignorant of Marxism.[29]

Meanwhile, on the outbreak of the Sino-Japanese War in 1895, Limanowski became hopeful that Russia would also become involved. He suggested to Jedrzejowski that in the event of such a development, it might be possible to enlist the United States as an ally in the Polish struggle against Russia. Namely, he proposed the formation of an armed force made up of Polish exiles and financed by American capitalists; it would attempt to separate Siberia from Russia and set up the former as an independent state. Limanowski felt certain of moral backing for this venture from the American people; he was hopeful, too, of material support from American capitalists, whom he believed to be economically interested in Siberia. He felt that Polish socialists ought not to refuse financial aid from American capitalists in support of this project, if offered; he identified the cause of the independence of Siberia with the Polish cause and the cause of freedom everywhere. Moreover, he had heard that the American critic of Tsarist Russia, George Kennan, was agitating for Siberia's independence and had already discussed this problem with some Polish and Russian exiles.[30]

However, Jedrzejowski was completely out of sympathy with "non-party" projects like that proposed by Limanowski; the former argued that only socialist activity might deploy Polish manpower resources effectively at this time. Jedrzejowski himself doubted, quite sensibly, the existence of any separatist movement in Siberia and he dismissed the possibility of Russia's involvement in the Asian conflict. Also, he was sceptical of the benefit to the Polish cause from this venture, and of securing Polish volunteers. Most important, as Jedrzejowski pointed out, the ZZSP was not now in a position to sponsor a venture of this sort; it did not even dispose of a sum of several hundred francs required to rent a room for a single agitator! London felt that any démarche on their part in this matter would only make the ZZSP appear ridiculous.[31]

In reply, Limanowski explained that the idea of setting up an independent Siberia had been suggested to him on a number of occasions; and he himself had been thinking it over for a long time. Limanowski continued to believe that many Poles would be only too willing to fight against Russia in Asia, just as some of them were then, in 1895, engaged in fighting against Spanish rule in Cuba. Furthermore, he reassured London that he was not contemplating any kind of Siberian expedition at the moment but, aware that a member of the ZZSP, Stanislaw Grabski, was about to go to the United States, he felt that London ought to instruct him to sound out Kennan, as well as representatives of the *émigré* Russians in North America, for their opinion in the matter. The idea of this Siberian venture seems preposterous today; it was certainly unrealistic of Limanowski to expect that even Russians themselves might be willing to bring about a dismemberment of their country.[32]

Nevertheless, Limanowski might usually count on support among those exiled Poles who were members of, or allied with, the ZZSP or the PPS. For instance, in 1895 a Polish socialist student organization in Paris, the "Union (*Spojnia*)," arranged a solemn celebration of Limanowski's thirty-fifth anniversary as a writer: on this occasion, they praised him for his perseverance in the two causes of independence and socialism. However, when in the same year he spoke before the Polish *émigrés* in Paris at large, at a commemoration of an anniversary of the 1830 uprising, he incurred the displeasure of some of his audience by deploring that the ideological heirs of the Democratic Society, the National Democrats, were now hostile to Polish leftist parties like the PPS and Wyslouch's newly founded Peasant Party (SL).[33]

Having thus antagonized some of the non-socialist exiles on his right, Limanowski was soon to earn the enmity of Polish socialists on his left as well. When the Paris Branch of the ZZSP commemorated on January 28, 1896, the tenth anniversary of the execution of the four leaders of the Proletariat (Kunicki, Bardovsky, Ossowski and Pietrusinski), Limanowski came under attack from a representative of the SDKP in Paris, Adolph Warski [Warszawski]. During his speech given before an audience composed of members of the Paris Branch and their guests, both Poles and Frenchmen, Limanowski in surveying the development of socialism in Poland ignored the existence of the SDKP. Thereupon, Warski protested, arguing that the PPS had no right to consider itself heir to the Proletariat, for the PPS was merely a patriotic organization which used socialism as a means to attract the workers.

Although Limanowski was not ashamed of his patriotism, he did not reply to Warski. He was aware that its patriotism had made the PPS suspect in the eyes of the French socialists and he despaired of SDKP was ideologically closer to the Proletariat than the PPS. Thus, he did not trouble to reply to Warski's arguments. But Limanowski himself failed to perceive that Warski's charge had a basis in fact;

there were many within the PPS who considered socialism a mere means to independence. His usual tolerance did not extend to those who opposed Polish patriotism. As a result of this controversy, Warski, who had helped Limanowski to secure his post at the New York Life, ceased to have any dealings with members of the ZZSP. Both the PPS and the SDKP now challenged each other's claims to act as the representatives of Polish socialism.[34]

Just prior to the next international socialist congress which was held in London in July 1896, Rosa Luxemburg published in *La petite République* a resolution renouncing the struggle for Polish independence, which she planned to introduce at the congress. Limanowski hastened to inform the Centralization of her intention, urging that London prepare a note for publication also in *La petite République*, denying that the PPS had had any part in drafting such an anti-patriotic motion. He suggested also that the ZZSP provide each delegate to the forthcoming congress in London with a detailed critique of Luxemburg's resolution, which would at the same time deny that the PPS wished to resurrect a "bourgeois" Poland while asserting that the goal of Polish independence was less Utopian than Luxemburg's expectation of a constitution in Russia.[35]

Limanowski could not attend the congress, because he was not given leave of absence from his office. However, he wrote an Open Letter in French to the delegates. It informed them that the PPS aimed in the long run at a socialist Poland, but that its immediate aim was "national unity" for the sake of regaining independence. According to Limanowski, this was a moral imperative, in view of the repressive regimes in both Prussian and Russian Poland, which stifled all manifestations of Polish life—culture, intellectual development and material well-being as well. Limanowski pointed out (as he had done at the congress of 1889) that in the eighteenth century the Polish-Lithuanian Commonwealth (what would become Prussian Poland) possessed more industry than other regions of the country. Annexed to the more economically advanced—but socially conservative and chauvinist—Prussia, it became the least radically inclined area in Poland and it proved, too, least conducive to socialist activity. Due to the province's separation by political boundaries from Austrian and Russian Poland, the unemployed proletariat in Prussian Poland could not migrate to either of these two regions and, instead, had been forced to seek work in Western Germany (only part of this proletariat found employment in Prussian Upper Silesia); there Polish workers helped to lower the local wages as well as to reinforce the strength of German conservatism by virtue of their political ignorance. In Galicia, despite its very dense population and rich natural resources, industrialization had proceeded very slowly, due to the competition from goods manufactured in the more advanced parts of Austria, the customs barrier with Russian Poland, and the restricted character of the local market. In the most industrially advanced area, Russian Poland, manufacturers were now facing stiff competition from their counterparts in Russia; Polish goods were subjected to discriminatory railway tariffs demanded by the Russian industrialists. Nevertheless, Limanowski stressed that he attached less weight to the economic argument in support of Polish independence than to the moral imperative of doing away with the humiliating dismemberment of Poland. He argued above all that Polish socialists should not remain indifferent to the national tragedy of being partitioned among three hostile powers. [36]

Limanowski told the congress that the Polish tragedy was proving detrimental to the cause of progress everywhere, as Marx, Engels, Becker and Liebknecht had argued; the partitions had strengthened Russia and Germany, which were the mainstays of militarism and reaction in Europe. He stressed, moreover, that the cause of Poland was the cause of international peace and order; subjugated Poland would remain the powder-keg of Europe. [37] Actually, Limanowski now envisioned a resurrected Poland (even if a "bourgeois" Poland) as the *cordon sanitaire* separating Russia from Europe.

Limanowski explained, too, why he considered Polish independence a less Utopian idea than a constitution in Russia; he felt that her rigid bureaucracy and conservative society, which was imbued with an authoritarian spirit nurtured by the Orthodox Church, precluded liberalization in Russia. He dismissed as ridiculous the idea that Polish aspirations might impede the struggle of the Russians themselves with their autocratic regime, believing that the growth of Polish separatism would actually accelerate the social and political transformation in Russia by forcing the Tsar to make concessions. (This belief was contrary, however, to the opinion of National Democrats like Dmowski.) Limanowski countercharged that the Russian critics of Polish separatism in the press and in society, as well as the SDKP—in effect though not in intent—all were seconding the anti-Polish propaganda of Russian officialdom. Finally, he criticized the "economism" of the SDKP; he felt that its fatalism, i.e., an apparent reluctance to engage in a political struggle, was tantamount to a lack of revolutionary spirit. [38]

Limanowski stressed to the London congress—probably for tactical reasons—that independence was necessary for Poland as a goal in its own right. Earlier, at the Possibilists' international socialist congress of 1889 he had put emphasis on Polish independence as a means to socialism. Convinced that socialists must fight national oppression as well as social exploitation, he asked the delegates to vote for a resolution endorsing the idea of struggle for Polish independence. As he told them: "All peoples have the right to national self-determination and you ought to morally support all their efforts to this end."[39] Actually, the congress adopted a compromise solution, resolving that all nations were entitled to self-determination, but without reference to Poland as an especially deserving instance.[40]

Meanwhile, by 1896 Limanowski became a very popular figure with the Polish socialist youth in Galicia. In May 1896, one of Limanowski's correspondents in Cracow, Zygmunt Klemensiewicz, writing on behalf of a local student group at the Jagiellonian University of Cracow, had invited suggestions from Limanowski regarding a cycle of lectures which these young people contemplated giving before working class audiences in Cracow and its vicinity. In appealing to Limanowski for support, Klemensiewicz had referred to him as "Boleslaw Limanowski, that great and dependable friend of Polish youth."[41]

In December of that year and again in July 1897, Klemensiewicz asked Limanowski for contributions to the new socialist organ, the *Right of the People (Prawo Ludu)*, edited by him in Cracow, which was designed for peasant readers. This periodical was published by the new Polish Social Democratic Party of Galicia and Teschen Silesia (PPSD).[42] In reply to Klemensiewicz, Limanowski sent him two articles on the peasant problem both entitled "Peasants and Socialism," which the latter—for an undetermined reason—published in two other organs of the PPSD, *Forward (Naprzod)* and in *Workers' Calendar (Kalendarz Robotniczy)*.

As already mentioned, a Polish Peasant Party (SL) was formed in Galicia in 1895; it had elected a number of deputies to parliament, both peasants and intellectuals. In his first article Limanowski deplored that peasant deputies, like Bojko and Wojcik, were treated in parliament with contempt not only by conservatives but by some democratic deputies as well. And yet, as Limanowski pointed out, these very moderate peasants "resembled the Polish populist democrats of the 1830s." As a spokesman for the PPS, Limanowski felt—and this was the moral which he sought to bring home in this article—

that experience was bound to convince the more enlightened peasants in Poland that their only true allies were to be found in the socialist camp.[43]

However, the peasant deputy Wojcik, far from taking this view, had rejected socialism as contrary to peasant aspirations. In his second article Limanowski argued against Wojcik that peasant suspicion of socialism was based on sheer "superstition." He contended that the majority of peasants in Galicia, who were either landless or farmed dwarf holdings, would, once they recognized the facts, be eager to support socialization of agriculture. He ignored the opinion of substantial peasant proprietors like Bojko and Wojcik, because he believed they were "doomed to extinction" as an economic group. Limanowski reiterated in this article his old argument for socialization of land; he continued to feel that it was a means of equalizing income and increasing the productivity of the soil. Again, he justified his belief in agrarian socialism, by appeal to the native Polish tradition as transmitted by Mickiewicz and Lelewel as well as by presenting a Marxist analysis of the peasant problem on the model of Becker's *Manifesto to Farmers*, already discussed above.[44]

Notwithstanding his "Marxist" analysis of the peasant problem, Limanowski continued to struggle with the leader of Polish Marxist socialists, Rosa Luxemburg. In the *Week*, in the course of commenting on a collection of articles written by Liebknecht during the Russo-Turkish War of 1877-1878, Limanowski charged Luxemburg with Russophilism. Liebknecht had felt that those who sympathized with the plight of the Cretans and Armenians, but were indifferent to the plight of the Poles, were either confused, hypocrites, or hired agents of the Russian government. In Limanowski's view, the Polish socialist left, Russophile out of conviction, was more dangerous to the cause of Poland than those misguided or self-interested individuals, to whom Liebknecht had alluded. And Rosa Luxemburg, in advocating the "organic incorporation," unintentionally became an ally of the Polish conservatives. In the guise of the then "fashionable" historical materialism, she had "maneuvered" her theory of capitalist development to suit her *ad hoc* purpose. He pointed out her inconsistency in arguing, on the one hand, for integration of each section of Poland with the respective partitioning power and, on the other, championing the completion of national self-determination in the Balkans. According to Limanowski, Luxemburg, when told that this last might facilitate the aims of Russian imperialism, had contended that the current process of industrialization in Russia was not

conducive to adventurism in foreign policy. Limanowski regarded
Luxemburg's belief in the peaceful intentions of the Russian capitalists as entirely fallacious; he agreed with Liebknecht, for instance,
that the capitalists in Germany had supported Bismarck's aggressive
foreign policy. (Limanowski was unaware that, in attributing aggressive intentions to an industrializing Russia, he had contradicted his
earlier belief—in the Saint-Simon tradition—in industrialization as a
means to lasting peace.) A greater Russophobe now than he had
been twenty-six years earlier in Lvov, Limanowski preferred the
Turkish yoke to Russian rule in the Balkans. He favored national
self-determination for the remaining unliberated Balkan areas only
because, mindful of the example of Bulgaria, he was now certain of
their "ingratitude" toward their potential "liberator," Russia.[45]

In 1898, to mark the fiftieth anniversary of the European movements for national and social liberation of 1848 and 1849, known
collectively as the "Spring of Nations," the Centralization had issued
a special Polish edition of Marx's *Revolution and Counter-Revolution
in Germany*. At Jedrzejowski's request, Limanowski contributed
inter alia a brief account of events in Poland from 1846 to 1849, to
be appended to this pamphlet. In addition, Leon Wasilewski, the
editor of the ZZSP organ in London, *L'Aurore*, asked him to prepare
a special article for this paper, commemorating the revolutions of
1848. (Limanowski had met Wasilewski at the Wyslouchs in Lvov in
1893.)[46] Limanowski entitled this article "1848-1898." In it he
concentrated on events in France. He pointed out that it was due to
the February revolution of 1848 that France became the first country in Europe to adopt a universal system of male suffrage. But he
considered the bloody defeat of the French workers in June 1848
more significant than the achievements of the February revolution;
he felt that the former had contributed to the re-awakening of class
consciousness in Europe, and had initiated the process of transformation of the proletariat into an autonomous political force. According to Limanowski, thus had begun a new " scientific" era in the
history of socialism.

And now, Limanowski in his prediction of the future course of
struggle for socialism in Europe, became ostensibly even more radical than Engels. Limanowski disagreed with Engels that civilian
struggles on the barricades were rendered obsolete by advancement
in military technology. He felt that a passive general strike was ineffective as a means ot winning power for the proletariat in those
countries in Europe which still lacked universal manhood suffrage,
and that the proletariat might not preserve its present gains, nor

secure new concessions in the future, unless the danger of the struggle on the barricades continued to menace the ruling classes in Europe. He argued that, though the broad width of modern city streets complicated such struggles, there was plenty of new material to be used for building barricades, like overturned streetcars. He believed that a barricade had become as much a moral barrier as a physical one and that modern armies, especially in the democracies, were not likely to become blind instruments in the hands of the ruling classes. More radical when discussing socialism in Europe rather than specifically in Poland, Limanowski nevertheless appeared to oscillate here between the notion of political democracy as an end in itself and the notion of political democracy as a means of creating democratic socialism.

Limanowski went on to argue that the gulf between the army and the civilian population was no longer as great as it was in the past; the proletarian was a demobilized soldier and, conversely, the army consisted mainly of workers and peasants. He felt that the proletarian soldier was bound to become sympathetic to socialism, but he believed that in the Continental armies (especially in Eastern Europe) the peasantry was still in the majority. He feared the proprietor peasant as a counter-revolutionary force; yet he expected an ever-accelerating process of peasant proletarianization which, in his opinion, boded well for the victory of socialism. It was the peasant then, according to Limanowski, who held the key to the future of socialism.[47]

In 1898, in addition to commemorating the fiftieth anniversary of the "Spring of Nations," the Poles solemnly celebrated the one-hundredth anniversary of Mickiewicz's birth. At the request of Maria Wyslouch, Limanowski wrote an article about Mickiewicz for the *Week*.[48] Limanowski argued here that Mickiewicz, in spite of being a poet and perhaps for this very reason, was able to predict the revolution of 1848 in France in his two fragments known as "History of the Future." (This was fictional historiography, rooted in the eighteenth century rationalism, which must be distinguished from the poet's romantic, mystical prophetic writings.) Like Mickiewicz, Limanowski awaited a European war that would result in expulsion from Poland of the foreign invader and of the "loyalist worshippers of the Tsar."[49]

Yet, many *émigré* Poles seemed indifferent to the fate of the fatherland and socialism. The ideology of the Paris Branch of the ZZSP had a very limited appeal to the Polish community in Paris; since 1894 there were few new arrivals from Russian Poland, while Poles of the earlier emigrations were usually "careerist." In addition

to publishing the *Bulletin Officiel* for the French, the Paris Branch
arranged lectures and commemorated socialist and patriotic anniversaries. For instance, in 1895 it sponsored a lecture before members of a Polish organization sympathetic to socialism, the so-called
Association of the Working Poles, and arranged a May Day celebration, for which it provided a speaker. Actually, it was Krauz who
became the chief figure of the Paris Branch in its first stage of existence from 1894-1899. Krauz wished it to give an ideological lead to
both Prussian and Russian Poland, because he believed the freedom
enjoyed in emigration facilitated ideological training of party members and their detachment from the pettiness of local concerns.
However, he did not wish to subordinate the home organization to
the ZZSP; he felt that close collaboration with the party at home
would in fact simplify the task abroad.[50]

From the beginning of the existence of the Branch, Limanowski
opposed its program in a pamphlet form, prepared for publication
by Krauz in 1894. It was entitled the *Class Content of Our Program
(Klasowosc naszego programu)* and laid down two policy guidelines:
1) commitment of the PPS to the struggle for independence; 2) rejection of alliance with non-socialist parties. This reflected Krauz's
fear that the PPS might become dominated by people who were
mere patriots. He believed that the PPS program of 1893 was completely Utopian. In making a struggle with Russia their immediate
aim, its authors had failed to foresee that this was bound to attract
non-socialist elements to the PPS. Instead, Krauz advocated common
tactics to serve both the ends of independence and socialism. His
motto was "Independent Poland for the Proletariat, and not the Proletariat for Independent Poland." Krauz predicted that non-socialists
within the party would use it as a stepping stone to a career in the
"national government," in the "national universities," or in the "national press" of the future Polish state. Despite his strong criticism
of the SDKP, he felt that the PPS had more in common with the latter than with its own members who were merely "pseudo-socialists."
Krauz was even prepared, if absolutely necessary, to abandon the
slogan of independence to preserve socialist purity within the PPS.
Limanowski objected; he believed that publication of Krauz's pamphlet would prevent the PPS from becoming a strong party. Apparently, however, Limanowski was in a minority within the Paris Branch,
for he received only four votes while Krauz received eight.[51]

Again, in 1896 Limanowski clashed with the Paris Branch (i.e.,
Krauz) regarding admission of the delegates of the SDKP to the

forthcoming international socialist congress which, as mentioned above, took place in London in July of that year. Unlike Limanowski and the Centralization in London, Krauz believed that the PPS should welcome the representatives of the SDKP to the Polish socialist delegation, provided they had valid mandates from home.[52]

Moreover, Limanowski dissented from the opinion of the Branch, too, on the occasion of the student riots which took place at the University of Warsaw in 1899. Wladyslaw Studnicki-Gizbert, the assistant to the editor of *L'Aurore*, had written then an article entitled "Votum separatum" which opposed any manifestations of solidarity with the current unrest among Russian students at the University in Warsaw. The Paris Branch took exception to Studnicki's article for two reasons. They felt that his Polish "separatism," with its strongly anti-Russian coloring, was contrary to the principle of socialist internationalism, and that it placed the leftist PPS in the same political camp as the right-wing National Democrats, who had condemned these student riots too. Yet Limanowski disassociated himself immediately from this protest by the Paris Branch against the article of Studnicki.[53]

Almost from the beginning of its existence, Limanowski desired to transform the ZZSP into an organization for the whole of Poland, rather than being representative of only Russian Poland (as it in fact was). He felt that this change would enhance the prestige of the ZZSP among its own members as well as among outsiders. He suggested a motion—to be presented at the annual congress of the ZZSP in December 1895—whereby the validity of resolutions passed at the annual congresses of the ZZSP (always held in London) depended on attendance of representatives from all three parts of Poland. Should some of the delegates from the home country fail to arrive, he proposed that special mandates be given to members of the ZZSP to act on behalf of these absent delegates. Since it was difficult to secure regular representation from Poland, Limanowski's proposal seems quite unrealistic.[54]

In fact, a reorganization of the PPS did take place in 1899; the ZZSP and its branches then became an integral part of the home organization, subject to control by the Central Workers' Committee in Warsaw. The Centralization became the Foreign Committee (*Komitet Zagraniczny*) of the PPS; it was now appointed in Warsaw. Initially, the Paris Branch opposed this change; it objected to the loss of autonomy by the ZZSP which became the Foreign Department of the PPS (*Oddzial Zagraniczny PPS*).[55]

Although both Krauz and Limanowski opposed this reorganization, they did so for divergent reasons. Krauz desired decentralization to stimulate the intellectual life of the party by means of exchange of ideas; to foster individual initiative and activism; and to democratize the party on the model of some of the foreign socialist parties. On the other hand, Limanowski represented the centralizing tendency within the PPS. He wished to transform the ZZSP into an organization which would embrace the whole of Poland as well as the Polish socialist movement in the United States. Once the reorganization was agreed upon in 1899 by a majority of the ZZSP, he reluctantly accepted its verdict. Yet in May 1900, he told London that he still hoped for organizational unity, which he felt might now come about by means of the PPS in Russian Poland absorbing the separate organizations in Poznania and Galicia, and that he wished to raise the issue again that year at the forthcoming international socialist congress in Paris. But London were of the same opinion that unity might be maintained only by the three separate Polish delegations acting in solidarity at international socialist congresses. For practical reasons, the Foreign Committee did not pursue his suggestions for further reorganization. Since almost all of the members of the ZZSP were refugees from Russian Poland, therefore they already belonged to the PPS. In these circumstances, London considered the new definition of the relationship between the home and *émigré* organizations, supported as it was by a majority of the latter, to be in a sense an acknowledgement of the status quo.[56]

Ostensibly the reorganization of 1899 aimed at removing grounds for conflict between the activists in Poland and those abroad, in the cause of efficiency. It seems that it was principally due to the chronic shortage of funds in London; in exchange for being relieved of financial obligations, the *émigrés* agreed to subordination to the PPS in Russian Poland. At the decisive ZZSP congress of January 1899, at which time the majority voted for the motion of reorganization as sponsored by the home party, the delegates from Paris (with Limanowski, as usual, absent), now aware of the situation, unanimously supported the verdict of the majority.[57]

Between 1889 and 1900 Limanowski could not secure leave of
absence from his office, in order to participate in socialist gatherings abroad. However, in 1900 he was able to attend a socialist congress in Paris and was chosen chairman of the Polish delegation. Just
before the congress began to assemble, Rosa Luxemburg in an article
published in *Vorwärts* attacked the PPS for sending an allegedly false
report to that paper on the party's activity in Poland during the past
five years; she was also to argue with the delegates of the PPS, in
person, at the congress. To retaliate against Luxemburg's attack in
Vorwärts, some members of the Polish delegation, like Daszynski
and Limanowski, signed a resolution declaring Rosa Luxemburg
"not worthy of associating with Polish socialists, pending her public
apology for slandering them." As a result, Luxemburg was excluded
from the Polish delegation. She participated in the deliberations of
the congress only because the German socialists invited her to join
them.[1]

Around the time of the congress, the Polish delegation issued a
declaration drafted in French by Limanowski. In it he paid homage
to all those members of the PPS who had suffered a martyr's death
in Russian Poland. His intent was to make foreign socialists aware
that in Russian Poland, unlike in the West, the penalty for political
activism was often death; he told the congress on September 22 (the
day before its deliberations began) that six of his comrades in the
PPS had been condemned to death by a Russian court martial in
Warsaw.[2]

In January 1901, a few months after the congress, Rosa Luxemburg's followers in Paris commemorated the fifteenth anniversary of
the execution of four leaders of the Proletariat (Kunicki, the Russian
jurist Bardovsky, Pietrusinski and Ossowski), hanged by the Russian
authorities in Warsaw. (The SDKP had almost ceased to exist in
1896; it was reconstituted around 1900 by the future Bolshevik

Feliks Dzierzynski as the Social Democracy of the Kingdom of Poland and Lithuania—SDKPiL.) On the occasion of this anniversary,
the French Marxist leader Guesde wrote a letter to the organizers
of the commemoration (i.e., members of the SDKPiL), which
glorified the "martyrs"; he referred in it to the Proletariat as an
affiliate of the Russian *Narodnaia Volia* and criticized the PPS for
being a nationalist rather than a socialist party. His letter appeared
in the socialist periodical *Le petit sou*, published by Guesdists and
Vaillantists in Paris. Limanowski, angered by these remarks, replied
to Guesde in an Open Letter and communicated it to several socialist papers and personalities in Paris, as well as to the International
Socialist Bureau itself. Among other things, Limanowski told Guesde:
"Celui qui a exprimé la fameuse opinion que 'Guillaume ou Loubet
[i.e., the German Kaiser or the French President], empire ou république, c'est tout un pour le proletariat,' comprend bien la lutte des
classes à la manière de Bakounine, mais jamais à la manière de
Marx."[3]

Meanwhile, back in February 1900, the Paris Branch of the PPS
had formed the Jubilee Committee to arrange a solemn celebration
of Limanowski's sixty-fifth birthday, as well as of the fortieth anniversary of his career as a political writer. The Committee, in its first
appeal to compatriots for support, had referred to Limanowski—
somewhat exaggeratedly—as "the Nestor of Polish socialism, a stead-
fast patriot and democrat, as well as a distinguished sociologist and
philosopher."[4]

Around the time of the Paris congress, on September 29, 1900,
the Jubilee Committee had arranged a banquet in honor of Limanowski, to which they invited representatives of all local Polish organizations. After all the congratulatory messages had been read—
these came chiefly from socialist organizations sympathetic to the
PPS—Limanowski was honored in a number of speeches. The most
noteworthy was given by Krauz. In commenting on the erstwhile
estrangement of Limanowski from Polish socialism, Krauz analyzed
the development of the movement in terms of the Hegelian dialectic;
he pointed out that in order to assert itself as an independent political force Polish socialism had to attack its ideological "parent"—the
erstwhile democratic movement for social justice and independence.
Krauz felt that by 1892 Polish socialism, now a confident and autonomous political force, was in a position to reclaim the legacy of
Polish democracy; by synthesizing socialism with the old democratic
ideology, the PPS had thereby advanced onto a higher ideological
plateau and acquired the right to aspire to the leadership of the entire

nation, under the slogan of independence. And, according to Krauz, Limanowski himself played a vital role in this dialectical process of development of Polish socialism. Turning to Limanowski, he said: "But you, comrade, nothwithstanding all your criticism [of Polish socialism in the past] were always so understanding. . . . that when the right moment arrived. . . you were ready for unconditional collaboration with us, without any reservations whatsoever! And that is why you yourself constitute an example to all Polish patriots, an example of how to be democratic and patriotic. . . . and how to gain the workers' trust. . . ." Krauz went on to deal with the rather unique personal characteristics of Limanowski: "It is [relatively] easy to be a thinker, [relatively] easy to be a scholar and writer, but to demand no reward for oneself, to live one's entire life dedicated to an ideal, without any personal ambitions and so completely disinterested financially—as is Limanowski—this is most unusual!"[5]

In his speech delivered at the banquet, Limanowski deplored the split within the Union of Polish Students Abroad, which had taken place back in December 1899, at the XIII congress of the Union. There, the non-socialist majority had been hostile to anti-government student demonstrations in Warsaw(which, as mentioned above, had occurred that same year). Consequently, the socialist minority had seceded from the Union and formed their own Union of Progressive Youth. Nevertheless, both student Unions continued to honor Limanowski; and their respective representatives came to his Jubilee banquet.[6]

On this occasion, Limanowski was impressed with his growing popularity at home as well, even among non-socialists, due to the many tributes and gifts which he received from Poland. It is noteworthy that Limanowski's writings had a special appeal to women and young people. He treated them as equals, sympathizing with their aspirations.[7]

The Jubilee Committee in Paris were eventually to publish a special book in Limanowski's honor, entitled *Socialism-Democracy-Patriotism*; it netted him nearly 1000 francs in several installments, which was equivalent to about four months of the wages he then earned in Paris as an insurance clerk.[8]

In this book Limanowski included an autobiographical sketch, entitled "How I Became a Socialist," which he had written at the request of a socialist high school student group in Cracow, known as "Radiant Ones *(Promienisci)*." Many of these young people, for whom Limanowski's ideology became a source of inspiration along with the example of Kosciuszko and the legacy of the Democratic

Society, were eventually to play an important role in the Polish socialist movement (notably the historian, biographer and admirer of
Limanowski, Adam Prochnik).[9]

Abroad, around the time of Limanowski's Jubilee banquet, the
PPS organ *Light (Swiatlo)* in commenting on his career as a sociologist, stated that "the Polish socialist camp was proud of this their
veteran still so fully of energy and enthusiasm," adding that Limanowski was a man of a "crystal clear" character as well as a conscientious party worker, whose mental powers had remained unim-
paired despite the difficult conditions of his existence in exile.

At home, the Galician PPSD had honored Limanowski on the
occasion of his Jubilee by arranging a special issue of its *Workers'
Calendar*.[10] A biographical sketch also appeared anonymously in
the *Week*; in it the author referred to Limanowski as "one of the
most distinguished and most sympathetic of our scholars and publicists," and he especially praised Limanowski as a political opponent
of the conservative Cracow school of Polish historiography and as
one who had re-awakened pride of their past among all progressive
Poles.[11]

Except for this eulogy in the *Week*, none of the Polish non-socialist
papers, or major political organizations at home or abroad, remembered Limanowski at this time. True, the Union of Polish Exiles
in Paris had honored him on this occasion of his Jubilee by greeting
him as their former member and a popular historian, but they were
silent about his socialist activity. All National Democratic organizations, including the National Treasury and the Council of the Polish
Museum at Rapperswil, ignored Limanowski, thereby snubbing the
PPS as well. This upset its leaders who had hoped that the occasion
would help to effect a detente with the National Democrats. In fact,
Limanowski's Jubilee resulted in an even greater estrangement between the PPS and the National Democrats than had existed thereto.[12]

Limanowski himself was soon to experience the enmity of the
National Democrats. In 1896 the Union of Polish Student Societies
Abroad had decided to finance the publication of his *History of
Polish Democracy*. This book went to press in 1897. However, the
National Democrats soon began to gain the upper hand in the student
Union; they resented the idea that Limanowski's work was being
printed by the PPS in London. Even prior to the split which, as al-
ready mentioned above, took place within the Union at the end of
1899, they had begun to delay the printing by withholding the
funds to cover its cost. Moreover, the National Democrats in the
Union compelled Limanowski to omit his introduction, because in it

he had criticized the Viceroy Badeni, with whom their leaders were now on excellent terms. Also, they had demanded of Limanowski that the place of publication be shown as Zurich rather than "socialist" London. After the split, both student organizations began to argue over their respective right to Limanowski's *History*; this caused a further delay in completing the printing. In addition to contending with the Union of Polish Student Societies Abroad, the PPS also wrangled on Limanowski's behalf with the bookbinder who failed to fulfill the terms of his contract. Finally, after many vicissitudes, in 1901 the *History* was ready for distribution. But the Union was now reluctant either itself to promote it, or to release a few copies to be reviewed by outsiders. Six months after the book was ready, the entire edition still lay untouched in a warehouse in London. Being angry that partisanship had thus impeded the common cause, as well as inconveniencing him personally, Limanowski in a letter to Wyslouch of July 22, 1901 was asking him to admonish the leaders of the Union in his *Lvov Messenger*, in order that they might be shamed into releasing his book.[13] However, as late as 1903, when the literary critic Wilhelm Feldman wished to review it in his *Criticism (Krytyka)* in Cracow, the book was still unavailable in Galicia.[14]

The reason for this was its lack of appeal to the non-socialist student Union. A scholarly version of the popular *One Hundred Years of Struggle for Polish Independence* (1894), this book was based on Limanowski's lectures in Zurich in 1887-1889 and on his subsequent studies in Paris. In it Limanowski concentrated on the activities of the Democratic Society; he emphasized here its struggle for independence, which accorded with his belief that independence was the primary aim for every Pole, irrespective of his social creed.[15]

His emphasis on the national issue is not surprising in view of his growing Russophobia, manifesting itself especially when he discussed the problem of the national minorities in Russia. In October 1901 Limanowski was approached by Jedrzejowski to write propaganda pamphlets for translation into Yiddish, in order to awaken patriotism among Polish Jews. He wholeheartedly approved of the idea, provided—to enhance their effectiveness—they were to be written by a Jew. He then argued that Polish Jews would be better off in an independent Poland than under Russian rule; he assumed that those Jews who had participated in the anti-Russian national movement in Warsaw in 1860-1862 were well aware of this. A protagonist of Jewish assimilation in the past, Limanowski was now in favor of granting the Polish Jews the right to full self-determination. Though he condemned contemporary Zionism as a reactionary political trend,

Limanowski believed that eventually a progressive patriotic movement would also arise among Polish Jews; its leaders, in his opinion,
would in all likelihood come to terms with the PPS. He preferred to
deal with Jewish patriots, rather than with Russified Jews.[16]

In April 1902, Limanowski complained to Jedrzejowski that the
Paris Branch had invited a Russian socialist to speak to them in Russian during the forthcoming May Day celebration. And, the following March, he indignantly told London that he had just been snub-
bed by the leader of Proletariat III in Russian Poland, Ludwik Kulczycki, who invited a number of Russian *émigrés* to his Paris lectures
on European socialism, but initially omitted including Limanowski.
(The Proletariat III, which seceded from the PPS in 1900, was a
splinter group aiming at constitutional transformation in St. Petersburg as a means to social revolution both in Poland and in Russia.)
However, when Kulczycki eventually did extend an invitation to
Limanowski, the latter decided not to go.[17]

Limanowski was now convinced that the Russian administration
was more oppressive than German rule in Poland; and he believed
that a constitution in Russia—if it came—was bound to be inferior
to that now in existence in Germany. Writing in the *Week*, back in
February 1900, Limanowski had argued that under the milder German rule censorship was less strict than in the Russian parts of Poland. In Germany there was a legal socialist political party; Poles
could allegedly still speak in their native language in public everywhere, and they were permitted to buy land, which the Russian
authorities forbade them to do in the Lithuanian and Ukrainian
provinces.

Limanowski himself was to prove his contention that Polish might
be spoken freely in the areas under German rule to be erroneous, for
instance when he was dealing with the problem of Mazurians who inhabited East Prussia, as will be shown below. In addition, he now
implicitly belied his earlier belief that Poles under German rule faced
a serious threat of eviction from the land; he had once agreed with
the Polish philosopher Libelt that in Prussian Poland it was imperative for Polish peasants and large landowners to collaborate, in order
to resist jointly expropriation by powerful German capital, which
was backed by the German state.[18]

In the German-occupied parts of Poland the native element was
Polish, or of kindred Western Slav stock, of whom the majority were
Roman Catholics. On the other hand, in the eastern borderlands, the
majority of the population differed, both ethnically and in religion,
from the Polish gentry settled there. Hence, Limanowski sometimes

misunderstood the aspirations of the borderland peoples. When late in 1902 the PPS began issuing a socialist paper—the *Little Farm (Hutorka)*—in the Cyrillic alphabet (*grazhdanka*), as requested by the Belorussian Department of the party, Limanowski strongly opposed this in a letter to London. He felt that most Belorussians preferred the Latin alphabet, even if they were Orthodox, because among the latter those who were former members of the Uniate church still secretly favored Catholicism. He believed that the Latin alphabet symbolized the anti-Russian feeling existing in Belorussia. However, the Foreign Committee in London contended that they had decided to use the Cyrillic alphabet, arguing that this was actually the best means of appealing to the Orthodox among the Belorussians, whose approval the PPS was especially seeking.

A few months later, in March 1903, Limanowski objected to a pamphlet written by a leader of the Ukrainian Social Democrats in Galicia, Vityk, in which the latter condemned oppression of Ukrainians by the Poles in the past. Limanowski fully expected a Russian to attack the Polish gentry element in the eastern provinces, but he resented the idea that a Ukrainian, too, might do so![19] Nevertheless, Limanowski himself criticized the chauvinistic policy pursued by the National Democrats around this time toward the Ruthenians in Galicia; he considered this to be contrary to both Polish democratic traditions and the idea of progress professed in contemporary Europe.[20]

He continued to clash with the National Democrats on his right and with Rosa Luxemburg on his left, to whom he referred as "the evil spirit of Poland." When in the fall of 1903 he had learned that the PPS *Workers' Gazette*, then appearing in Katowice in Upper Silesia, was on the verge of bankruptcy, Limanowski hastened to tell London that, in his opinion, the demise of this paper would constitute a triumph for Luxemburg; he urged that everything possible be done to prevent this disaster. Early in 1904, when he had been informed of Rosa Luxemburg's attempts to bar the PPS from attending the forthcoming international congress, Limanowski urged London to retaliate in an Open Letter addressed to the International Socialist Bureau.[21]

Around 1903, the Foreign Department of the PPS began transferring its headquarters from London to Galicia. In a letter to its executive, the Foreign Committee, Limanowski again wished to make it autonomous, as a means of uniting Polish socialism at home and abroad. But the Committee spokesman, Aleksander Malinowski, contended that the Foreign Department of the PPS was but an auxiliary organization which lacked authority in Poland, especially since

it consisted mainly of students. The only practicable way of achieving unity in the Polish socialist movement was an alliance of execu-
tives of the PPS in Russian Poland, the PPS in Prussian Poland, and
the PPSD in Galicia. Limanowski replied that the Foreign Department could act effectively only so long as it was independent of the
conspiratorial type of organization in Russian Poland, which was in-
appropriate for *émigré* groups. He urged the need for an *émigré* political body to act as liaison between the socialist parties in Poland
and in the United States, as well as with the International Socialist
Bureau.

As this discussion with the Foreign Committee shows, Limanowski's main concern was national unity rather than practical party
politics. Early in 1904, his proposal to convene a special congress
was approved by one other member of the Paris Branch. The Foreign
Committee, (i.e., the *émigré* PPS executive) considered it an act of
insubordination vis-à-vis the party's executive in Warsaw, and regretted the omission by Limanowski of representatives of the home PPS
from the proposed congress which was to discuss issues of vital importance to them as well. The Foreign Committee considered such
omission improper, because the *émigré* organization trained individuals for socialist activism in Poland and thus had to act in concert
with the executive at home. Limanowski's proposal was rejected and
his discussion with the Foreign Committee was adjourned *sine die*.[22]

While Limanowski remained in faraway Paris, the center of PPS
activity outside Russian Poland was shifting to Galicia. From 1901
on he attempted to persuade the Galician authorities to let him settle in Cracow; he gave as extenuating circumstances for his request—
his advancing years, the changed political situation in Galicia, as well
as his desire to be near his third son whom he wished to enroll in a
Galician high school. However, the Chief of Police in Lvov, Schaetzel,
was convinced that Limanowski was still a radical socialist. He had
shown to Viceroy Badeni certain passages from Limanowski's *Socialism* of 1879 ("a revolutionary upheaval is unavoidable in countries
deprived of political freedom"; "we sympathize with those who are
hungry, rather than with the well-fed," to cite two examples). In
addition, he pointed to Limanowski's continued appeal to student
radicals. He also listed the names—some of them suspect—of persons
and organizations wishing Limanowski well on the occasion of his
Jubilee in 1900. In short, Schaetzel reacted to Limanowski's petition
by cautioning Badeni against admitting the former to Galicia.[23]

However, Limanowski still contemplated naturalization as a
French citizen. Notwithstanding his age and political radicalism, the

management in his firm recognized his diligence by granting him regular salary increases. He was now very glad to have secured this position rather than employment at Rapperswil, being certain that he would have been eventually forced out from the latter post by his erstwhile political friends, the National Democrats.[24]

As early as the summer of 1902, Limanowski in a letter to Lucjan expressed his conviction that mankind was approaching a turning point in its history. He felt somewhat troubled: "I am so depressed by sad thoughts. . . . The world to which we have become accustomed, had learned to love, is about to collapse; and we are about to perish with it. However, we should look to history for consolation. That which is vigorous seldom perishes; rather, it merely changes its form while adapting to new conditions."[25]

Russia's forthcoming defeat by Japan in 1905 represented the beginning of the reassertion of Asia against the dominance of the great powers of Europe. On the outbreak of Russo-Japanese hostilities, Limanowski became hopeful that this war would soon lead to Polish independence. He was now anxiously awaiting the potential involvement of the United States in the hostilities, on the side of Japan; in a letter written in March 1904, he urged the Foreign Committee of the PPS to agitate among Polish Americans in order to influence their government to take part in the war. He also told the Committee that it was necessary to prepare for the eventuality of an armed uprising at home.[26]

Around this time, Limanowski wrote an article devoted to the problem of planning an insurrection, which was published in the *Workers' Calendar for 1904*. In his article he argued that erstwhile Polish democrats had laid a basis for independence by forcing both Austrian and Russian authorities to emancipate the peasants in their respective parts of Poland. He believed that in Russian Poland after 1863 peasants suffered more from national oppression than from dependence on landlords; hence, they were bound to become patriotic, and would be prepared to fight for Polish independence. However, Limanowski cautioned against rising prematurely; this, he believed, had contributed significantly to defeat, both in 1846 and in 1863. In his opinion, in order to succeed the Poles must rid themselves of their two chief political vices—an impulsiveness to fight without adequate preparation, and inability to persist in the struggle, once they had begun it. It was his conviction that only a *levée en masse* against Russia, which was recruited from well-disciplined individuals, fully conscious of their political aim, would result in winning independence. Limanowski felt that a guerilla army might well paralyze enemy forces by cutting them off from their supply bases. But he warned realistically that "anarchistic guerillas, inspired by poetic manifestos, were no match for regular enemy armies, for enemy cannons, and for enemy fortresses."[27]

On May Day 1904, Limanowski, speaking before an audience composed of Russians, Ukrainians, Lithuanians and Poles, at a meeting sponsored by the Paris Branch of the PPS, urged the overthrow of the Tsarist Empire both because, in his opinion, it constituted a "prison" for its constituent nationalities and impeded propaganda for socialism. Though Limanowski considered war barbarous and an anachronism in the twentieth century, he nevertheless felt that it was a lesser evil than national oppression. He pointed to war as an instrument of liberation for Poland in the past. He argued that it was the Napoleonic wars which resulted in the partial re-establishment of Poland as the Duchy of Warsaw in the period between 1807 and 1815, and the Russo-Polish struggle of 1863 paved the way for the final act of peasant emancipation in Poland. He spoke in a similar vein at many other *émigré* gatherings.[28]

Limanowski engaged in the activities of the Paris Branch in so far as he had time to do so. In May 1904, he told Lucjan that the Paris Branch agitated against the imminent granting of a French loan to Russia, commented on Polish affairs in the local press, and protested expulsion from Paris of the Russian Socialist Revolutionary Burtsev, which was due to pressure exerted on the French authorities by the Russian ambassador Nelidov.[29]

Meanwhile, shortly after the outbreak of war, Limanowski asked Aleksander Debski of the Foreign Committee, who was influential with Polish socialists in the United States, to seek their assistance in organizing a Polish legion, to be equipped by the Japanese and based on Formosa. Limanowski suggested that in return for fighting in the Japanese cause, the Poles ought to be granted this island as a colony for settlement. This fantastic proposal was, of course, contrary to his professed aim of full self-determination of all nationalities everywhere. Subsequently—in August 1904—after Limanowski had arrived in Galicia on his way to lecture in the resort town of Zakopane, he approached Jedrzejowski and Wasilewski with this proposal, at the latter's summer cottage near Cracow. Thereupon, Wasilewski reassured Limanowski that Jozef Pilsudski (then a leader of the home PPS) was already looking into the matter.[30]

This trip to Zakopane Limanowski had undertaken during his two-week vacation in August 1904. In the previous February Limanowski had been invited by Krauz, on behalf of the newly formed Society for Sponsoring Advanced Summer Courses (*Towarzystwo Wyzszych Kursow Wakacyjnych*) in Zakopane, to take part in its program for 1904 at the Society's expense. This project was modelled on similar institutions in the West; it was conceived by Krauz and its chief organizer was the literary critic and PPS sympathizer, Wilhelm Feldman. Their aim was to gather progressive Polish intelligentsia from Poland and abroad, especially from Russian Poland—where there was severe censorship impeding free exchange of ideas—in order to hear a series of lectures, mainly in the social sciences and the humanities, on topics which were relevant to the Polish situation. The intent of the series was, on the one hand, to propagate socialism and, on the other, to stimulate independent research in Poland, as well as to acquaint the audience with progress made in the natural and social sciences abroad. These lectures proved to be both popular and intellectually stimulating; and by rallying intelligentsia from the whole of Poland, they helped to enhance national unity.[31]

Since once again Limanowski failed in his attempt to secure a permit to stay in Galicia, he had no choice—after completing the series of four lectures on Polish history—but to return to Paris. A farewell party was arranged in his honor just prior to his departure; at the party his sympathizers, using rather flowery language, even hailed him as a "true poet." Limanowski took advantage of the presence there of representatives of Polish left-wing socialism to urge reunification of the movement, and as the administrator of the so-called "Red Cross" (see Chapter VII, note 17), he unrealistically proposed that this—no longer politically neutral—organization might become the means of uniting socialism in Poland. As a result, he was ridiculed by his political antagonists, whom Krauz chastized severely. Notwithstanding this incident, Limanowski considered his trip to Zakopane another modest personal triumph for him; he was welcomed everywhere and was well looked after in Galicia. He returned to Paris feeling refreshed and younger in spirit.[32]

Subsequently, in 1906, the PPS published his four lectures in pamphlet form, under the title *The Development of Democratic Convictions among the Polish People*; it sold quickly and is still considered to be not without value as a work of scholarship.[33] In it Limanowski defined the Polish people as the vanguard both of progress and Western civilization in Eastern Europe.[34] In addition to tracing historically the development of democratic ideas in Poland, in his pamphlet Limanowski speculated on the origin of socialism in general. He explained socialism as an attempt on the part of scholars to define the vague aspirations of the common people to traditional freedoms in terms of social science (i.e., sociology) and history. Li-manowski felt that the Manifesto of the Prague Slav Congress of 1848, which had proclaimed the principle of national self-determin-ation, was as much of a milestone on the national issue as the con-temporary Communist Manifesto was on the social one.[35] He con-sidered these issues inter-related, arguing that the ideal multi-national state of the future would ensure the well-being of both individuals and peoples, on the basis of voluntary contract in all political relation-ships—whether local, national or international.[36]

In another treatise, entitled *Nationality and State*, published also in 1906, Limanowski was to expand his views on "international so-cialism."[37] To prevent small nations from being exploited by the powers, he recommended a World Commonwealth, with a central in-ternational legislature and executive, or at least a European concert of nations as, for instance, proposed by Gladstone. Mindful of the contemporary Swiss model as well as of the sixteenth-century Union of Lublin, he envisioned a federation of Poland-Lithuania-Latvia-Ukraine as a means to the proposed World Commonwealth; the federa-tion would act as a counterweight against Germany and Russia. His idyllic vision was marred by the reality of power politics in Europe.[38]

Yet there was much sense in his ideas. He knew that fortune was changeable; the oppressed of today, once free, might well become the oppressor of tomorrow. Therefore—viewing the state both as an effect of international anarchy and its cause—he urged that all nations should have equal rights, in order that they might combine in what-ever political and economic unions they deemed to be most conven-ient and profitable to themselves.[39]

Back in Paris, Limanowski became so busy with frequent speak-ing, chairing of meetings and preparing various reports that he found little time to undertake any serious writng. In October 1904, he told Lucjan that news of the war was now his primary concern: "I am be-coming almost a Japanese, taking each of Japan's setbacks so much to heart as if they affected my own fatherland."[40] The anti-govern-ment patriotic manifestations which the PPS began to organize in Russian Poland in the fall of 1904 reminded Limanowski of 1861, but now the future appeared to him even brighter than the situation in Poland had been on the eve of the uprising of 1863.[41]

In 1904 the PPS had kept aloof from economic strikes in Russian Poland, and its demonstrations were usually of a purely patriotic character. The leaders of the PPS mistrusted then the revolutionary movement in Russia and treated it with condescension. For instance, Pilsudski sought an alliance with non-Russian nationality parties rather than with the Russian opposition. Demanding that the Russian parties support his postulate of complete independence for Poland within the boundaries of 1772, he forbade them at the same time to encroach on his territory, i.e., Lithuania and Russian Poland. (However, there were leading members in the PPS, who were reconciled to the idea of a purely ethnic Poland, notably Wladyslaw Gumplowicz.)[42]

Upon the outbreak of the Revolution of 1905 in Russia, a rift soon developed within the PPS between the "old guard" (i.e., the right wing) and the "youngsters" (i.e., the left wing), who now flocked en masse to the party. The latter insisted on solidarity with the revolutionary movement in Russia and were even willing to abandon the slogan of independence. At the VII congress of the PPS held in March 1905, a compromise was reached: the entire body agreed to demand a Constituent Assembly in Warsaw. However, the very concept of "Constituent Assembly in Warsaw" was ambiguous. The "old guard" in the PPS, including Limanowski, interpreted it as a means of securing complete independence, whereas the "youngsters" considered it to be merely a guarantee of an autonomous status *within* a constitutional Russia. Moreover, the left and the right wing in the PPS were eventually, by the end of 1905, to clash on tactics. Because the "old guard" subordinated the class struggle to the struggle for national liberation, the logic of their position led them, by the beginning of 1906, to oppose strikes not only because they were proving detrimental to the worker, but also as being harmful to the Polish cause. This made the SDKPiL charge the PPS right wing with unwittingly aiding the Polish counter-revolution.

Nevertheless, during 1905 the PPS remained united and all Polish socialist groups usually attempted to act in solidarity with the Russian revolutionary movement. It was the National Democratic Party which became the real counter-revolutionary force in Russian Poland. By the end of 1905 its opponents labelled it the "Polish Black Hundreds," due to its policy of fratricidal assassinations. This party was now essentially loyalist toward the Tsarist authorities, and it combatted the revolution both verbally and by armed struggle.

Shortly after the outbreak of the Revolution, the Paris Branch arranged a meeting of representatives of all Polish socialist organizations in Paris. When Limanowski began to advocate national unity, discordant voices made themselves heard in reply to him. A certain Ulanowski made a speech which was too "anti-bourgeois" for Limanowski's taste. An acquaintance of Limanowski, Dr. Jozef Zielinski, who was a left wing member of the PPS, attacked his party for neglecting the cause of international revolution. Nevertheless, the meeting ended in collecting a considerable sum for revolutionary purposes.

The economic situation in Russian Poland did not appeal to Limanowski. There the population was suffering materially as a result of a protracted wave of industrial strikes. Reluctant to expose his family to the hardships of Russian Poland, he decided to remain in Paris, awaiting the outbreak of an armed insurrection at home.[43]

Limanowski spoke at the May Day celebration of 1905, to which the Paris Branch had invited the Paris representatives of other socialist parties active in the territories of the Russian Empire. As was his wont at such gatherings, Limanowski began by reviewing the achievements of the socialist movement in Europe during the past year and by outlining future goals. But he concentrated on events in Russian Poland. He contended that it was the strike movement there during the previous winter and spring, involving almost half a million Poles, which had influenced Russian liberals like Petr Struve to demand a return to the 1815 status in Russian Poland. Limanowski expected that Polish factory workers would remain in the vanguard of the Revolution; but he was uncertain of the peasants, though he hoped that they would also join in the struggle.

In the course of his speech he referred to a recently concluded agreement between some of the Socialist Revolutionary nationality parties in the Russian Empire and the PPS, somewhat exaggerating its significance however.[44] On the one hand, Limanowski now advocated an alliance with all oppositional parties in the Russian Empire (as in the 1893 PPS program) and, on the other, he urged reunification of all Polish socialist organizations in a common front against the Polish bourgeoisie (as in the resolution agreed to at the international socialist congress held in Amsterdam in 1904). He was especially concerned with effecting unity in the socialist movement in Russian Poland. In his opinion, a recent compromise reached at the VII PPS congress in March, i.e., the new demand for a Constituent Assembly in Warsaw (in addition to one in St. Petersburg), went a long way to meet the present objective of the SDKPiL—a Constituent Assembly in the Russian capital, as a means of effecting social revolution in the Tsarist empire. Limanowski mistrusted the idea of one Constituent Assembly in St. Petersburg for the whole of Russia, where Polish deputies would be in a minority.[45]

From the beginning of the Revolution, the existing feud between the *émigré* leaders of the PPS and the SDKPiL was aggravated by their bickering over the funds collected in France for Russian Poland, which Limanowski deplored in his capacity as chief cashier of the "Red Cross." (The PPS, unlike the SDKPiL, had no easy access to the French socialist organizations, and to the French press in general. On the other hand, the PPS was able to obtain some financial aid from the British.)[46]

By October 1905, he became sceptical whether the Russian revolution would actually succeed; nevertheless, in the event of a total collapse of Tsarist authority, he anticipated the need for quickly organizing an insurrection in Russian Poland. To this end, he had sent to Jedrzejowski in Cracow a copy of Kamienski's *People's War* (written in the period of the "Great Emigration") for perusal by Pilsudski, who had just then become the leader of the PPS "Militant Organization" in Russian Poland.[47]

After the Tsar had issued his October Manifesto, granting a limited constitution, the *émigré* "old guard" of the PPS decided to transfer its headquarters from Cracow to Warsaw. However, Limanowski was not in favor of this move. He believed that, when strengthened, the autocracy would hasten to return to the *status quo ante*. He himself was reluctant to go to Warsaw, as he was urged to do. If he were single, he might have returned secretly, to struggle as a conspirator. But to arrive there openly, he felt, would be understood as coming to terms with the hated Tsarist regime; it would be a humiliating denial of his entire past.[48]

Around this time, Lucjan informed him of peasant disturbances in the borderlands. This news upset Limanowski. In reply to Lucjan he warned him that the "storm had only just begun," that the agrarian revolution would be violent and perhaps not entirely to the liking of those who had incited it. However, Limanowski considered it a necessary evil and felt that in the long run it would be beneficial to the Polish cause and to the cause of progress in general. In another letter, Limanowski told Lucjan that he was convinced that "we have sown well" and he now hopefully awaited the "ripening of the harvest."[49]

The "conscious revolutionary element" in Russia, drawn from workers, intelligentsia and a small section of landowners, was, in his opinion, but "a drop in the vast peasant sea." He felt that peasant revolution would be the best means of overthrowing the autocracy in Russia. He now credited the Russian Social Revolutionary Party with success in overcoming the apathy of the Russian peasant, doing away with his fear of bureaucracy, and he praised the Social Revolutionaries for teaching the peasant the meaning of self-esteem. According to Limanowski, it was the Socialist Revolutionaries who had reminded the Russian peasant of his ancient communal freedoms and communal equality. Limanowski believed that the contemporary *mir*, corrupted by Tsardom into an instrument of tax collection and for the maintenance of the political status quo, was hated by the peasants. And he prophesized that should a new Tsar appear in Moscow, in order to oppose the "German Tsar" in St. Petersburg under the slogan "down with the German rule," undoubtedly the Russian peasand would rally under his banner.[50]

In December 1905, Limanowski eagerly followed news of the armed uprising in Moscow which, however, was soon suppressed by the autocracy. Subsequently, in his May Day speech of 1906, he described the situation in Russia as being unlike that in France during the Revolution; rather, he said, it resembled the Roman Empire on the eve of its collapse, when only brutal military despotism prevented its imminent disintegration. He considered the despotism in Russia to have been even more cruel and unpopular than the oppressive rule of the later Roman Empire.[51]

For news of the situation in his native north-eastern borderlands, Limanowski now turned to the conservative Polish periodical *Home Country (Kraj)*, published in St. Petersburg. This paper was the organ of the Polish Realist Party, which had its support in the upper bourgeoisie, aristocracy and episcopate of Russian Poland. Previously, he had never read it, but now he found it more congenial than the National Democratic *All Polish Review*.[52]

In an article on Mickiewicz as a publicist, written late in 1906 for the PPS *Tribune*, Limanowski described National Democracy as a party which "hid its commitment to vested interest and exploitation under the cloak of religion and patriotism," somewhat resembling the "Orleanism" of Mickiewicz's era. He was convinced that if Mickiewicz were alive, he would not have called Pilsudski's militants "bandits," as the writer Henryk Sienkiewicz had done. Like the National Democrats, the latter condemned outright the revolution in Poland. The National Democrats regarded it, Limanowski argued, as "total anarchy which threatened the very fabric of social order" in Poland. And he unequivocally rejected the politics of National Democracy in the revolutionary years of 1905 and 1906 as both counter-revolutionary and anti-democratic.[53]

Meanwhile, by the end of 1905 the PPS and SDKPiL in Russian Poland drew closer ideologically and tactically. The "old guard" in the PPS became dissatisfied with the evolution of the majority in their party at home toward the left, and with the tactic—no longer economically (or politically) effective—of industrial strikes. It was the Galician socialist Daszynski who dared to express openly the dissent of the "old guard" in his Open Letter of January 1906, which he addressed to the Central Workers' Committee of the PPS (CKR) in Warsaw. In it he deplored that the left majority in the PPS, led by Max Horwitz, had now abandoned the goal of an independent Polish republic. Daszynski wished to aim at more than "this or that Duma in St. Petersburg" and attacked the then "fashionable concept of a Russian republic extending from Lodz to Kamchatka." He felt that the PPS should not confine its appeal to factory workers; it ought to also rally intelligentsia and the petite bourgeoisie in the cause of regaining Polish independence. Daszynski doubted that a "bourgeois" Russian republic would be established in the foreseeable future, due to the present "uncoordinated revolutionary tactics" in Russia. He believed that both Luxemburg and Horwitz were under the spell of the anarchistic Russian masses as well as the idea of an economic union with Russia and that, moreover, the latest general strikes had brought nothing but hunger and demoralization to Russian Poland. Daszynski's "counter-revolutionary" démarche made him for a time very unpopular with Polish socialists. Only Limanowski wrote him a friendly letter from Paris, in which he thanked Daszynski for expressing his very own thoughts—clearly, concisely and convincingly.[54]

Thus, the PPS had to contend with an armed counter-revolution on its right, with verbal battles with its foes on the left, and was itself rent by an internal conflict. As a result of the revolution the PPS had become a mass party; and its new cadres—as already mentioned above—were willing to abandon the slogan of independence. This "left" orientation had won at the VIII party congress which met in February 1906. The new left wing majority aimed merely at a federal Russian republic with a Constituent Assembly in Warsaw and relegated independence to a very distant future.

The final split, which took place at the IX party congress held in Vienna in November 1906, was only partly due to ideological differences. A personal struggle for power also ensued between Pilsudski as the leader of the "old guard," who still believed in an armed insurrection as a means of securing Polish independence, and the new Central Committee led by Horwitz, who now considered an insurrection in Poland a Utopian idea. The right wing broke away from the party majority; henceforth it was to be known as the PPS "Revolutionary Fraction." The Fraction adopted a new program in March 1907, and the Left PPS followed suit in December of that same year. Both resembled their predecessor of 1893, since both were based on the Erfurt Program. Unlike the somewhat vaguely formulated social program of the Fraction (based on both the Gotha and Erfurt models), the minimum demands of the Left PPS were conceived only as a *means* to social revolution. On the national question the Fraction aimed at the establishment of a Constituent Assembly in Warsaw as the next best strategy in the struggle for independence. The Left PPS, having abandoned the slogan of independence altogether, called for broad autonomy for Russian Poland, with a legislative Polish diet, while omitting the former demand of the still united PPS for a Constituent Assembly in Warsaw. Its position on the national question now resembled that of the SDKPiL, which had relegated solution of the national question to the post-capitalist era. (The social program of the SDKPiL was modelled on that of the Russian Social Democratic Labor Party.) Whereas the Left PPS now comprised the majority of members of the former PPS, the minority Fraction proved to be the more dynamic group. Limanowski learned of the split within the PPS from *L'Aurore*; he deplored the shattering of party unity, but of course, he favored the right wing orientation now that the split had become a *fait accompli.* [55]

Meanwhile, Limanowski's second wife had died on September 19, 1906. His two older sons were now financially independent adults, whereas the youngest boy Stanislaw was attending a Polish boarding school at Batignolles, a suburb of Paris, on a scholarship. Limanowski, having reached the age of seventy-two, felt lonely and more than ever frustrated by his office work. He decided to resign and to return to Cracow where he thought he could now support himself by writing. If he were again expelled from Galicia, he was prepared to take up a clerical position at the International Socialist Bureau, which since 1900 had been based in Brussels. His two older sons urged him to go to Galicia; Zygmunt even wrote a letter to Daszynski, presently a socialist deputy in Vienna, to intervene on his father's account. In return, Daszynski informed the senior Limanowski that he had spoken about his return to the Minister for Galician Affairs in Vienna, Wojciech Dzieduszycki, who—in turn—discussed the matter with the Galician Viceroy, Andrzej Potocki. Daszynski advised Limanowski to send immediately a petitition to Potocki, motivating his request to stay in Galicia by personal, rather than political considerations. [56]

Limanowski did not expect to receive an answer from the Viceroy to his petition, a copy of which he had enclosed with his reply to Daszynski, until the new system of universal manhood suffrage for the central parliament became law. Pending this, the struggle for the new electoral law was absorbing the energy of his political friends, like Daszynski, placing them in opposition to the ruling conservatives, on whose decision depended Limanowski's return to Galicia. However, on January 23, 1907, the Chief of Police in Lvov agreed to admit Limanowski, assuming that he was now too old for socialist activism and the Cracow police would watch him closely. In February Limanowski was informed that he might reside in Galicia until the end of the year; he was notified of this via the Austro-Hungarian embassy in Paris.[57]

However, Limanowski had to delay his return for a few months, in order to take care of expenses incurred by him due to the illness and funeral of his wife. After he had discharged his financial obligations in Paris, he planned to travel to Crakow where he was to manage an apartment building for his brother Lucjan and, in exchange for his services, was to be given rent-free accommodation. In April 1907, Limanowski was invited by his other brother Jozef's widow to stay permanently with her on the family estate in Jassy; she intended thereby to provide her brother-in-law with security and peace in his old age. But he as yet had no intention of retiring to a country estate in distant Belorussia; he wished to be at the center of political activity in Poland at this time, in Cracow.

Limanowski decided to take advantage of his last few months in the West, in order to collect further materials for his projected biography of the early Polish socialist Worcell. He planned to do research in Brussels, London and Rapperswil. While in London he intended to inquire from Voynich (an acquaintance, who was a Polish archeologist and permanently resided in London), whether Voynich's project to open a bookstore in Galicia, to be managed by Limanowski, was likely to materialize or not.[58]

In the spring of 1907 he spent his annual vacation in Brussels and in London. He arrived in Brussels on May 21, stayed there for three days—during which time he met Camile Huysmans, the secretary of the International Socialist Bureau—and spent the remainder of his free time, until June 4, in London.[59] In London Limanowski stayed at the home of the Voynichs. (Mrs. Voynich, née Ethel Lillian Boole, was the author of a popular novel *The Gadfly*.) Since the Voynichs had to be away from London at the time, Limanowski was entertained by friends of theirs. He was to enjoy himself immensely in London; he found the Londoners very friendly and eager to help him, and not at all reserved toward strangers, as he had expected them to be—with the exception of Bernard Shaw! Limanowski again met Max Hyndman, the leader of the British Social Democratic Federation, with whom he had already got acquainted at the Possibilists' congress in 1889. Hyndman received him with open arms and even arranged special access to the reading room of the British Museum, which was then closed to the public for alterations, so that he could work on his projected biography of Worcell.[60]

Just before his arrival in London, the Russian Social Democratic Labor Party (including the affiliated SDKPiL, Jewish Bund and Latvian Social Democrats) assembled there for its fifth congress. The 180 delegates from Great Russian organizations of the party were almost evenly divided—91 Bolsheviks; 89 Mensheviks. And the two factions failed to arrive at a joint appraisal of the existing situation in Russia. In 1905-1906 members of the SDKPiL had agreed with the Bolsheviks to boycott the new Duma. On the other hand, they now opposed Bolshevik terrorism and expropriations, along with the Mensheviks, the Jewish Bund and with some of the Latvian Social Democrats. The SDKPiL delegates charged that preparations for an armed insurrection, and tactical terror, tended to demoralize a proletarian party; such militancy "was pushing the petite bourgeoisie into the arms of the reaction."

Limanowski wished to attend the congress, but was prevented from doing so by British police because, according to Hyndman, the meeting hall was already dangerously overcrowded. In reporting the proceedings of the congress in *Forward*, Limanowski based himself on the account in Hyndman's socialist organ, *Justice*. In his report he stated his belief that the party militant and the working-class organizer, who concentrated on economic aims, had each a role to play in the common struggle. He did not share the reservations of the Mensheviks and of the SDKPiL concerning the application of terror, feeling that even "banditry at its worst was less demoralizing than cowardice and servility" toward the oppressor. In Poland confiscation of Tsarist funds and executions of Tsarist officials in the course of the revolution in his opinion were an inevitable end product of the Tsarist foreign yoke.[61] In this connection, Limanowski pointed to an ideological inconsistency on the part of the SDKPiL delegates. At the congress they argued that militancy, including militancy in the cause of Polish independence, "was pushing the petite bourgeoisie into the arms of reaction"; whereas, previously they had considered terrorism and insurrectionary tactics to be manifestations of petite-bourgeois behavior in Poland.[62]

The SDKPiL delegates to the congress had arranged a meeting for all Polish socialists then in London, which was held at the local headquarters of the German Social Democrats. On this occasion, two of the SDKPiL delegates vehemently criticized the PPS Fraction. Limanowski replied. He was convinced that revolution in Russia would result in the disintegration of the Russian Empire and that this outcome was desirable. One of his antagonists castigated him for this "cruel" statement. Limanowski was somewhat astonished that the SDKPiL apparently identified the Russian state with the Russian people. He upheld his idea that the disintegration of Tsardom was the only means of liberating the non-Russian nationalities of the Empire. He believed that only an immediately victorious revolution might have resulted in a transformation of the present Russian state into a federal union, but now everything pointed to a long and stubborn struggle between an intransigent autocracy and the discontented nationalities, even though these were unfortunately not able to agree on a joint course of action.[63] Limanowski maintained that one could not compare contemporary Russia with France during the Revolution; in France the bourgeoisie was ethnically homogenous and was both a revolutionary and an integrating force. Thus it was empowered, as it were, to act on behalf of the French people as a whole. In contrast, in Russia there was no common revolutionary denominator, only separate national movements; hence, disintegration of the Empire was inevitable. It is evident that Limanowski underestimated in his speech the social content of the revolutionary movement in Russia and exaggerated the national one; he had overlooked the Bolsheviks as a potential instrument of "revolutionary centralism" in the Russian Empire.[64]

Limanowski, nevertheless, attributed to all Russian opposition parties, apart from the Socialist Revolutionary Party and the handful of anarchist followers of Prince Kropotkin, a "centralist outlook." He argued that conscious socio-political attitudes, like the ideas of autocracy and centralism, might, like instincts, be inherited from generation to generation; this was in accordance with his concept of sociology as a biological science. He believed, too, that the allegedly "Russian" tradition of centralism, and the worship of autocracy, was perpetuated by Tsarist schools, the church, the arts and literature, resulting in an indoctrination of the broad masses of the Russian people. To Limanowski Russian history represented but a record of "Basils and Ivans, more or less terrible." The Russian Tsardom, in his opinion, was modeled on the erstwhile tsardom of the Tatars in Russia; and the Russians had always dreamed the imperialist dream that "all Slav rivers converged to form a single Russian sea."[65]

Now, however, according to Limanowski, Russia as a power would never recover from the defeat inflicted on it by the Japanese. He expected that Russian imperialism in Europe would soon be checked, because Russia would be forced to strengthen her defences in Asia against a potentially powerful modern China. He predicted that, with the rise of national and social consciousness in China, Russia would find it very difficult to maintain the status quo in the Far East.[66] Meanwhile, Russia's "grand" politics were exhausting the country's resources. He pointed out that to pay for her immense army, navy, and military railroads, Russia had been compelled to contract loans abroad, the interest on which was payable in gold; and Russia was obtaining this gold by means of her exports of grain which was needed to feed the hungry at home. He sympathized with the Russian people for being exploited economically twice over; they paid a heavy tax to the state and, at the same time, were at the mercy of private capital subsidized by the state. Limanowski agreed that there was an urgent need for reform in Russia, but he was sceptical of the idea of active Polish collaboration with Russian liberals or radicals to this end, without simultaneously aiming at Polish independence, and he criticized the new group of Progressive Democrats in Russian Poland as well as Polish left socialists for allegedly wishing to do so. He demanded equal status with Russia for Poland in the proposed Slav federation.[67]

It should be emphasized that Limanowski tended to idealize "federalism" as a peculiarly "Polish tradition," embodied in the erstwhile Union of Lublin. He contrasted Polish federalism with the principle of centralism as embodied in "half-Asiatic Russia," rationalizing the Polish conquest of Moscovy during the Time of Troubles as a means of achieving federation in Eastern Europe. In regard to the period following the partitions, Limanowski argued that the Polish revolutionary democratic ideology was based on two principles: republicanism and federation. He cited with approval the Russian Novosiltsev's remark that, in the cause of national liberation, "Poles had absorbed 'Jacobinism' with their mothers' milk." But he forgot that Jacobins were in fact centralists!

Convinced that the principle of "centralism" had an almost metaphysical hold on Russia, Limanowski in fact dismissed the idea of a federation based on Russia as nothing but wishful thinking on the part of Polish left socialists and liberals. It was in Poland, not in

Russia, that "federation" remained a realizable ideal. Lithuania and Latvia, he argued, were too weak to stand on their own; their history as well as their Western European culture drew them into the "Polish orbit." He assumed that they would be better off in a federal union with Poland than remaining within even a democratized Russia which, he felt, would at best grant them only a very limited degree of autonomy. He continued to believe that Jews would enjoy full civil and political rights in a future federal Poland; they were less likely to obtain this in a centralist Russia, which was historically alien to them. He was certain that the Belorussians, many of whom were Roman Catholics, would naturally gravitate toward a federal Poland; on the other hand, he feared the "historical prejudices and hatreds" of the Ukrainian patriots, which he believed clouded their judgment in the matter of union with Poland. The future might face the Ukrainian people with three alternatives: union with Russia, union with Poland, or full independence. In Limanowski's opinion, irrespective of their future choice, it was in their interest in the present to cooperate with the Poles against the Russian autocracy.[68]

Back again in Paris, Limanowski continued working in the office for a few more weeks. He departed for Galicia toward the end of July 1907. On his way back to his native land, he stopped for a fortnight at the Polish Museum at Rapperswil to collect more material for his projected biography of Worcell. Finally, after twenty-nine years of exile, he arrived in Galicia around the middle of August. Invited by a friend of his son Mieczyslaw to spend a vacation on her landed estate near Kurzany in Eastern Galicia, Limanowski rested there for three weeks prior to settling in Cracow.[69]

Limanowski came to Cracow at the beginning of September 1907.
On arrival in Galicia, he was assured by the Viceroy, Andrzej Potocki,
that the authorities would not bother him, provided he refrained
from political activism.[1] Nevertheless, during his Cracow years he
did take an active interest in politics, *inter alia* attending party congresses and lecturing on Polish history to intellectuals as well as workers.[2] In addition, he wrote a great deal. Hired by the editor of *Forward*, Emil Haecker, on October 1, 1907, Limanowski henceforth
drew a regular salary as a reporter and critic on this paper.[3]

I

In the period of seven years preceding the outbreak of the World
War, Limanowski was to contribute numerous articles on contemporary politics both at home and abroad, as well as reviews of recent
Polish historical writings. Of special interest to us are the articles
which dealt with women's rights, Russian affairs, the "Rosa Luxemburg problem," and peasants—all published in *Forward* in 1907 and
1908, as well as two brief treatises on historical materialism and the
nationality question, respectively, both appearing in Feldman's *Criticism (Krytyka)* in Cracow, in 1908.

Concerning women's rights we find his standpoint expressed in a
review of an article by a German feminist Lily Braun, entitled "Die
Frauen und Politik" (it had appeared in the Berlin *Vorwärts* in 1903.)[4]
In it Limanowski agreed with the author that Social Democratic
parties—the parties of all the disinherited and oppressed—were the
only true champions of women's rights. In his opinion, women's
emancipation was not just the cause of liberating women; feminists
were bound to become socialists and would bring up their children to
be socialists as well. Thus, liberation of women would considerably
augment the ranks of socialist parties. He considered feminism a decisive factor in the victory of socialism. Apparently, Limanowski
underestimated the social conservatism among women.

Limanowski noted that in Western Europe lawyers and politicians now agreed with the idea of granting women civil rights, but they still as a rule objected to women in politics. Influenced by John Stuart Mill, he opposed this prejudice on various grounds. He believed that participation of women in politics was absolutely essential to guarantee them their rights. Moreover, he was convinced that female politicians would perform at least as well as men. Finally, he assigned to women a vital role to play in the struggle for universal brotherhood (which he considered the ultimate goal of socialism). In his opinion, the ideal of brotherhood was grounded in emotion rather than in thought—and women were more emotional than men! He did not mean that women were irrational, however; he attributed to them a certain propensity toward altruism, which he felt men often lacked. He noted that there were many dedicated female activists in the PPS. And, long ago when Christianity was still a religion of the meek and oppressed, women were among its most ardent apostles; it was, too, female rulers like the Czech Dubravka, the Russian Olga (in fact, a blood-thirsty tyrant!), and the Polish Jadwiga, who spread Christian civilization in Eastern Europe. However, Limanowski assumed that, so long as Poland was subjugated, women's emancipation was but a theoretical problem. Therefore, he urged Polish women to concentrate on independence as their primary aim.[5]

Limanowski again anticipated an eventual revolution in Russia, which he felt would help to achieve Polish independence; hence, his preoccupation with Russian affairs.[6] In 1908 he reviewed in *Forward* a recent history of the Russian revolutionary movement, by a Polish author Feliks Kon.[7] In this review Limanowski predicted that a contemporary lull in revolutionary activism in Russia was not likely to last. He pointed out that the revolutionary years of 1905-1906 had produced the first, albeit only temporary, proletarian authority—the Soviet of Peasant and Workers' Deputies in St. Petersburg. Henceforth, Limanowski noted with satisfaction, the idea of an armed insurrection was on the Russian revolutionary agenda. He explained why the Soviet failed in its attempt to arm the Russian masses in 1905-1906; it had had no access to state arsenals. Unrest in the Russian army in 1905 prompted Limanowski to hope that it would eventually come down on the side of the revolting masses. Therefore, he urged agitation for a general mutiny in the army as the only sure means to a victorious proletarian uprising in Russia.

Mistrustful of the Russian premier Stolypin's reforms (1906-1911), Limanowski was contemptuous of the Duma as well. He felt that new revolutionary struggles were imminent in Russia, as was another onslaught of reaction in response thereto, assuming that there was now no alternative to a violent upheaval as a means to a social transformation of that country.

Limanowski speculated that a victorious revolution in Russia would have a tremendous impact on European politics. He expected the downfall of Russian Tsardom to result in the demise of "Prussian Caesarism" as well—both were the mainstays of reaction in Europe—assuming that a reactionary Germany was unlikely to survive in a progressive Europe without support from its conservative Russian ally. Therefore, Limanowski again attacked the National Democrats for their loyalism toward the Tsarist regime; he believed that they not only impeded a social revolution in Russia by this loyalism, but also the cause of freedom throughout Europe.[8]

Yet, while commenting earlier on the Anglo-Russian Convention of August 31, 1907, Limanowski implicitly agreed with the National Democrats that Germany was then the principal foe of Poland; since the turn of the century Germany had pursued a ruthless policy of Germanization in the Polish lands which it occupied. And he was even ready to admit that Germany, rather than Russia, had become the greatest obstacle to national self-determination in Europe. In his opinion, Germany, being dominated as she was by the reactionary Prussia, was bound to intervene on the side of the autocracy in any reactionary upheaval in Russia.

Limanowski praised the Anglo-Russian Convention as a means of loosening the ties binding Russia to Germany. He felt that the Anglo-Russian *rapprochement* of 1907 was popular in the latter country, arguing against a German social democrat, M. Beer, who criticized the Convention in *Die Neue Zeit*, that this realignment of the powers was unlikely to impede the cause of national self-determination in Europe, as Beer had feared.[9]

In 1908 Limanowski had also resumed his argument with Rosa Luxemburg while reviewing a recent pamphlet written by Wladyslaw Gumplowicz (the son of the famous sociologist) in response to Luxemburg's various declarations on the national issue. Disdainful of idealism, Rosa Luxemburg professed "the pessimism of historical necessity." After many years of political campaigning, she had produced in 1908 a final synthesis of her views on the national question in "The Question of Nationality and Autonomy," which she published in her organ the *Social Democratic Review (Przeglad Socjal-Demokratyczny)*.[10] In his pamphlet entitled *The Polish Question and Socialism*, Gumplowicz had attacked Luxemburg's views—first from an idealistic premise and, secondly, by means of a counter-economic argument influenced by the late Krauz. Limanowski agreed with Gumplowicz that political democracy merely "extended a promise of equality;" socialism—economic democracy—would fulfill this promise. He asserted, as Gumplowicz had done, that a prerequisite for establishing political democracy, as a means of creating socialism in Poland, was abolition of the foreign yoke; true socialist internationalism was possible only for those parties which functioned in independent states. Limanowski felt that Luxemburg might well reject this reasoning, because it assumed the primacy of "superstructure," but that she ought not to ignore Gumplowicz's economic argument in favor of independence. Gumplowicz had pointed to an "inconsistency" in her theory of development of capitalism in Poland. According to Luxemburg, Prussian Poland was to provide a market for German manufactured goods whereas, conversely, it was Russia which was to be the market for Russian Poland! Limanowski fully agreed with Gumplowicz in urging the need for independence as a means of securing a permanent and fair access to the Russian (as well as to other markets); of ensuring expansion of the internal market; and of providing cheaper food for the city worker, cheaper implements and fertilizer for agriculture, as well as cheaper primary products for industry in Poland. He believed that Gumplowicz's pamphlet had succeeded in demolishing Luxemburg's "theory of organic incorporation."

In opposing this theory, Limanowski assumed that the political connection with Russia (and with the other partitioning powers as well) impeded development of capitalism in Poland. He implied that there was a reciprocal relationship between development of capitalism in Poland and freedom; industrialization was to be a means of providing the proletarian cadres for the struggle for independence and, in turn, independence was to result in further industrialization which would lead to socialism.

Nothwithstanding Limanowski's preoccupation with the agrarian problem—the fourth topic under discussion here—he never idealized the peasant, but he continued to idealize the worker. He wished, as we have seen, to ameliorate life in the village not merely for the sake of the peasant himself, but also in order to expand the internal market as a means to industrialization. Limanowski believed that it was the industrial proletariat, rather than the peasant, which was the mainstay of socialism; he considered the former the vanguard of progress, in virtue of its alleged disinterestedness, its propensity toward socialism and organization, and its mobility.[11]

Limanowski gradually changed his mind about the peasant farmer, assuming that his thrift, combined with inability to take a long-term view, and his intellectual myopia, obstructed progress. The original program of the PPS, published in 1893, contemplated no distribution of land to peasants for individual family use, whether on temporary or permanent tenure. This was in accordance with Limanowski's views on the question. However, at the turn of the century a number of young social democrats like the German Eduard David, or the Pole (and member of the PPS) Wladyslaw Gumplowicz, had dissented from the agrarian "orthodoxy" of the Marxist "old guard." In contrast to Limanowski, Gumplowicz argued that the relationship between a large landed estate and a small land holding was unlike the relationship between a factory and an artisan shop. He considered the large estate to be a remnant of feudalism, rather than a model for the future. And the agrarian situation in Galicia was, in his opinion, quite untypical (i.e., dwarf holdings, illiteracy and lack of co-operatives); it was, therefore, illogical for Polish socialists to dismiss peasant proprietorship—or tenant farming—as doomed, on the basis of the backward agriculture in Galicia. In fact, in championing the idea of peasant proprietorship or tenancy for Poland, Gumplowicz was influenced by the agrarian situation in the West. He warned, however, that the Polish peasant must have an adequate allotment of land for efficient farming. To this end, he advocated confiscation by the government of large estates; this land was to be leased to the peasants.[12]

Gumplowicz's ideas were embodied in the first PPS agrarian program of 1905, which represented a compromise between Krauz's "othodoxy" and Gumplowicz's "revisionism." It was influenced by the peasant platform of the Russian Socialist Revolutionary Party and treated the theory of capitalist concentration in agriculture as unproven, at least regarding Poland, but it nevertheless deplored the peasants' propensity toward individual land ownership as contrary to the spirit of socialism. The program of 1905 consisted of a preamble and two parts: i.e., the minimum and maximum demands. It envisioned confiscation by the state of private land, to be leased to the poor on long-term tenure, but this was considered as a means of creating socialism rather than a final goal.[13] Incidentally, both the Left PPS and SDKPiL dismissed the farmer as a socially conservative force and therefore, like socialist parties in the West, did not bother to draft peasant programs as such.

After the PPS had split, in 1907 the Revolutionary Fraction prepared a new peasant program—which remained in effect thirty years—

surviving Limanowski himself. Drafted by Gumplowicz, Perl and Jodko-Narkiewicz,[14] it incorporated the minimum demands of 1905, but it relegated socialism to a more distant future than its predecessor had done. Moreover, it introduced a new principle of compensation for the expropriated landowner, both in the form of rent and as a charge against the general tax fund. (Limanowski advocated only the latter.) Under the maximum provisions of 1907, expropriation was to be limited to estates of 100-120 morgs each of good land (3.5 morgs = 5 acres). And the state was to lease the expropriated land to peasants on fifty-year tenure. It was expected that, after the initial leases had lapsed, a rise in productivity during this period would allow for a redistribution of a given area of land to a larger number of tenants than hitherto. Although much land was to be acquired by the state, private ownership would remain and the agrarian reform was to be implemented prior to the general socialization of the means of production, that is, within the framework of a capitalist economy.[15]

Limanowski criticized this program of the PPS, arguing against the belief of its co-author Gumplowicz that the peasant was unlikely to work well on large collective or state farms. He maintained that the lot of contemporary farmers in the West (for instance, in France) was grim. In an attempt to prove that only large socialized farms might deploy labor efficiently and without exhausting the agricultural worker, Limanowski cited some examples of contemporary communal large-scale farming in the United States. Also, he reiterated his argument in favor of socialist estates, which assumed that only large-scale farming would employ the best methods (i.e., rotation of crops, widespread use of artifical fertilizer and mechanization of agriculture); he dismissed as impractical Gumplowicz's suggestion that a small farmer might well rent the necessary machinery. And he foolishly repeated Fourier's alleged dictum: "better one good barn than several hundred poor barns." Limanowski proved impervious to Gumplowicz's argument in favor of small farms for Poland, which would represent a compromise between the backward dwarf holdings in Galicia and the huge capitalist landed estates in the West. Yet, the small farm had proven viable in Western Europe, as Gumplowicz noted. In addition, Limanowski seems to have been somewhat insensitive at this time to the peasants' ambition to possess land, though earlier in his *History of Polish Democracy* he had praised the priest Sciegienny's agrarian program of 1844 in Russian Poland as one which "reconciled peasants' propensity toward individual possession of land . . . with public interest and the communal ideal." While accept-

ing the PPS program of 1907 for the present as the practical policy,
i.e., a tactical compromise aimed at enlisting both gentry and peasant in the struggle for independence, Limanowski remained unrecon-
ciled in the long run to the agrarian "revisionism" of the right-wing
PPS.[16]

Though Limanowski fully agreed with Krauz's "orthodoxy" on the
agrarian question, he opposed the latter's historical materialism. In
1908 a collection of Krauz's writings entitled *Economic Materialism:
Studies and Sketches (Materializm Ekonomiczny: Studia i Szkice)*
appeared posthumously; in it he admitted an interaction between
the "base" and "superstructure."[17] Shortly after its publication, Limanowski reviewed the volume in his "Monism in History" which he
published in Feldman's *Criticism.*

In his review Limanowski commended Krauz for avoiding "a narrow doctrinaire approach."[18] Nevertheless, he attached Krauz's philosophical monism, arguing that it was inapplicable to a concrete empirical discipline like history. He felt that historical materialism was
a retrogressive approach to history; it was a legacy of a theological
interpretation which originated with St. Augustine and assumed that
Providence guided human actions. Limanowski noted that Hegel had
substituted Idea for God, and Marx had reversed Hegel's theory by
enthroning Matter. In Limanowski's opinion, in their respective inter-
pretations of the historical process both Hegel and Marx merely substituted metaphysics for theology. In accordance with Comte's law
of the Three Stages, Limanowski postulated an empirical history,
employing the positivist method.[19]

Yet, notwithstanding his profession of an "objective" approach
to history, Limanowski adopted a partisan stand in rejecting the
Marxian concept of ideology as a reflection of the contemporary
material conditions of life, which Krauz espoused; he actually attributed primacy to the "superstructure," arguing that it was applied
knowledge (i.e., technology) which determined economic development! And he insisted that Marxists themselves, in stressing the importance of their ideology, belied their belief in the primacy of the
economic "base." Limanowski maintained that progress was accomplished also by the moral and emotional faculties—including religion—
and not exclusively by intellect, as Buckle had argued. Yet Limanowski agreed with Buckle (whose works he had read while in exile
in North Russia and shortly after his arrival in Russian Poland) that
the primitive man was fully dependent on his external environment,
but in more advanced societies it was his own nature which was primarily responsible for man's development.[20]

Limanowski again attacked the Marxian idea of the primacy of the economic "base" in an article entitled "Nationality, State and Internationalism," also published in *Criticism* in 1908. Economic relations, he argued here, were not always the determinant factor of socialism; that is, socialism might well develop in countries which lacked modern industry and a numerous factory proletariat. To prove the point he cited the case of Galicia where, notwithstanding its economic backwardness, there was a well-organized socialist movement, whereas in the industrialized Upper Silesia the movement had made little headway. In his assessment of socialism in Galicia, however, Limanowski overlooked the fact that the PPSD was principally a party of the petite bourgeoisie as well as a rallying point for the socially radical and patriotic intelligentsia in Galicia; this last group "substituted" itself for the liberal bourgeoisie in the Western countries.[21]

In "Nationality, State and Internationalism," he was reacting to Kautsky's *Nationalität und Internationalität* (Stuttgart, 1908), as well as to a study by Otto Bauer (who, along with Karl Renner, was a prominent theoretician on the national question in the Austrian Social Democratic Party) entitled *Die Nationalitätenfrage und die Sozialdemocratie* (Vienna, 1907). Limanowski argued that, because Kautsky had overlooked the antagonism inherent in the relationship between nationality and the (alien) state, he included Prussian Poland within the German nationality. And Limanowski disagreed with Kautsky that there was a single European culture which was international in its scope; rather, he supported Bauer's thesis, i.e., each nationality produced a unique culture.[22]

Limanowski went on to argue, as Bauer had done, that socialism by emancipating the victims of capitalism (i.e., creating a truly "participatory democracy") should lead to the realization of the postulate of national-cultural self-determination, and that under a socialist world-government international boundaries would be readjusted in the interest of all. In cases of conflicting national interests due to an intermixture of ethnic groups, Limanowski recommended a cultural-national autonomy for each minority group. In effect, his program on the nationality issue resembled somewhat the resolutions adopted at the Brno (Bruenn) congress of September 1899, which ralled Social Democratic parties of the Austro-Hungarian Empire. The nationality program agreed upon at the Brno congress represented a compromise between a concept of national-cultural administrative autonomy defined in terms of the territorial principle and a concept of national-cultural autonomy based on the linguistic principle, with a leaning toward the former type of autonomy. It was, however, the extra-territorial type of national cultural autonomy which was to be adopted and further developed by Renner and Bauer as a solution to the nationality problem in the Austro-Hungarian Empire.[23]

In the article under discussion, Limanowski, in briefly surveying (and in effect much oversimplifying) the past, assumed that the era of feudalism in Western Europe was characterized by "inter-state" relationships, i.e., relations which were based on a brute force. He argued that the capitalist West had identified the concept of nationality with the concept of state; the European liberal "bourgeois" regimes championed "national rights" in theory, but in fact they usually appealed to the "state law," that is, to coercion, in order to maintain the political status quo. Thus, Limanowski attributed to the era of capitalism in Europe a policy of "half-measures" on the nationality issue. He insisted that in future the transformed, socialist state would only admit the rule of the "law of nationality," that is, self-determination on the basis of an impartial plebiscite. (In distinguishing between the state as a "thing," and the nationality as a "collective person," Limanowski opposed what he thought were two basic principles of human behavior, that is, coercion and voluntary association.) He believed that socialism represents the only means of gradually creating a truly democratic world commonwealth. And he now assumed that the Austrian state might become a model for such an international *res publica.*[24]

II

Two years after his arrival in Cracow, Limanowski completed the research for his projected biography of Worcell. The book came off the press in April 1910. Both as a history of the first Polish People and a biography of Worcell it remains unsurpassed. It was Limanowski who first drew attention to Worcell as a pioneer Polish socialist. [25] It is interesting to note that Limanowski had modelled to some extent his own political activity on that of Worcell, whether consciously or unconsciously. For instance, when Limanowski broke away from the socialist "sect" *Equality* eventually to join a "national" organization, the Polish League, his action was not unlike Worcell's erstwhile departure from the Polish People and eventual participation in politics of the Democratic Society, a predecessor of the League. In his biography, Limanowski characterized Worcell thus:

> Due to his commitment to the cause of Poland, Worcell did not remain within the narrow circle of his socialist group. . . . He felt an urgent

need to act, and to organize an armed struggle against Muscovy, because Poland epitomized for him the common people, while Muscovy represented the principles of aristocracy, monarchy, tyranny . . . . And when the Democratic Society had summoned [compatriots] to battle, Worcell humbled himself before the Society, and recognized it as the rightful leader of the nation, rallying all democrats in the common [Polish] cause.[26] . . . As a pioneer of socialism in Poland . . . Worcell represented the best example of how to reconcile patriotism with loving mankind as a whole.[27] . . . The personality of Worcell became like a lodestar for all those Polish socialists who valued the democratic past of their forefathers and did not wish to abandon the . . . struggle for independence . . . . His name became a symbol for them.[28]

Limanowski's political friends, unlike the conservative *Time*, were enthusiastic about his new book. For instance, Emil Haecker considered it Limanowski's best work so far. However, the criticism of the Galician conservatives was directed against Worcell himself, rather than Limanowski, though subsequently they modified their stand—perhaps because they became mindful of Worcell's aristocratic origin. Limanowski entered his book in a contest sponsored by the conservative Academy of Learning in Cracow, but he was not awarded a prize.[29]

He had contemplated writing a biography of Worcell for twenty years, but could only begin the task after he had perused sources available in various libraries in Zurich, Rapperswil, Paris, Brussels. London and Vienna. While composing his brief biographical sketch in 1892 (mentioned above), Limanowski decided that in his projected monograph he should append Worcell's two extant theoretical works (*On Property* and *On the Natural and the Social Associations*). These treatises served Limanowski as the point of departure for composing two theoretical chapters on socialism and patriotism, which were only indirectly connected with the subject matter of the rest of the book. In doing this, Limanowski wished to remedy "the too German . . . character" of contemporary Polish socialism and to provide it with a native root (as he had confided to Maria Wyslouch). Limanowski attributed a dual role to his study on Worcell. On the one hand, it was to be a contribution to modern Polish historiography, specifically to the history of the Polish Great Emigration and, on the other, it was to synthesize Limanowski's ideology as he had developed it to date. In the pages that will follow we shall discuss only the theoretical chapters which constitute a direct contribution to Limanowski's political ideology.[30]

Worcell in his essay *On Property* considered private ownership socially determined and a historically conditioned (i.e., a relativist) concept which was not based on a natural right. This was acceptable to

Limanowski. But he opposed Worcell's ideology in so far as the latter ignored the "economic" roots of socialism, that is, drew his inspiration for socialism from Christianity rather than from the myth of the primitive Slav communism. Limanowski argued that the Polish People might have been successful in overcoming the Lelewelian aversion to Christianity on the part of the Democratic Society, had the Polish People asserted the idea of conquest by a foreign ethnic group as explanation of the antagonism between Polish gentry and peasant.[31] Unlike Lelewel, Limanowski believed that Christianity—which, along with feudalism, Lelewel considered one of the two alien principles which destroyed the primitive Slav democracy—was not to blame as such for the demise of the ancient Slav socio-political system. He maintained that Christianity had merely served as an instrument of the alien state: "The very fact that Christianity became a deadly weapon in the hands of the state against the Slav communal custom, convinced me that this state was . . . of alien origin."[32]

Back in 1876 Limanowski had argued that it was impossible for the decentralized system of self-governing tribal communes to survive the challenge of an armed conflict imposed from without; like Bagehot, Limanowski considered the state a necessary outcome of war. Nevertheless, he was convinced that modern ideology was reverting to the ancient concepts of internal consensus and external federation, as war-oriented "world-society" was allegedly giving way to "world-society" based on peaceful coexistence of divergent ethnic groups. Limanowski believed that family, tribe and nationality had developed naturally and spontaneously; and he distinguished these "natural" social groupings from "artificial" political institutions like the commune or the state. He argued that natural evolution led from family to tribe and from tribe to nationality. On the other hand, in his progressive sequence of principal political forms—commune, state and federation—only the state had allegedly come into being by violent and essentially "unnatural" means; for both the primitive commune of antiquity and the federation of the future were based on the twin principles of consensus and voluntary association.[33]

Limanowski objected to the "original," i.e., "feudal" predatory state (he seldom referred to the pre-feudal states in antiquity), because it gave rise to social inequality. True, he rejected Rousseau's idealization of primitive man in the pre-agricultural stage of human development and continued to consider enlightenment one of the most potent factors of progress, assuming that early agricultural societies were subject to crude political organization, to the rule of violence, and to cruel customs. Man in these societies suffered both from discomforts due to the harsh external environment as well as from the man-made calamity of constant warfare. Yet such primitive agricultural societies were attractive to Limanowski, because he believed that they were "natural," that is, had come into being by means of an internal consensus and external federation, rather than by a force applied from without. He argued: "It is true that this democracy was raw, egoistic, based on the law of the fist rather than the brain but, in the final analysis, therein all were equal, and all equally shared material goods."[34]

In a sense Linanowski's concept of progress was incompatible with democracy; his ideal democracy was in the past and he attributed progress to the future. He reconciled this inconsistency by looking simultaneously backward to a golden age in the past (a characteristic of Romanticism, e.g., Rousseau), as well as to a golden age in the future (like the *philosophes* of the Enlightenment, e.g., Condorcet). Akin to anarchism, Limanowski's ideology was "like Mohammed's coffin suspended between the lodestones of an idealized future and an idealized past."[35]

He had combined two divergent concepts of nature: Hobbes' "state of nature," wherein the life of man was "solitary, poor, nasty, brutish and short,"and Locke's (and Rousseau's) primitive condition of man, characterized by peace and mutual aid. Hobbes had conceived nature as amoral, and his outlook was conducive to formulating objective natural laws, that is, scientific generalizations. On the other hand, influenced by the Stoics, Locke's view had identified nature with the idea of reason and with right morality. Limanowski considered federation "natural" according to this Stoic concept of nature, asssuming that federation was the ideal principle of ordering human affairs, whether in the past or in the future. To this concept of "federation" which he believed was a "popular" idea, he opposed the idea of a "state" as embodiment of the principle of "coercion," which was "natural" only in the sense attributed to nature by Darwin or Hobbes. Limanowski asserted that every state had begun as a highly centralized unit, gradually undergoing transformation in the direction of political decentralization and social liberalization. In defense of this thesis, he argued that modern capitalist concentration was compatible with preserving local autonomy. For instance he pointed to Great Britain which was more industrialized, but at the same time more decentralized, than was France. Limanowski did not realize that this was so only because in these countries the economy was not owned by the state. He went on to defend decentralization on political grounds, too, rejecting Wladyslaw Gumplowicz's argument directed against the political regime in Switzerland. Gumplowicz had felt that federation was a less efficient political system than the unitary state. He argued that in Switzerland, due to a potential predominance of the Germans, the other constituent nationalities resisted extension of power at the center; this Gumplowicz believed was harmful to them all. In reply, Limanowski pointed out that highly centralized France lagged in social progress behind federal Switzerland.[36]

These ideas of Limanowski did not stem from a systematic empirical investigation, but were due in part to his observation of the contemporary political scene in Western Europe and in part to his studies of the European past, especially the histories of England and France. On the other hand, his ideology owed much to his speculations on the relationship between Polish nationality and the alien political regimes imposed on Poland by the partitioning powers, especially Russia. We may also note the influence of Lelewel as well as of the Russian Slavophiles.[37] The latter influence is clearly apparent when Limanowski discusses the origins of nationality and of the state, for the Slavophiles believed that Western societies were formed by conquest and maintained by coercion, whereas in contrast Russian society was allegedly a voluntary association originally ruled on the basis of full consensus. At times Limanowski partly agreed with the Slavophiles by implicitly substituting the Polish nationality for the Slavophiles' ideal, that is, the early Great Russian state.

Under the influence of the French anthropologist Topinard (who, like many of his contemporaries, e.g., Gobineau, had claimed that "long-headed" racial types in Europe were nobility, whereas "round heads" characterized the European common people, that is nobility and peasants originated from divergent ethnic stocks), Limanowski argued that tribes differed from nationalities in that the former were racially homogenous, while the latter were ethnically mixed, being composed of both descendents of the original rulers and of the ruled. And, therefore, nationalities were more "spiritual," and less "racial," social entities than were tribes. Limanowski went on to draw an analogy between individual human beings and nationalities that was completely "spiritual," that is, he now claimed that a person was a member of a given nationality solely by virtue of his cultural affiliation with it. This was in contrast with his earlier attempt to reconcile a mechanistic sociological organicism—for instance, as professed by Albert Schäffle—with a psychological, "spiritual," or even mystical conception of nationality, which was in fact a legacy of Romanticism.[38]

Limanowski felt that without patriotism modern man would sink to the moral level of "tribal barbarism"; yet he distinguished "state patriotism" (i.e., chauvinism) from "people's patriotism" which he defined as "tribal solidarity," manifesting itself best in France, Italy and Poland. (Simultaneously, he implicitly defined modern nationality as less barbarian but less moral than mankind allegedly had been in the tribal stage of development!) He sensibly argued that to appropriate enemy territory of a developed nationality was self-defeating; for one thing, it weakened the national unity of the aggressor nation, and loyalty of subjects was a better guarantee of both internal and external security than armaments were.[39]

Limanowski had by this time come to consider religion a closer national bond than common race or even speech. For instance, he pointed to the eastern borderlands; here, Poles were closer to the ethnically alien Lithuanians than they were to Ukrainians who resembled Poles both ethnically and linguistically. Also, mindful of the Irish, Limanowski argued that, though a language was the usual attribute of a nationality, it was not necessarily its true indicator.[40] Nevertheless, like Otto Bauer he considered speech a unique expression of a national culture; hence his belief that schools in an alien language, imposed on a conquered nation by an invader, impeded its intellectual development.[41]

Limanowski firmly believed that Western European nationalities resulted from an ethnic conquest, but was somewhat uncertain whether this was so in Poland. As noted above, he at times defined a nationality as a racially heterogeneous group formed from forced intermingling of two or more ethnic stocks. But on other occasions, like Lelewel he maintained that nationality developed in a "natural" manner, a product of voluntary union of kindred tribes federating in reaction to a common danger imposed from without. On the one hand, he felt that nationality came into being prior to the state, as he had argued in his *Nationality and State* (1906) and, on the other, he at times assumed that nationality was a by-product of the state. He admitted in his *Sociology*, composed earlier than *Worcell*, that the state might provide its citizens with external and internal security, arguing that it became a means of advancing civilization and of creating a homogenous nationality "where the ruling class stood above conflicting interests of the constituent clans and tribes."[42]

This concept of a ruling class as a neutral political force was incompatible with his socialist belief, being more in line with the socially non-radical and paternalistic political philosophy of Comte.

By the beginning of the twentieth century, Limanowski leaned more and more toward an evolutionary socialist creed, and opposed violent social revolution. His ideology was spiritually akin to the "integral socialism" of Benoît Malon. In his *Sociology*, Limanowski had defined "integral socialism" thus: "An expression of humanitarianism and altruism, and a by-product of progress in morality and scientific research, as well as of development in economic, social and political relations, 'integral socialism' does not confine itself to church, school, economics or politics, but it penetrates into all areas of human experience."[43]

Both Malon and Limanowski opposed the "mechanistic determinism" of "German" socialism. Limanowski felt that it was "necessary" only in the sense that in time the majority were bound to become converted to socialism. He opposed the fatalism of "German" socialism as an impediment to political struggle, including the struggle for independence. Both Limanowski and Malon had a profound belief in the moral progress of mankind and stressed the ethical aspect of socialism more than Marx had done. And both appealed especially to the intelligentsia, due to their humanitarian and democratic approach to socialism.[44]

In his biography of Worcell Limanowski defined socialism as "primitive communism adapted to the complicated forms of modern social life."[45] He distinguished three aspects of "integral" socialism: 1) idealistic; 2) economic-deterministic, i.e., Marxist; and 3) positivistic or sociological.[46] Limanowski defined the idealist aspect of socialism in terms of a religious experience. He believed that the essence of religion—which he considered a "culture in the emotional realm of life"—was not theology but ethics, that is, its purpose was human brotherhood, the ultimate goal of socialism too. Limanowski agreed with Comte that in the past human conduct had been influenced by emotions more than was presently the case, and that intellect would become the determinant factor of human actions in the long run. He even assumed that man's rationality had already gained the upper hand over his emotions! Though he admitted that Worcell's socialism was still rooted in religion and emotion, rather than in intellect (i.e., "science"), as was the entire Utopian socialist movement prior to 1848, he believed that since that date Marxist economic determinism—according to his definition the second of the three aspects of socialism—had been gradually gaining the upper hand in European socialism.[47]

Finally, Limanowski explained that he had already defined the positivistic or sociological approach to socialism in his *Nationality and State* (1906). There he had argued that socialism, "rather than being a product of an economic change, had originated as a result of an evolution of the concept of the 'state.' "[48]

Limanowski felt that the contemporary capitalist state in Europe was clearly the transformed "original" feudal state which was based on a brute force. In the course of time, he argued, racial divergencies eventually became attenuated and ethnic strife was superseded by the less acute antagonism of classes. And this, in his opinion, was a result of the rise of a middle class which was intermediate racially between the conquerors and the vanquished within the original state. Styling itself the "Third Estate" and, as such, representative of the entire common people, this enlightened new class appealed to national unity and used the speech of the lower classes—the vernacular—which it transformed into a national language. Though the bourgeoisie had succeeded in abolishing despotism and privilege based on birth, the transformation of the state which it effected was incomplete and the idealism of the French Revolution—"liberty, equality and fraternity"—was realized only partially. Today, Limanowski maintained, the bourgeois state was allegedly ruled by the people, but in practice it was in the hands of the owners of the means of production, though this was concealed behind the facade of indirect popular rule (parliamentarism) and indirect taxation. He was convinced that such a state was characterized by a policy of "half-measures"; i.e., indirect democracy, political freedom combined with economic coercion, and national self-determination in theory only. As an example of the last characteristic, Limanowski pointed to the annexation by Austria of Bosnia and Herzegovina in 1908. However, he overlooked the realities of the European situation, that is, that even "bourgeois" democracies (and Austria was not even a "bourgeois" democracy, strictly speaking) might pursue an aggressive foreign policy (actually, he had attributed aggressive intentions to both the German and the Russian bourgeoisie), that external relations might well escape control of the people in parliamentary democracies, and that the people themselves were not likely to be more fair and reasonable than were their governments.

Limanowski went on to argue that the middle class, on achieving political equality with the nobles (i.e., alleged descendants of the conqueror tribe), had isolated itself from the proletariat (i.e., alleged descendants of the vanquished people) due to its economic privilege, that is, control over the means of production. But the proletariat longed for a revival of the old communal benefits: self-government, economic equality, and social security. Socialists, mindful of the complexities of modern life, were attempting to devise a new and better social system which would be based on the old communal principles, that is, they were striving to create a democratic socialist republic, in which the people would be truly "master in their own house."[49]

Limanowski's eclectic ideology lacked the logical coherence of Marxism as well as its uncompromising and ruthless passion for making a revolution, for it was based on a belief that class struggle diminished in the course of history.[50] And Emil Haecker, in attributing Limanowski's political determinism to such factors as his conversion to socialism by Lassalle rather than by Marx or to his interest in history rather than economics, was only partially right; this political determinism was mainly due to Limanowski's commitment to the idea of struggle for independence. In attempting to give Polish socialism the sanction of a native root, Limanowski did not clearly distinguish between what was common to socialism in most European countries (i.e., what was a product of modern industrialization) and what was the result of specifically Polish traditions.[51]

III

In addition to contributing *inter alia* to *Forward* and to *Criticism*, and lecturing on Polish history before intellectuals and workers (and also pursuing scholarly research), Limanowski while in Cracow was frequently in demand as either chairman or speaker during commemorations of patriotic anniversaries, arranged by socialists or democrats.[52]

He was now very popular with the PPSD, and regularly attended its congresses as a guest. For instance, he was present at the XI congress of this party, which was held in Cracow in early June 1908. The ovation given Limanowski at this gathering by both students and older socialists was publicized in all Galician dailies, including the conservative *Time*. Yet the authorities did not bother him.[53]

From 1907 the right-wing PPS held its congresses semi-legally in Vienna, where the political atmosphere was more favorable toward the PPS than in Galicia. Limanowski was invited to come to the XI congress of this party, held in late August 1909. It discussed *inter alia* the agrarian problem in Poland; the proposed incorporation into Russia proper of the Kholm district in south-eastern Russian Poland, due to its large population of Uniates; the nature of the relations which the PPS was to assume with Russian oppositional parties; and the international situation. Enjoying his stay in Vienna, Limanowski now speculated that Austria, since becoming the Dual Monarchy in 1867, had been gradually losing its "predatory German character"; perhaps some day Vienna might even become the capital of Slavdom. He was very favorably impressed by the kindliness and integrity of the Austrian workers. The congress of the PPS Fraction was held in the magnificent Workers' Hall located in the outskirts of Vienna. In addition to being shown the building and its facilities, Limanowski visited a model bakery run by the Austrian socialists in Schönbrunn, near the capital. And his remaining spare time Limanowski devoted to completing the research on Worcell in the University of Vienna library.[54]

In 1901—in addition to the appearance of *Worcell*—two important events took place in Limanowski's life; he was chosen a delegate to the international socialist congress which was held in Copenhagen in late August, and subsequently the solemn Jubilee of his seventy-fifth birthday and of his fiftieth anniversary as a political writer took place both in Cracow and Lvov in the fall of that year.

Shortly after his arrival in Copenhagen, Limanowski attended a reception arranged for the delegates at the City Hall, and he was introduced to the incumbent mayor, Knudsen. In Copenhagen, Limanowski observed jubilant crowds in the streets of working class districts; there, women and children threw flowers from windows and balconies to the delegates. And the congress itself became significant in that it acquainted foreign socialists with the splendid progress of cooperatives in Denmark. Limanowski considered his sojourn in Copenhagen—where he met the Russian Marxist Plekhanov—a memorable experience; yet, due to his deafness, he did not take an active part and this time made no speeches.[55]

In the fall of 1910 the PPSD planned a celebration in Cracow of the Golden Jubilee of Limanowski's public career. However, the local Chief of Police Flattau, mindful of the arrest of several refugees from Russian Poland on the night of September 30 (an indication, incidentally, of collaboration between the Russian and Austrian police) and of the recent assassination of an *agent provocateur* in Cracow, feared that festivities in Limanowski's honor might become the pretext for radical elements to incite a riot. Early in November, two weeks before the scheduled event, Limanowski was summoned before the police and warned that he must prevail upon his friends to cancel the celebrations, otherwise the authorities would expel him again from Galicia. This ultimatum resulted in a veritable storm of protests both in Cracow and Lvov. Almost all Galician newspapers, with the sole exception of the conservative *Time*, championed the cause of Limanowski's Jubilee. By 1910 he was famous outside of socialist circles in Galicia; in a sense he had become a "living legend" and, for many patriots, a symbol of faith in the resurrection of Poland. In Cracow students and workers staged a joint street manifestation, protesting both the arrests, mentioned above, and Limanowski's intimidation by the police. In Lvov, in addition to students and workers, progressive democrats also actively championed his cause. Moreover, two Polish socialist deputies, Diamand and Hudec, interpolated on his behalf both the Prime Minister and the Minister of the Interior in Vienna. The authorities in the Austrian capital put pressure on Flattau and he had to withdraw his ban.[56]

The festivities took place as scheduled on November 10, 1910: a matinee concert for over 2000 persons and a banquet in the evening for about 100 guests. The audience at the concert was representative of various political orientations. Among the speakers paying tribute to Limanowski on this occasion were: Dr. E. Bobrowski, the former editor of the PPS *Tribune,* whose speech honoring Limanowski resembled a religious litany; Feliks Perl, who commented objectively on Limanowski's career as a political activist and socialist writer; Jozef Pilsudski, who spoke on behalf of industrial workers in Russian Poland; two workers from Austrian and German Silesia, respectively, Porebski and Stec; representatives of students from secondary and higher schools; and, finally, two Ukrainians, Tennitsky and Hankevich. Even the police were satisfied that Limanowski's festivities had been conducted in a solemn and dignified manner![57]

Limanowski received many congratulatory messages for the occasion from abroad. *Inter alia*, tributes came from: a group of prominent Polish writers in exile in Paris; the International Socialist Bureau, signed by Anseele, Furnemont, Vandervelde and Huysmans; the German Social Democratic Party, signed by Hermann Müller; the Russian periodical *Byloe*, signed by Burtsev; the Russian Socialist Revolutionary Party, signed by Chernov; the Lithuanian and Latvian Social Democrats; and from a Czech admirer, the lawyer Frantisek Soukup. There were articles about Limanowski in Vienna's *Der Kampf*, Copenhagen's *Social Demokraten* and in the *Schlesisches Tageblatt*. Limanowski also received a telegram from Daszynski, who at that time was touring Polish working class communities in the United States.[58]

In Lvov, Limanowski's Jubilee was celebrated on December 4. A democratic journal described it enthusiastically as a "universal holiday for Polish democracy." Limanowski was overwhelmed by the reception; and yet this was the same Lvov, from which he had been expelled thirty-two years ago.[59]

During the festivities in Cracow, Limanowski had been given 1400 crowns toward publication of his memoirs which he was to begin writing two years later, in 1912. Following the Jubilee celebrations, almost all Polish papers published a biographical sketch of Limanowski, including of course the socialist *Forward* and *L'Aurore*. Writing in *Forward*, Haecker referred to Limanowski as the "symbol of the PPS," epitomizing its traditions, development and idealism. And Perl, nevertheless compared the latter to Worcell, whom he considered a "saint." Writing in the democratic journal *Life (Zycie)*, the young Marian Kukiel paid a tribute to Limanowski for his pioneering historical writings. Finally, Feldman also commented favorably on Limanowski's career in his *Criticism.*[60]

Though now popular with a broad spectrum of centrist opinion in Poland, Limanowski was taken to task on the occasion of his Jubilee by his old Paris antagonist from the SDKPiL, Warski. (The extreme left—and the extreme right—did not share the current enthusiasm for Limanowski.) Warski underrated Limanowski's influence both within and outside the Polish socialist camp. In an article written around this time, Warski denied that Limanowski's Jubilee had any historical significance whatsoever; to him this occasion appeared like a retirement of a diligent and respectable banker. Warski, in addition to ridiculing Limanowski for his habit of citing a long string of both Polish and foreign names as his authorities, poured scorn on his idealistic non-historical socialism which, like that of Mickiewicz, was rooted in "religious feeling and in patriotism." Warski maintained that, unlike the French Utopians, whose concrete criticism of contemporary society was valid and had influenced Marx himself, Limanowski in harking back to the Polish "abstract" Utopian socialism of the 1830s produced a socially reactionary ideology which was intellectually reagressive as well.

Warski further argued that Limanowski's "metaphysical idealism" was indeed consonant with his "mechanical" understanding of history. He criticized Limanowski's stress on organization as a means of a political struggle. In one of his histories of the uprising of 1863 Limanowski had stated "that it is most important for a subjugated nation to have a vigilant conspiracy which is always ready to act."[61] This statement he backed up by a number of historical examples. Warski attacked it, opposing the "voluntarism" of the right-wing PPS which, like Limanowski, considered conspiratorial organizations a means to a successful revolution. (It should be remembered that Limanowski, in turn, ridiculed the SDKPiL for its cult of spontaneity and its fatalism, due to its belief in economic determinism.) Warski's final criticism was perhaps more valid than the earlier one. He recalled an old childhood pledge of Limanowski to follow in the footsteps of the Irishman who planned to avenge his country by blowing up the Bank of England. (This was a fictitious character in a fantastic novel by the French author Paul Féval, *Les mystères de Londres*, see Chapter II.) "Behold," Warski wrote sarcastically, "the vigilant conspiratorial organization of his party makes his childhood dream come true . . . by robbing [Russian] state monopolies and post offices."[62]

During 1911 and 1912 Limanowski travelled much in Galicia and Austrian Silesia, lecturing on Polish history—it was a time of intensive patriotic propaganda in these provinces—as well as campaigning on behalf of the PPSD and agitating for reform of the electoral law in Galicia. He also concerned himself with the plight of Polish political prisoners. On July 21, 1911, for instance, he spoke before the XI congress of Polish physicians and biologists, held in the salt-mining town of Wieliczka; their wives had organized a fund-raising campaign to aid imprisoned patriots in Russian Poland and made him an honorary chairman of this fund.[63]

By 1912 Limanowski had received much evidence of his popularity among Poles outside the socialist camp. He now enjoyed considerable prestige abroad as well.[64]

Around that time he prepared two major historical works for the press, dealing with Polish struggles for independence in the past. The first of these books was a collection of biographical sketches of distinguished but little known Poles, most of whom had died in emigration. Limanowski had begun writing these sketches back in 1889 on gaining access to the Polish library in Paris; since that time, he published a number of them in various periodicals. In 1911 he revised them for publication in book form; the collection was designed to present in chronological order various aspects of the Polish struggles for independence and social justice from 1831 to 1871. He entitled it *Champions of Freedom* (1911).[65]

Among these sketches Limanowski included a biography of Wiktor Heltman, a leading member of the Democratic Society, who considered socialism to be the antithesis of individual freedom. Limanowski argued against Heltman that, just as without equality there could be no true freedom, so without freedom there could be no real equality.

In his view, socialism would eliminate political corruption. By abolishing the private means of production, it would put an end to intimidation at the polls of the worker by his employer. Thus a socialist parliament would truly register "the national will."[66]

Limanowski's second work, *A History of the Revolutionary Movement in 1846* (1913), was based on lectures given during 1912 at the School of Political and Social Sciences in Cracow. Writing in 1912-1913, Limanowski stressed that it was not his intent to revive old animosities in Galicia, such as had resulted from the massacre by peasants of Polish landlords in Galicia in February 1846 (discussed above). Actually, he was convinced that the Galician Poles as a rule were now favorably inclined toward Austria. The only lesson which he wished to bring home at this time was that "one should prepare scrupulously for an uprising which, once begun, should be persevered in—until victory or complete exhaustion of resources."[67]

Meanwhile, Limanowski had been himself actively participating in the preparations for a struggle for independence, which started in Galicia in the wake of the Bosnian crisis of 1908-1909. This event had reawakened Polish hopes for a European war in the near future. After the revolution of 1905-1906, the right-wing PPS transferred from Russian Poland to Galicia; here, in its patriotic activity the PPS could count on support from the PPSD, but it was soon to lose even an appearance of a working class party. Around the time of the Bosnian crisis, Pilsudski founded in Galicia the Union of Active Resistance (*Zwiazek Walki Czynnej*) as a political arm for the paramilitary units which he planned to form in Cracow and in Lvov. (The Union's social program was confined to a vague formula: "social reforms to guarantee the right to work and food for all.") The authorities in Galicia supported Pilsudski's activity in exchange for his intelligence on Russia. Moreover, in the event of war they expected his paramilitary units to become a means of enlisting Poles from Russian Poland on the side of Austria.

On the eve of the first Balkan war, Limanowski attended the so-called first Congress of Polish Irredentists which met in Zakopane on August 25-26, 1912, rallying activists from Galicia and Russian Poland (i.e., delegates from the PPSD, the right-wing PPS, democratic and populist groups, as well as from anti-Russian "rebel" National Democratic organizations). They all agreed with Pilsudski's plan for an eventual insurrection in Russian Poland. On Daszynski's motion, the congress adopted a joint political declaration and founded the Polish Military Treasury to finance training of officer cadres. At the congress Limanowski, who was chosen the chief administrator of the fund, declared himself a strong protagonist of collaboration with non-socialist parties in the cause of Polish independence.[68]

Nevertheless, Limanowski in his speech at the congress had stressed that he approached the question of struggling for independence from a "socialist standpoint." He noted that after 1863 there were two dominant anti-insurrectionary political trends in Poland at the two opposite poles of the political spectrum: conservatism and internationalist socialism. Limanowski went on to deplore the lack of recognition on the part of the Polish international socialists that primacy must be given to the struggle for independence, which was both a means of achieving socialism and a moral imperative in its own right. He charged that the exclusive preoccupation of the left-wing Polish socialists with the social issue had been an important factor in the defeat of the Revolution of 1905 in Poland. In his opinion, the revolutionaries' lack of arms and ammunition was another significant cause of this defeat. Therefore, as Limanowski told the congress, he wholeheartedly supported all efforts aimed at military preparedness for the revolutionary struggle which he anticipated in the near future.[69]

The Zakopane congress of August 1912, gave birth to the idea of the Polish *Legiony* which fought against Russia in the First World War. In 1912, among Poles expectation of an imminent war was general; and the paramilitary units led by Pilsudski were now ready immediately to march into Russian Poland. But no general war resulted from the complications in the Balkans in 1912-1913. Therefore Limanowski felt disappointed; he believed that the Poles had missed an opportune moment to organize an insurrection, as neither Russia nor Austria—in his opinion—were then ready for war.[70]

On November 10, 1912, Limanowski attended another Congress of Polish Irredentists which met in Vienna, gathering delegates from seven independence parties from Galicia and Russian Poland. Here, Pilsudski formed a new political arm for his paramilitary units, the so-called Temporary Commission of Confederated Independence Parties (TKSSN); it held its initial meeting on December 1, 1912. However, some sincere socialists within the right-wing PPS deplored the idea of a united front with non-socialist (i.e., democrat and populist) political groups; they charged that this tactic, in effect, resulted in abandonment of socialism by the PPS Revolutionary Fraction. In 1912, Perl led a rebel group which seceded from the right-wing PPS, the so-called "PPS-Opposition"; however, the following year on the outbreak of the war this splinter group reunited with the parent party.[71]

Limanowski remained with the PPS Revolutionary Fraction, opposing the defection of Perl. However, nothwithstanding his tactical compromise, he did not at any rate in theory divorce socialism from the Polish cause. For instance, while inspecting a military school shortly before the war, Limanowski *inter alia* told the officer cadets that independence would pave the way for socialism in Poland. Thereupon, they informed him that he was the first person to speak to them openly and unambiguously about socialism.[72] In the years 1912-1914, Limanowski preached socialism at gatherings of the intelligentsia, while at the same time he sought to arouse patriotic feelings among the workers. Around this time, he lectured frequently in a working class district of Cracow, Podgorz, where a school for officers functioned in the local Workers' Hall.[73]

Abroad, Polish patriots had come to consider Limanowski somewhat of an authority on Polish affairs. As an instance of this, we might point to an inquiry addressed to him in November 1912, by the secretary of the non-party Independence Committee, which the foreign department of the PPS Fraction—the Union of Polish Socialists—had organized in the United States. The Committee was representative of several American Polish religious, patriotic and socialist organizations. Limanowski was asked the following questions: 1) whether there was a possibility of an Austro-Russian war issuing from the complications in the Balkans; 2) in the event of such a war, would the Poles rise against Russia; 3) whether an armed struggle for independence would be in the Polish interest; 4) which of the political parties in Poland sincerely aimed at independence and would likely implement their independence program; and 5) what was his opinion, in general, of the present situation and of future prospects of independence. Limanowski replied (though in his *Memoirs* he expressed some uncertainty as to exactly what he wrote) that war was indeed possible and that, in the event of war, the Poles should utilize the occasion for organizing a *levée en masse* to achieve both political independence and social transformation of the country.[74]

In Galicia, in the years immediately preceding the outbreak of the war, Limanowski noted a change in attitude toward him, as a socialist, in non-socialist circles, In 1913, on the occasion of the fiftieth anniversary of the 1863 uprising, he was invited by a group of student democrats to lecture at the Jagiellonian University of Cracow. While the conservative professor Stanislaw Tarnowski vehemently objected to this, his colleagues Ignacy Chrzanowski and Waclaw Tokarz intervened on Limanowski's behalf. The rector refused at first to allow suitable accommodation at the university for the proposed meeting, should Limanowski in fact speak. The students prevailed upon the rector to permit Limanowski to share in the festivity as an honorary chairman; however, he insisted that there should be three other honorary chairmen in addition to Limanowski, including a Roman Catholic priest. Thus, the students scored a partial victory. After leaving the university festivities, Limanowski attended a reception given in honor of surviving participants in the patriotic activities of the 1860s, this time as a guest of the City Council.[75]

Notwithstanding modest personal triumphs like the above, by 1914 Limanowski had become depressed. Due to his deafness and advanced age, he ceased—for the time being—to take an active part in politics. Bothered by loneliness and idleness, he again found himself in financial difficulties. Hitherto, the pressure of preparing books for the press had kept him contented and busy. Having finished all his major projects, including his *Sociology*, he now planned to write with the sole purpose of earning a living. But such work was hard to find. He had become accustomed to contributing exclusively to socialist papers; due to his radical views, it was still difficult for him to find non-socialist publishers for his articles and pamphlets.[76]

However, the outbreak of the war cheered him up and he became politically active again. Immediately after the German ultimatum to Russia had expired on August 2, 1914, Pilsudski sent the first of his paramilitary detachments into Russian Poland; it proclaimed the formation of an in fact non-existent "National Government" in Warsaw, which supposedly had nominated Pilsudski the Polish military commander-in-chief. By means of this mystification (in which he was aided by the "old guard," in the PPS, including Limanowski), Pilsudski aimed at securing a "national" saction for recruitment of the masses in Russian Poland, which might not heed an appeal solely in the name of his party or even that of the TKSSN. Actually this deception enabled Pilsudski to exert pressure on Galician politicians who were outside of his political camp (i.e., the TKSSN). He proceeded to create local conspiratorial organs of civil administration in Russian Poland, the so-called Commissariats of the National Government. To this task, which Pilsudski considered very important, he assigned the "old guard" members of the PPS, forming the Citizens' Committee with Limanowski as its vice-chairman. The Committee was to aid the Polish Military Treasury and the Commissariats. However, Pilsudski soon withdrew both his military contingents and civilian agents, in part due to the anti-insurrectionary mood encountered in Russian Poland and also because he was ordered to do so by the

Austrian General Staff—after having scored, however, a moral victory.[77]

The Austrian General Staff compelled Pilsudski to subordinate himself to the transformed TKSSN; on August 16, 1914, the latter was renamed the Supreme National Committee (NKN). Four days later, the NKN began recruiting the famous Polish *Legiony*, the first brigade of which was commanded by Pilsudski. The latter was thus given an opportunity to distinguish himself as a military leader and to gain popularity in the whole of Poland. Based in Cracow, the NKN represented the idea that Galicia was destined to become the "Piedmont of Poland"; it aimed at creating an autonomous Polish state comprising Galicia and Russian Poland, ruled by an Austrian archduke. However, this project held many dangers for the Polish cause, for Austria was the weaker partner in the Austro-German alliance and was bound to comply with Germany's wishes on the Polish issue. The NKN came to include all anti-Russian Polish political groupings, from conservatives to right-wing socialists.[78]

Limanowski collaborated with the press bureau of the NKN by writing articles on historical and political themes for home and foreign consumption, aimed at defending Poland's right to independence.[79] The Left of the NKN, which included Limanowski, in acquiescing in alliance with Austria was going against the Polish democratic tradition of seeking allies among "peoples" rather than "governments" in Europe, whereas Limanowski's cult of the original Polish *Legiony* led by General Henryk Dabrowski of the Napoleonic era, which fought against Austria, was incompatible with the conservative politics of the NKN's right wing. Thus, when Limanowski sent an article about these *Legiony* to the conservative *Polish News (Wiadomosci Polskie)*, its editors rejected it. This was scarcely surprising since in his article he had underlined the "revolutionary spirit" of the *Legiony*, which had experienced brutality and treachery on the part of the Austrian troops during the capitulation of Mantua in July 1798, and back in 1797 had aided Napoleon to set up his "Roman Republic" and expel the incumbent Pope, Pius VI, the ally of the Habsburgs, thereby antagonizing both the Pope and Austria.[80]

Nevertheless, Limanowski himself was now in favor of an alliance with Austria. In his *One Hundred and Twenty Years of Struggle by the Polish People for Independence* (1916) he justified this alliance both in terms of a common political interest as well as common ideology aimed at "freedom and progress." He believed that Austria was now evolving toward democracy and federation. (In contrast, he considered "Prussia," i.e., Germany—Austria's ally—to be still the old type of a predatory and centralist state.)

In most of his writings during the war Limanowski continued to underline the connection between the struggle for national liberation and the quest for social justice in Poland. This social radicalism distinguished Limanowski from conservative writers in the NKN's political camp.[81] Nevertheless, the NKN did tolerate Limanowski; it could not dispense with his services, due to the great prestige which he now enjoyed among the democratic and socialist youth in Galicia. But the NKN liked to refer to Limanowski as a veteran of the uprising of 1863, rather than as a creator of nationalistic socialism in Poland.[82]

The entire Polish "activist" political camp (i.e., the protagonists of collaboration with the Central Powers) publicized the Union of Lublin to justify its own aim of restoring a Poland-oriented commonwealth within boundaries of 1772. In January 1917, the "activists" brought out a new edition of Limanowski's brief *History of Lithuania*, first published in 1895, in which he had attempted to counteract the Russian viewpoint of the Polish-Lithuanian connection, that is, that the historic lands of Lithuania had closer ties with Russia than with Poland. Once more he asserted that "it was in the interest of these nationalities [i.e., Lithuanian, Latvian, Ukrainian and Belorussian] to maintain the old union . . . which developed naturally in the course of history."[83]

He argued that the Union of Lublin had endowed the Lithuanian nobility with civil and political rights equal to those of the Polish nobility, as well as introducing Lithuania to Western European civilization. The Union had saved the Lithuanian nation from foreign conquest, first by the Teutonic Knights and secondly by Muscovy. On the other hand, it enabled the Lithuanian nobility to escape the despotism of their princes. To this end, the former chose Polonization. He pointed out, however, that this political and cultural assimilation of the Lithuanians took place at a time when the official language in Poland was Latin, prior to the rise of modern nationalism. Limanowski went on to argue that joint struggles for independence by Lithuania and Poland, often led or inspired by Polonized Lithuanians, like Kosciuszko, Mickiewicz, Emilia Plater, and Romuald Traugutt (the last two were insurrectionary leaders in 1830-1831 and 1863-1864, respectively), only strengthened the bond already existing between the two nations. And he stressed an important legacy bequeathed by the last Polish insurrection to future Polish generations: the insurgent "National Government" had issued in 1863 a decree recognizing Lithuania and Ukraine as equal partners in the future commonwealth.[84]

Limanowski wanted to safeguard the borderland nationalities against Germany and, above all, against Russia. He assumed that there was a basic cultural, and also partly a racial, gulf existing between Russia and Poland, which precluded their coexistence within one state. He continued to think of Polish struggles against Russia in the oversimplified terms of a "struggle between Asiatic despotism and European liberalism."[85] In reference to the Decembrist "Jacobin" Pestel, who was an avowed Great-Russian centralist, Limanowski now wrote: "of German origin and upbringing, he was not a protagonist of the Russian [type of] predatory imperialism."[86] When he learnt of the fatal assassination attempt at Sarajevo, he attributed it to "Russian work"![87] In 1916, Limanowski immediately reacted to a suggestion put forward in *Die Neue Zeit by an Austrian Social Demo*crat Engelbert Pernerstorfer (who was friendly to the Polish cause) that Poles collaborate with the other nationalities of the Russian Empire, in order to transform it into a democratic federation. In reply, Limanowski sent an article to Feldman, who from October 1, 1915, edited in Berlin a periodical for the NKN Department of War, entitled *Polnische Blätter*. Feldman published his article in his paper. In it Limanowski argued that Poles must not wait passively for a revolution in Russia but, rather, they ought to separate from her at the first opportune moment. He was now suspicious even of a potential revolutionary regime in Russia, arguing that all attempts at a revolutionary Russo-Polish alliance, which he traced back to 1863, had so far proved a disappointment to the Poles.[88]

Yet in March 1917, Limanowski was astonished by "the mildness and progressiveness" of the Russian revolution. Now that the Russian Tsardom had ceased to exist, he believed that with it there had disappeared the greatest obstacle to freedom in Europe. He hoped that the war had taught European nations that collaboration, rather than enmity, ought to become the basis for international relations and would best serve their respective national interests. And, in reaction to the Treaty of Brest-Litovsk, Limanowski—apparently quite out of touch with reality—wrote in March 1918: "Today, upon the complete disappearance of the threat [presented in the past] by [Russian] Panslavism, the Slav aspirations—one must assume—will intensify among the Polish people . . . in the direction of a Slav federation based on the principles of communal rule."[89]

In the previous October the Bolshevik revolution had taken place. Limanowski's reaction to the consequences of this revolution will be discussed in the next chapter.

CHAPTER X

The Poland sanctioned by the Treaties of Versailles (June 28, 1919) and Riga (March 18, 1921)—this last ended the Polish war against the Bolshevik Russia—represented a victory for the "centralist" doctrine of Dmowski and a defeat for the federalist idea of restoring the old Jagiellonian Commonwealth of Nations as advocated by Pilsudski (or Limanowski).[1]

In addition to shaping nationality policy, the Polish right also affected social policy in interwar Poland. Social change from below in some continental European countries—for instance, France, Prussia and Austria—had preceded the calling of a Constituent Assembly (e.g., in France in 1798; in Prussia and Austria in 1848). There, the bourgeois aided the peasant in liquidating feudalism, or at least remained neutral. The Assembly legalized the changes from below, moderating them, but nevertheless significantly altering the power relationship between classes. In contrast, popular movements in partitioned Poland usually were too weak to do this, repressed as they were by both the alien regimes and a native reaction. In interwar Poland, the Sejm (i.e., the lower house of parliament), dominated by parties of the right, could not become an effective instrument of social change.

Another important reason for the preponderance of the right in the Sejm was that representatives of the national minorities—who usually were liberal nationalists or peasant radicals and constituted at least one-fifth of the number of deputies in the Sejm—were extralegally excluded from influencing the formation of governments. True, the Polish moderate left (i.e., the PPS and radical peasant groups) was only slightly weaker than the Polish right. But, prior to Pilsudski's *coup d'état* in 1926, the most important party of the center (i.e., the peasant party "Piast") usually collaborated with the right.

The situation in Poland after 1918 was peculiarly difficult; in addition to the problem of the industrial workers there was also an unsolved agrarian issue. Concerning the latter, socialists as well as Communists postulated "socialization" of the land and, on the other hand, the right-wing peasantists demanded division of estates into private allotments. A coalition of all peasant groups and socialists could have been possible only if the latter were to have adopted an agrarian program approved by all peasants, as Lenin in Russia had done. In the existing circumstances, a large number of non-radical peasants, especially from former Prussian and Russian Poland, voted for the right-wing candidates. The very idea of socialization of the land—that is, acquisition of the expropriated land by the state (prior to dividing it into allotments for individual peasant family use), rather than handing the land over to the peasant for ownership *de jure*—repelled the richer peasants in Poland. Thus, before 1926 the radical agrarian program of the PPS accounted in part for the dominance of the right in the Sejm.

For this reason, and also as a result of the partial exclusion of the minorities, until 1926 (with the exception of the brief rule by the leftist Daszynski and Moraczewski cabinets from November 7, 1918, to January 16, 1919) Poland was in fact governed by coalitions of the center-right. After December 1922, following the assassination by a National Democrat of the newly elected first President, Gabriel Narutowicz, the moderate left was actually precluded from taking power. (The national minorities had voted for Narutowicz, along with the moderate left.) As a result of virtual monopoly of the right, parliamentary democracy in interwar Poland was defective, even prior to Pilsudski's coup in 1926.[2]

According to a special bill promulgated on February 20, 1919, by the Constituent Sejm (1919-1922), the Sejm was the sovereign of the state. The first Constitution (which was influenced by the contemporary French model) of the reborn Polish state went into effect on March 17, 1921. Henceforth both houses of parliament were to be chosen for a five year term, on the basis of equal, direct, secret, proportional and universal suffrage. Cabinets would be responsible to the Sejm. On November 14, 1918, Pilsudski had become the Chief of State (a title once borne by Kosciuszko), pending election of the President. The President, who was to be chosen by both houses of parliament for a period of seven years, would be merely an honorary figure. (The National Democrats insisted on weakening the Presidency, in case Pilsudski were elected to this office.) The franchise proved somewhat complicated for the average voter; this enabled a small group of parliamentarians in the Sejm to acquire a monopoly of power and to distribute among themselves positions both in the Government and the Opposition, as had also been the case in France. Proportional representation resulted in a multi-party system; this, in turn, produced shaky governmental coalitions.[3]

From 1923 to 1926, during the pure parliamentary rule in Poland, an almost continuous state of political emergency indicated, in Pilsudski's opinion, that the Sejm could not cope with both political and economic crises in the country; and he also alleged corruption in Polish life, due to the preponderance of the Sejm. However, one must not underestimate the tremendous task which faced the Polish parliament in these first few years of independent existence of the reborn Polish state, emerging from more than a century of subjection to the three alien powers. The real reason for Pilsudski's overt contempt for the parliamentary rule in Poland was that he disliked the idea of placing the Polish army, which he considered the sole guarantor of Polish independence, under control of the civilian authority.[4]

After his coup of May 1926, in the new government led by his nominee Professor Kazimierz Bartel, Pilsudski took over the post of Minister of War, which became the basis of his "moral dictatorship." He entrusted the Presidency to an old acquaintance and former member of the PPS, Professor Ignacy Moscicki, who was elected by votes of both center and left. The new regime effected an amendment to the Constitution on August 2, 1926, which was passed by the votes of the right. Henceforth, the President could dissolve parliament and issue decrees between sessions, subject to later ratification by parliament. In addition, the government could now spend sums of money equivalent to the previous year's budget, if parliament failed to ratify the current budget in time. The aim of this constitutional amendment was to provide continuity and stability of rule in the country. Pilsudski himself felt that he had effectively done away with "interregna" in Polish politics; he had accomplished a "unique historical experiment." (But the Russian premier Stolypin had issued decrees between sessions, too, subject to later ratification in the Duma!) Having gained power with the aid of the left, Pilsudski, like Millerand, proceeded to rule with the support of the right.[5]

When the rigged elections of November 1930 gave Pilsudski a majority in parliament, he considered himself empowered to draft a new Constitution. The amendment of August 2, 1926, left parliament with the power to overthrow governments and, in theory, to to form new cabinets as well. Pilsudski objected to this; he felt that the Sejm should merely act as a check on the government (as had been the case with the Russian Duma). The new Constitution of April 23, 1935, prepared by a group of Pilsudski's political friends under the direction of Colonel Walery Slawek, in many ways could have provided a model for de Gaulle in 1958. Whereas the chief defect of the 1921 Constitution was the weakening of the Presidency by the right in order to thwart Pilsudski, the Constitution of 1935 transformed the Presidency (which was meant for Pilsudski) into a virtual dictatorship; the most grievous fault of the new Constitution, however, was that it allowed, after Pilsudski's death on May 12, 1935, the evolution of the regime toward authoritarian rule, freed from control by society.[6]

The main theme of this last period in Limanowski's life was his uncompromising struggle against the authoritarianism of Pilsudski. Limanowski still pursued the aims of economic democracy and civil rights for all, becoming a champion of political prisoners in Poland, as well as of the non-Polish minorities, especially the kindred Slavs. In addition, he continued as a spokesman for the Polish *irredenta* in the west and in the northwest. (This last aspect of Limanowski's activity will be discussed in the next chapter.)

In Limanowski's opinion, one of the most vital problems in Europe both during the war and after the war was the agrarian question. On the basis of war-time experience, he argued that self-sufficiency in foodstuffs was indispensable to victory for a belligerent power. He believed that improvement in agricultural production would be one of the most important tasks confronting European countries in peace time, too.[7]

In Poland, he feared an elemental and uncoordinated movement in the Polish village directed against landlords, a result of the influence of the Bolshevik experiment. He wished to counteract such unrest by an agrarian reform from above. In his program Limanowski argued against the platform of the Polish peasant parties, on both ethical and economic grounds.[8]

Limanowski now based his economic argument in favor of socialization of farming in Poland on two assumptions: 1) small peasant holdings in Poland would be unproductive and would lead to dependence of the poorer peasants on the richer farmers; and 2) without intensive industrialization there was not enough land for distribution to each peasant family in Poland. But he admitted that peasant farming had proven viable in the West. Yet, he felt that small farming was "unthinkable" in Poland; it was bound to be backward, to waste labor-time, and to unduly exploit the rural workers. He reiterated his belief that, in general, only large farms had access to credit, to machinery, and to other means of improving agricultural production; such farms were situated on better soils. In Tsarist Russia, Limanowski maintained, it was the gentry who exported grain, rather than the peasants. He still detected a historical trend to ever larger political as well as economic units.[9]

In theory, none of Limanowski's arguments against small farming in general seem convincing. Peasants might well rent machinery and form credit unions and cooperatives, as Limanowski himself had shown in his writings discussed in previous chapters. And why must small farms be situated on poorer soils than large estates? In Tsarist Russia, peasant farming was indeed as backward as in Galicia; it was not fair for Limanowski to use Russia as an example of poor productivity of peasant farms. However, (as he had already pointed out), without extensive and costly industrialization there was not enough land in Poland to assuage peasants' land hunger, and this had been the case in Russia as well. Limanowski continued to believe that only large state farms, administered directly by the peasants themselves in their own interest, would lead to general democratization in Poland as well as to improvement in agricultural methods. He argued that, in order to counteract the influence of the Bolshevik example, the Polish parliament ought to declare immediately "the right of each worker to share profits in the [agrarian] enterprise in which he was employed, according to the worker's contribution."[10]

Limanowski assumed that the peasants' quest for land was not an "innate" characteristic, because so many of them were landless and as serfs they did not own land. He seems to have underestimated the longing of the agrarian proletariat to own land and lost sight of the fact that serfs had considered the allotments which they tilled for themselves as their own. He continued to bolster his argument in favor of agrarian collectivism by appealing to the traditions of the primitive Slav commune, of the Polish People of the 1830s and 1840s, and of the Sciegienny conspiracy in Russian Poland in 1844. (Yet the rank and file of the Polish People, for many years divorced from the land, existed under abnormal *émigré* conditions; they espoused a very unrealistic brand of socialism, while Sciegienny had contemplated individual, rather than collective, possession of the land, as had been the case in the Russian *mir*.)[11]

However, Limanowski admitted that peasants were anxious to possess land, for the sake of economic security. Therefore, he also advocated now granting peasants state land on long-term tenure, in order to reconcile the general interest with peasants' self-interest. Yet this was the intent of the PPS agrarian program of 1907, which he had then rejected. In addition, he was now prepared to leave the existing small farms in Poland intact, while ruling out an excessive subdivision of peasant land.[12]

Limanowski based his agrarian program on the following principles: 1) productivity of the land to be increased by the peasants themselves. 2) The land, a means of production, should be accessible to all, i.e., nationalized. He felt that the Constituent Sejm in Poland should be the final authority in the matter of nationalization, with provision for a possible referendum preceding the calling of this assembly. He believed that the expropriated must be assured an alternate means of earning their livelihood. 3) The Ministry of Agriculture ought to be the central authority administering agricultural production in the country. In addition, Limanowski envisaged a Supreme Planning Agrarian Council of experts, to be chosen by autonomous County Councils, in turn chosen by local Commune Councils directly elected by the people. (It is not clear how he related the Ministry to the Planning Council and how he reconciled expertise with the elective principle.) 4) Both temporary and permanent rural workers were to share in the profits of their collective enterprises and were to be paid by Communal Councils. 5) Inheritance in land was to be abolished. However, in the interest of economic stability, a local Council might leave a given farm in the hands of the heirs of the deceased.[13] Limanowski seems to have attempted to adopt all possible solutions to the agrarian problem. He left small farming intact where practicable; he combined a collective possession of land (Point 4) with the principle of an individual possession (Point 5).[14]

Limanowski was aware that, prior to the coup in 1926, it was the peasants who held the balance of power in the Polish parliament; he even assumed that this was also the case in contemporary Germany, France and even in Great Britain! In 1921, he contributed an article to the PPS *Tribune*, entitled "One of the most important tasks of socialism in Poland," apparently his last treatment of the agrarian problem. In the article he argued that in highly civilized countries in Europe peasants allegedly tended toward socialism; conversely, in backward states they were ignorant and supported the right. (Actually, however, socialists in the West produced no radical peasant programs involving compulsory expropriation.) Limanowski deplored the fact that in Poland the conservatism of the peasant deputies was hampering the business of the Sejm, to the detriment of both the rural poor and of the cause of general progress. Nevertheless, he reiterated his belief in the efficacy of socialist propaganda as a means of overcoming what he thought was a "narrow" egoism and sheer ignorance on the part of the peasant; and he urged that Polish socialists must especially appeal to the poorer peasants, to ensure that the Sejm indeed became a means of social change.[15]

A leading member of the PPS until almost the final year of his life, Limanowski nevertheless was not this party's official spokesman on the agrarian issue. He evolved an eclectic agrarian program of his own, which was inspired by both Marxism and the tradition of the ancient Slav commune. An advocate of a compulsory expropriation of gentry land, the mature Limanowski envisioned conversion of gentry estates into cooperative farms, for the benefit of the agrarian proletariat, with some of this land being leased by the state to individual peasant families on long-term tenure. Adopting a pragmatic approach toward the issue of peasant land holdings, in his program Limanowski advocated a combination of voluntary collectivism and individual possession of the leased peasant land. Where practicable, he was even willing to leave peasant proprietorship intact. It seems that Limanowski was an inconsistent ideologist. In part, this followed from his eclecticism and a faulty synthesis of the divergent theories to which he had been exposed. On the other hand, his inconsistencies may well be explained in terms of his pragmatism. He considered his ideology not as a dogma but—rather—as a flexible guide to action, to be tested and shaped by actual experience. It would be unfair to dismiss his agrarian platform as being impractical per se.

In the period from March 12, 1914, to almost the end of 1928 Limanowski kept a diary, in which he recorded the main incidents in his public life. On August 29, 1919, to cite just one example, he notes in this diary that he was elected to two committees that day; the first was charged with placing a memorial plaque on the building on Smolna Street in Warsaw, in which the last "dictator" of the 1863 uprising, Romuald Traugutt had resided, and the other was to aid Poles in "Prussian" Silesia (where a plebiscite took place in 1921). Limanowski attended sessions of this last committee, but due to his deafness he heard nothing! On that same day, August 29, 1919, he was chosen a member of a citizens' delegation and in that capacity interviewed the Chief of State Pilsudski and the Premier Paderewski.[16]

On November 14, 1920, after the Polish victory over the Bolsheviks, which was mainly due to the military genius of Pilsudski and his charismatic influence with the people, Limanowski chaired a meeting in the City Hall, honoring the Chief of State; Pilsudski had earlier that day, during a solemn ceremony received a mace (symbol of his new office as the Marshal of Poland).[17] However, Limanowski's enthusiasm—both for Pilsudski and for the regime in general, in his so-long awaited independent Poland—was short-lived. On January 24, 1921, he comments in his diary: "On Saturday in the bookstores of Goebethner and Hoesick the authorities conducted a stupid and brutal search for Bolshevist writings. In protest, today all bookstores [in Warsaw] are closed."[18] And on March 10, 1921, Limanowski spoke at a huge founding meeting of the League for Protection of Civil and Human Rights. On April 18 someone proposed him as president of this League; however, it was recognized that his advanced age and deafness would prevent him from effectively discharging this function.[19]

In 1921 Limanowski published a pamphlet entitled *The Bolshevist State in the Light of Science.*[20] In it he argued that the struggle of the Bolsheviks with the Whites actually benefited Poland; it had prevented the formation of a "Great Russia" allied with France. (In fact, Pilsudski's policy contributed significantly to the defeat of the Whites.) It was now Poland which concluded an alliance with France, becoming the *cordon sanitaire* separating Germany from Russia. Secondly, as a result of fear of Bolshevism in Poland, it was the PPS that had come to power in 1918. (Limanowski was referring here to the short-lived Daszynski and Moraczewski governments, mentioned above.) This fear had restrained the Polish right so far from openly bidding for power; rather than becoming a reactionary monarchy, Poland instead was reborn as a parliamentary republic. And, he went on, the Moraczewski government (in office from November 17, 1918, to January 16, 1919) instituted a working day of eight hours, laid a basis for a comprehensive system of social security, and contemplated an agrarian reform. Though the leftist Moraczewski government was soon forced to depart in favor of a centrist cabinet formed by the famous Paderewski, the latter did not dare to amend the existing legislation on social security.[21]

Limanowski maintained that, if the electoral law in Poland had been less progressive, the composition of the Sejm would have been more conservative and the project of the agrarian reform (in fact, as shown in note 10, quite inadequate in the long run) would have been less beneficial to the rural poor. In addition, the freedom of speech, press, assembly and association, which had been instituted in postwar Poland, aided in enlightening the masses; in the less than two years of existence of the reborn Polish state they had become sufficiently patriotic to fight against the Bolsheviks in July 1920.[22]

In 1921, prior to the signing of the Treaty of Riga, Limanowski still hoped that fear of Bolshevism would also have a beneficial effect on Polish foreign policy. He felt that Poland must secure itself against possible Russian attack. To this end, he argued for a resurrection of the Union of Lublin, but adapted to modern conditions. He had in mind creating a "natural" barrier composed of borderland nationalities, i.e., Lithuania, Belorussia, Ukraine, Latvia, Estonia and Finland, but he rejected the idea of conquest both in the north and in the east. He wished for a confederation of independent buffer states allied with Poland, but—at the same time—friendly toward Bolshevik Russia. In the existing circumstances, this program, especially as applied to Belorussia and Ukraine, was completely unrealistic.[23]

Limanowski continued in his role of patron of Polish radical youth. In May 1922, in a special address he welcomed the founding congress of the Union of Polish Socialist Youth, which effected a merger of all Polish socialist student organizations.[24] And in November 1922, he was elected Senator on the PPS ticket. (He usually acted as chairman of the annual party congresses.) At this election, the PPS secured approximately 10% of the seats in the Sejm and about 6% in the Senate.[25] During the electoral campaign, the longest-established organ of the PPS, the *Worker* (founded by Pilsudski in 1894) concluded a somewhat flowery tribute to Limanowski as follows: "Let us rally to him as if he were a great banner, representatives of all democratic thought in Poland. . . . Long live Limanowski for the sake of every worthy and peace-loving cause. . . . [He is our] light in the darkness, leader of the young, as well as the conscience and honor of the party, of the working class, and of the entire Polish nation!"[26]

As the senior member, Limanowski was named Chairman of the Senate and in that capacity had to make the inaugural speech. He began it by saying: "It is surely an important sign of the times when the first session of the Polish Senate is being opened by a socialist." Limanowski went on to appeal to his audience to implement the idea of brotherhood as preached by Mickiewicz, Lelewel and Worcell, an idea which, he believed, had been embodied in the Union of Lublin:

> In both the Senate and Sejm let us strive . . . for perfectibility . . . . for the good of all; let the people experience our love and become attached to their Commonwealth as to its best and most cherished of possessions.
>
> Let the Senate collaborate with the Sejm in a spirit of mutual trust, as older brothers should work with younger brothers. The Senate ought to become an amicable assistant to the Sejm, rather than its antagonist. Due to their seniority and experience . . . the Senators should try to moderated extreme partisanship in the lower house and to encourage

the embattled sides to arrive at a consensus, based on a mutal compromise. Let us remember, however, that all creative endeavors, including creative statesmanship, are indeed attributable to the young—those who are still in the possession of their full physical strength and of all their faculties. Let us not restrain the Sejm too much .... Let the words of the poet come true: ... "Everywhere [in our Commonwealth] there is enlightenment and justice!"[27]

Two weeks after Limanowski's inaugural address in the Senate, in the course of a reactionary demonstration at the Square of Three Crosses in Warsaw, in the vicinity of parliament, even Limanowski's own life was to be endangered by a mob incited by the National Democrats.[28] On December 11, 1922, the first President of the reborn Polish Commonwealth, Gabriel Narutowicz, was to be sworn in. The extreme right opposed the election of this President because (as already mentioned above) he owed it in part to the support of the national minorities. The right, therefore, attempted to disrupt the ceremony, thereby preventing the President from assuming office. Many politicians who had voted for Narutowicz (principally of the left and of the minorities) were indeed denied entry to parliament. Confronted by an excited mob composed of students, hoodlums as well as other right wing elements, Limanowski (accompanied by Daszynski) was forced to take shelter in a nearby apartment courtyard at the Square of Three Crosses. The President, however, was duly sworn in. But in less than a week, on December 16, 1922, he was assassinated by the right-wing fanatic Eligiusz Niewiadomski.[29]

During his first period in the Senate Limanowski continued to be greatly concerned with the country's minorities. In May 1924, he wrote a lenthy article for the *Worker*, in defence of minority rights. In it he commended the government for finally attempting to deal with this urgent matter. But he distrusted the official commission appointed for this purpose, because neither experts like Wyslouch, nor representatives of the minorities themselves, had been included in the commission. He argued that:

To arrive at a wise and just political solution [of the minority problem] one must ... turn for guidance to the best traditions of our past ... act in accordance with progressive aims of contemporary mankind; and take into consideration the lessons of history in Europe at large, of which Poland forms an integral part.

* * *

The policy of the present Polish government toward the national minorities is contrary to the contemporary idea of national self-determination. Hence, the unfriendliness toward us in Europe. We are considered as an oppressor, ungrateful for our liberation. We must not wonder, consequently ... why Europe judges us harshly. . . .

... In contravention of the promise, given by the insurgents in 1863 to the fraternal borderland nationalities, of equal rights in the future Commonwealth, today our policy toward them proceeds along the path trod by Bismarck and Stolypin. Thus, even the hitherto friendly Belorussians hate us now.[30]

On July 25, 1924, Limanowski delivered a lengthy speech in the Senate, commenting on the current government legislation regarding the Slav minorities. He considered this legislation to be inadequate and probably intended primarily for foreign consumption. He felt that the first two Bills dealing with use of minority language in courts and administration were worth voting for, but only because they represented a measure of progress in the matter. But he condemned the third Bill regarding minority schools as faulty from both the standpoint of the philosophy of education and that of practical politics. In advocating amendments to the Bill, as upheld by the progressive minority in the Sejm (and rejected by a majority), Limanowski demanded that, in the event of their (certain) defeat in the Senate as well, the minority in the upper house vote against the Bill in solidarity with its counterpart in the Sejm.

Limanowski argued on behalf of the minorities by pointing to the recent lessons of Polish history. In Austrian Galicia, he maintained, it had been possible for Poles to co-exist peacefully with Ukrainians prior to the appearance there of the "All-Polish politicians" (i.e., the National Democrats) in the 1890s. However, the latter had instigated a vicious struggle between the Ukrainians and the Poles, resulting in such deplorable incidents as the wrecking of Lvov University (due to clashes between Ukrainian and Polish students) and the assassination of the Galician Viceroy, Andrzej Potocki, in 1908. Once again, Limanowski insisted, a bellicose nationality policy on the part of Poles had been proven totally self-defeating.

Limanowski now was especially concerned with the Polish government's aim to Polonize Belorussians. He argued that:

Public opinion seems convinced that we must arrive at an understanding with the Ukrainians ... but that the case of the Belorussians is different.... I feel ... that an administrative authority [in Poland] must not attempt to determine a person's nationality, as had been the case under Tsarist rule, where the police might tell an individual to which nationality he must belong, whether Polish, Belorussian or Russian. ... And to claim that the Belorussians have not yet proven themselves a distinct nationality is to disregard reality. The Belorussians already possess a literature; they have poets, nevelists and publicists. At the same time, one must not lose sight of the fact that the Bolsheviks had created a Belorussian Republic ... indeed, this does appeal to the imagination of the Belorussians. ...

* * *

We might add that this [cultural] struggle with the Bolsheviks, which awaits our Commonwealth, is the legacy of the old Polish Commonwealth which had fought great battles with Muscovy in the past....

* * *

This Bill seems to be a product of misunderstanding. Thereby we aim at transforming the Belorussians. And this unfair policy is to be applied against a nationality which has always been so dear to us, which had given us leaders of the caliber of Kosciuszko, Mickiewicz, and the great Belorussian populist Konstanty Kalinowski who perished on the [Russian] gallows in 1863. It should be very easy to come to an understanding with this people but, to this end, Poles must be sincere and open-hearted when approaching the Belorussians.[31]

In 1925 Limanowski prepared a paper for the IV congress of Polish historians, held in Poznan in December of that year. He entitled it *Centralism and Federalism; the National State and the Multi-National State.*[32] In it he attempted once again to influence both government and public opinion in Poland in favor of a just policy toward the minorities by arguing from the historical record. He maintained that the Lithuanian-Polish Commonwealth had prospered only as long as it pursued a wise policy of "federalism" (and he stressed that this "federalism" had preceded the idealism of the French Revolution by more than two centuries.) When the egoistic magnates and gentry in the Commonwealth abandoned "federalism," having been influenced detrimentally by the narrow-minded Jesuits, the result had been alienation of the kindred Ukrainian people; this proved instrumental in causing the Commonwealth's fall.[33]

Back in December 1924, the PPS had solemnly celebrated the beginning of Limanowski's ninetieth year by sponsoring an amateur workers' concert, which took place in the afternoon of December 8; and in the evening of that day the party honored him at a special banquet. On this occasion, he received a great quantity of flowers and souvenirs. The *Worker* reported that "all Poland joined Warsaw in manifesting its love and admiration to Limanowski." Daszynski was the first to speak at the banquet. He emphasized that no other socialist party in Europe had in its ranks a veteran like Limanowski. (Kautsky, celebrating his seventieth birthday around that time, sent Limanowski his best regards, through Daszynski, "from a seventy-year old youngster.") According to Daszynski, Limanowski—quiet, personally unassuming and disinterested—had acted as mentor and example to each generation of the modern Polish socialist movement.

In a voice shaking with emotion Limanowski replied to Daszynski and the others who had spoken. He cordially thanked his audience

for the tributes which he had received that day, rewarding him generously for all the difficult moments in his life, notably for imprisonment and exile. In his speech Limanowski modestly assumed that
this celebration had actually been intended to honor all Polish socialists who had made sacrifices for the people, and not merely himself.
At the end of his speech he exclaimed optimistically: "I believe that
eventually a socialist Poland . . . is bound to become a reality!"[34]

When Limanowski reached his ninetieth year, he felt tired physically and exhausted mentally. He decided to curtail his activities and
to travel only on exceptional occasions. As to writing, he planned
only to complete a pamphlet entitled *The Development of Polish
Socialist Thought*, which was eventually published in 1929, and to
continue preparing his memoirs and diary. But he was constantly being asked to contribute articles on various subjects. On October 4,
1925, he attended a gathering of the Association of Former Political
Prisoners in Lodz; he chaired the meeting and spoke. The IV congress
of Polish historians, which met in Poznan in December 1925 (as
mentioned above) had appointed him chairman, but around this
time he had to undergo a major surgery and could not attend it in
person. Discharged from the hospital in the middle of December
1925, Limanowski soon began writing again.[35]

As a result of the *coup d'état* in May 1926, Limanowski abandoned
the idea of retiring from public life. He now felt duty-bound to
struggle against the authoritarianism of Pilsudski. In the past, it is
true, Limanowski had on a number of occasions supported Pilsudski's
non-socialist causes during crucial moments in the existence of the
PPS. For instance, he had backed the Temporary Commission of
Confederated Independence Parties (TKSSN), which from 1912 to
1914 was the political arm of Pilsudski's paramilitary organization,
and had thereby weakened—albeit unwittingly—the democratic "PPS
Opposition" led by Perl (as mentioned above), as well as the leftist
"fronde" from the PPSD led by Boleslaw Drobner. But this was in
accordance with Limanowski's insurrectionary ideology and with his
tactical subordination of socialism to the struggle for independence.[36]
Again, while residing in Galicia in the period just preceding the outbreak of the First World War, Limanowski became a close friend of
Jozef Pilsudski's younger brother, the ethnographer Bronislaw Pilsudski. However, this was not a political friendship; and Limanowski
had never been close to Jozef Pilsudski himself.[37] Already in 1915,
Limanowski began to be disillusioned with Pilsudski. For instance,
on July 10, 1915, after speaking to some of his associates, Limanowski wrote regretfully: "Almost nobody believes in a social revolution

any more."[38] And again on November 14, 1915, Limanowski noted in his diary: "Yesterday, I conversed with Wasilewski and Gumpolowicz; they have given me the impression that our socialists have moved considerably to the right."[39] In 1916, writing in his *Memoirs*, Limanowski attributed somewhat greater ability and a more pleasing personality to Pilsudski than to the latter's military rivals, Waldyslaw Sikorski and Jozef Haller. Yet to him Pilsudski seemed overly ambitious; Limanowski felt he overestimated his influence even among the nationalistic socialists, the mainstay of his political support among Poles.[40] He attributed to Pilsudski megalomania and snobbery,[41] supporting him only as long as the latter was, at least formally, a socialist. As Limanowski wrote in his *Memoirs*: "At first, it was Sikorski who epitomized sheer militarism, rather than Pilsudski.... But as soon as he began to shed his socialism, his militarism lost its 'civil features'."[42] Opposing Pilsudski's concept of the traditional, undemocratic army, he favored a citizens' militia, as advocated by the French socialist Jaurès. In a militia, there was no blind obedience; discipline was due to the consciousness of his civil duty on the part of the individual soldier. In an article which Limanowski sent to *Polish Culture (Kultura Polski)* in 1917 he had advocated, without mentioning Pilsudski, the formation of such a militia in Poland.[43]

On May 29, 1926, Limanowski noted in his diary that since his coup Pilsudski had been constantly attacking the idea of party politics. "But," he went on, "if Pilsudski had not belonged to a political party, he would not have gained the popularity which he now enjoys in Poland." And on May 30 Limanowski wrote: "Pilsudski's speech was not at all convincing."[44]

On June 4, 1926, Limanowski attended a meeting in the parliamentary socialist club. At the meeting, "the majority decided not to attend the inauguration of the President. . . considering the proposed constitutional amendment humiliating to the Sejm. Thus, I returned home."[45] And on July 8, 1926, he reported: "It was hot today, but I feel fine and I must reply to the silly professors attacking the parliamentary system in Poland." On November 9, 1926, he asked: "What is the matter with Pilsudski? More and more he is denying his past. Pilsudski's press law evoked much indignation in almost all quarters."[46]

Lacking a firm parliamentary basis for his regime, Pilsudski in January 1927 instructed his political friend Colonel Walery Slawek to form a government party. The "Non-party Bloc of Cooperation with the Government" (BBWR), created by Slawek, united politicians from all parties, albeit principally of the right. The Bloc produced no

definite program, apart from an almost mystical idea of serving Poland under the leadership of Pilsudski. The elite of the BBWR was
composed of gentry from the eastern borderlands. This was the beginning of a new period of "Lithuanian" supremacy in Polish politics.[47]

It is not surprising that Limanowski's attitude toward the BBWR
was completely hostile. On June 2, 1927, he wrote in his diary:
"The Czechs are surpassing Poles in every way. Masaryk had been
chosen President for the third time. Immediately, he granted a complete amnesty. . . . And today's *Worker* asks: 'What about amnesty
in Poland?' The new press law is a hundred times worse than the old
law in Austria used to be." Limanowski indeed may well have been
among the first to question Pilsudski's "legend."[48]

In January 1928, Limanowski agreed to stand for re-election to
the Senate for another five-year term. On February 8, 1928, he
writes: "I must survive until the opening of the Senate, in order to
tell them in my inaugural speech what is in my mind." On February
11, Limanowski praised Herman Lieberman, a prominent PPS deputy,
for attacking a new decree applicable to the judiciary in Poland. Limanowski branded this decree as "a new subversion of the constitution, an open attack on independence of the courts of law." He contended that the President "was becoming a puppet in the hands of
reaction."[49]

Between February 21 and March 14, 1928, Limanowski—due to
his preoccupation with the electoral campaign—had interrupted the
writing of his diary. Yet, at the request of a fellow socialist Senator,
Stanislaw Posner, he committed himself to preparing two lengthy
articles for the PPS *Worker*: the first was to deal with the minority
problem and the second to defend the rule of law in Poland. Due to
his advanced age, Limanowski had to limit his writing to two hours
daily. Nevertheless, he met the deadline, and his articles were well
received by his political friends.[50]

In the first article Limanowski reiterated the historical argument
in favor of a policy of "federalism" which, in his opinion, had transformed Poland-Lithuania into a major power in Europe. He believed
that "the democratic tradition of respecting the autonomy of provinces and peoples, which represents the only effective means of
unity, has been bequeathed to us by our history. Only federalism, as
opposed to a policy of oppression of the fraternal nationalities, would
lead us onto the path of an honest, just and wise solution of the minority problem [in contemporary Poland] ."[51]

Due to being personally involved in the issue, Limanowski became considerably irritated while writing the second article, in which he sought to defend Polish democracy. As a candidate to the Senate on the PPS ticket, he could not act as an impartial government critic. In the article he attacked Pilsudski and his political friends thus:

> Each thinking and self-respecting individual ought to have certain political convictions. . . always consonant with his conscience. . . . They say, acquisition of knowledge and experience might well change an individual's convictions. True . . . but the change is usually a gradual one and ought not to proceed in the opposite direction [to one's original position]. If conscience were to tell us that black is wrong and white is right, we might attempt to evolve toward white and away from black. There is something radically wrong with one's conscience when yesterday's black suddenly becomes white today.
>
> It has become fashionable to attack the system of party politics. . . . Where there are political convictions, there must be parties. To realize their convictions, individuals with similar views join together to form a political party. A multi-party system is indeed impeding the general cause, but an absence of parties, that is, of political convictions, is a hundred times more detrimental [to the political well-being of the country] than partisanship might be. [Without parties], . . . self-interest becomes the only motive force of political activity.
>
> Of all political parties in Poland, it is the PPS which has proven itself the most stable and uncompromising both as to principles and as to tactics.[52]

The PPS scored a great victory indeed in the spring of 1928 (in the last relatively unimpeded general elections in interwar Poland). In almost doubling the number of elected candidates, it was the most successful of all parties participating in the elections of 1922. This almost "dramatic" rise of the PPS in both the Senate and the Sejm has been often overlooked by scholars in the West. Nevertheless, contemporary evidence shows this success to be an incontrovertible fact. And it was happening, in spite of an attempt on the part of the government to induce the populace to vote for the BBWR; to this end, the government had spent millions![53]

On March 21, 1928, Limanowski wrote in his diary: "I am preparing mu speech. I am afraid, it is too bellicose; they cannot reply to my challenge." He was prevented by the government from making it, as if they suspected what he intended to say. He published it in the *Worker* on March 25. In this speech he protested the regime's treatment of the nation "as a slave, whom one orders what to do, according to the arbitrary notions of a handful of individuals."[54] Without hesitations, Limanowski vehemently condemned his former political friends. The ruling elite in Poland now indeed feared the old man. They had wished to transform him into a "venerable relic," but he, in command of a moral strength unusual in an individual of his age, refused to retire. Hence, in spite of his seniority, the government now deprived him of chairmanship in the Senate.[55]

Nevertheless, Limanowski remained cheerful. On March 29 he noted jubilantly: "Great victory in the [new] Sejm: Daszynski has been elected Speaker by an overwhelming majority of voters!" (This opening of the Sejm by Pilsudski was memorable, due to the expulsion and arrest of the seven Communist deputies, who had shouted: "Down with the fascist government of Pilsudski!") While accepting his post, Daszynski promised to "safeguard the rights and dignity of this High Chamber." He was to become the leader of the Sejm in the forthcoming struggle for parliamentary democracy in Poland.[56]

On May 5, 1928, Limanowski wrote: "Warsaw historians [The Society of 'Lovers of History' (TMH)] made Pilsudski an honorary member. They had brought to me their resolution to this effect for my signature, but I refused to sign." He himself now received wide recognition in Polish society at large. To cite one instance of this, on July 1, 1928, the city of Lodz created a university scholarship fund in his name.[57]

The writing of his diary and memoirs became more and more difficult for Limanowski. After his operation late in 1925, he was no longer in daily contact with the outside world. Henceforth, the marginal entries (i.e., his diary) were usually based on information culled from the *Worker*. Limanowski's memory began to fail. Yet he was reluctant to give up the task because, as he wrote:

> It represents to me one of my greatest pleasures. While I write, it seems to me that I am not yet completely cut off from public life and that someone might still glance at my diary and consider my opinions. For instance, today, on August 29, 1928, the *Worker* cites two speeches—one by Moscicki and the other by Masaryk. How completely different are these speeches! The first, given on the occasion of harvest festivities at Spala, attempts to instill in the peasants an attitude of idolatrous servility toward Pilsudski and his regime. On the other hand, the other speech shows wisdom; it is intelligent and progressive. In it Masaryk attempts to explain what is the essence of so-called "democracy"; it is not a rule by ignorant masses, but an effort to ensure the greatest possible equality for all. [Masaryk says], "It would be incorrect to suppose that we aim merely at economic democracy. We must ensure equality everywhere: in law, religion, morality and in the entire spiritual realm of life."[58]

Nevertheless, in his *History of Polish Socialist Thought* (1929) Limanowski arrived at the conclusion that Poland was better qualified than Czechoslovakia to "lead Slavdom." He admitted that the Czechs possessed a rich historical tradition and a unique culture; they had advanced far intellectually. Czechoslovakia was economically better off and more democratic in its social structure than Poland, and it was also more industrialized and possessed a larger, more enlightened, working class. Led by progressive statesmen like Benes and Masaryk, Czechoslovakia—with a majority in its parliament which was representative of the working people—pursued a more democratic internal policy than contemporary Poland had done. No wonder, argued Limanowski, that it was Czechoslovakia rather than Poland which appealed to European democrats and socialists. Yet, in his opinion, Czech nationalism threatened to split the country. At the same time, the Czechs constituted a minority in their own state. In addition, he was mindful of the Czech-Polish rivalry over Teschen. It would appear that Limanowski's antipathy toward the Czechs, however, was principally due to their tradition of friendship with Russia.

Limanowski believed that Poland surpassed Czechoslovakia in many areas of public life. Notably, Poland's territory was three times as big as Czechoslovakia's; it was twice as populous as the latter; and Poles were the majority nationality in their state. (But, we may interpose, size as such is not tantamount to merit!) Furthermore, he argued, Poles were more creative than were Czechs, and Polish industry had a better potential for expansion. But most important, in his opinion, Poland could boast of her "great historical tradition of 'federalism'."[59]

During the period of partitions, the Poles—by their constant struggle for independence—had proven their worth both as individuals and as a people; the idealism of the French Revolution had appealed to many of them. Limanowski believed that in his own day "the idealistic minority in Poland"—notwithstanding the current setback for democracy in his country, due in large measure to the ignorance and passivity of the common people exploited by the privileged classes— had been successful in laying a strong democratic foundation in the reborn state. Therefore, Limanowski was optimistic as to the future of Polish democracy: "Despite the . . . temporary preponderance [in Poland] of a coalition of clergy, landowners, industrialists, bankers, rich peasants, profiteers of all sorts, usurers and home-grown fascists, the minority, which desires to advance rather than regress, is strong enough to withstand [this onslaught of reaction]. Instead of succumbing to the reaction, Polish democratic forces will grow in both numbers and strength, aided by the liberal constitution and the spreading enlightenment among the masses. The future is bound to prove the ally of [Polish] democracy."[60]

Around the time this pamphlet went to press, however, a new attempt at subversion of democracy took place in Poland. In October 1928, Pilsudski's political friends proceeded to split the PPS. Thereupon, on October 20 Limanowski wrote an angry letter to the *Worker*. In it he implored the working class in Warsaw: "Defend the PPS, do not let it be broken up." Limanowski regretted that his advanced age prevented him from demonstrating with the workers on the street. A few days later, at the XXI congress of the PPS held at Sosnowiec an overwhelming majority condemned the disrupters. Limanowski had sent a message to the meeting admonishing participants: "Do not allow them [the authorities] to succeed."[61]

On October 24, Limanowski commented, "Much noise on account of my letter of October 20 [which appeared in the *Worker* on October 21]. I have just read *Le Neuf Thermidor* by Louis Barthou. I cannot help but to compare Robespierre to Pilsudski."[62] On October 30, Limanowski paid a visit to the Sejm, and was thanked by his friends for writing the letter. On November 9, he was invited by the President to the Royal Castle for a *soirée*, though it is unlikely that he accepted this invitation. On November 16, he commented: "I was in our [parliamentary] club early in the morning to underline by my presence there that I am protesting against the disruption of our party's unity. That same day at 4:00 p.m. I was in the Senate. An interesting thing . . . neither the President nor Pilsudski dared to appear in parliament." On November 20, he noted, "Government-inspired troublemakers create much confusion in our party."[63]

In a letter he wrote around this time to Emil Haecker Limanowski deplored "the betrayal, fortunately of only a handful of our comrades." Limanowski had played a key role in consolidating the PPS. During the fall of 1928, on festive occasions and during congresses he continued to speak before socialist and workers' organizations. He had not yet lost hope, he told Haecker, having attended in October a meeting of working class youth, whose enthusiasm convinced him that it represented "the young vanguard of freedom and democracy"; it would not permit in Poland, he believed, the degradation suffered by the Italian people—"the nation of Mazzini, Cavour and Garibaldi"—at the hands of the "base dictatorship of Mussolini."[64]

In the spring of 1929 the struggle between Pilsudski and the Sejm became intensified. On March 20, 1929 the Sejm, on the initiative of the PPS lawyer, Herman Lieberman, demanded the arraignment of the Minister of Finance, Gabriel Czechowicz, in the Tribunal of State. Authorized by Pilsudski, Czechowicz had spent public money on election propaganda in favor of the Government Bloc, the BBWR.

Pilsudski fully backed Czechowicz against the Sejm. On April 12, 1929 he formed a cabinet, led by Premier Kazimierz Switalski, in which half of the ministers were officers. This represented the origins of a "colonels' rule" in Poland. On June 24, 1929, the intimidated State Tribunal declared the Minister of Finance formally innocent; however, it insisted on parliament's right to control expenditure.

Meanwhile, Pilsudski subjected the Sejm to a forced vacation of six months. Thus, in the summer of 1929 a stalemate had ensued in the conflict between the Sejm and Pilsudski. The relative political calm of this summer, however, was interrupted by Limanowski, who on August 17, 1929, addressed a vehement letter to President Moscicki, openly branding the government as a corrupt "dictatorship." In his letter Limanowski condemned subversion of the rule of law in the country, *inter alia* arguing that the state budget was not the private purse of the Minister of Finance. The letter, which proved instrumental in consolidating the Polish center and left, was promptly confiscated by the authorities.[65]

On October 31, 1929, the day parliament reconvened, Pilsudski walked into the Sejm, accompanied by a number of armed officers. Thereupon, Daszynski refused to proceed with the business of the Sejm. Pilsudski gave in and departed with his officers, having ordered another adjournment of the parliament. In retaliation for this, Daszynski and the PPS stiffened their opposition. The next day an anti-government bloc, the Center-Left (*Centrolew*), came into being, led by Daszynski. It initially united three major parties of the left and three parties of the center: in addition to the PPS, it included the three peasant parties, one of them being the hitherto rather conservative "Piast," as well as the center National Workers' Party and the Christian Democratic Party (this last had moved from the right to the center).[66]

When the Sejm finally reconvened on December 5, 1929, the Center-Left overthrew Switalski. Confronted with the choice between making concessions either to Dmowski's political camp or to Daszynski's, Pilsudski chose the latter. On December 29, he appointed another moderate cabinet under Bartel, who attempted to come to terms with the opposition. However, early in February 1930, the PPS and the other parties of the Center-Left caused the demise of Bartel's government, too. This was the end of "Bartelism," that is, of the policy of governing with an intimidated Sejm. Now fully committed to authoritarianism, on March 29 Pilsudski entrusted Slawek with forming a cabinet composed solely of the military friends of the Marshal. Henceforth, Slawek ruled by repeatedly adjourning the Sejm. Yet the Center-

Left continued to defend the idea of sovereignty of the Polish parliament.[67]

The Center-Left opposition did not know whether to expect another election or another *coup d'état*. It decided to appeal to society directly, by extra-parliamentary means, in order to activate public opinion in Poland. The opposition faced the following dilemma: how to make the government show its hand, so that it would either restore democracy or would abandon its facade of democratic rule. The Center-Left aimed at forcing the regime either to stop threatening the Sejm or to fulfil its threats. The atmosphere of intimidation, that is, the political "cat-and-mouse" game played by Pilsudski with his political enemies, was becoming intolerable to the latter.

On June 29, 1930, numerous politicians of the Center-Left political camp assembled at the Old Theater in Cracow. This rally was held under the slogan of "Defense of the Rule of Law and Freedom." The two leaders of the Polish democracy, Daszynski and Limanowski, neither of whom could come, sent telegrams to the meeting. In his message, Limanowski summed up the purpose of this mass gathering as follows: "To shake up spiritually the nation, so that, disgusted, it would cease to be shamefully servile toward the dictatorship."[68] The congress represented a breakthrough in the tactics of the opposition. While pointing to the fact that democracy in Poland was merely a shadow, the Center-Left now referred openly to Pilsudski's rule as a "dictatorship" and challenged him to a direct struggle.[69]

Thereupon, Slawek accused the congress of attempting to organize a violent coup against the government. Pilsudski himself referred to the Center-Left politicians as "scoundrels." Early in September he ordered the arrest of about seventy leading members of the Center-Left camp, charging them with high treason. (Previously, on August 30, 1930, he had again dissolved both houses of parliament and ordered new elections, which were slated for November.) He soon removed the arrested politicians from the jurisdiction of civil authorities and sent them illegally to the fortress of Brest-Litovsk, where many were severely beaten and humiliated by the commandant of the fortress, Waclaw Kostek-Biernacki. They were to remain incarcerated thus for about two months, including the period during which the electoral campaign and the elections took place. This enabled Pilsudski to score a victory in his struggle with the opposition. The rigged elections of November 1930 finally resulted in a majority for the Government Bloc. Limanowski, however, remained in the Senate, where he represented the center opinion in the PPS.

He appended his signature to a document known as the "Brest Protest," a brave act since he thereby risked prosecution by the government.[70] Notwithstanding his great age, even he "went out on the street" to demonstrate against Pilsudski's rule. On September 14, 1930, during a huge protest rally held in Warsaw "Swiss Valley" Limanowski as chairman of the meeting and its first speaker severely chastised the regime for subverting the rule of law in Poland.[71]

On January 11, 1931, Limanowski in a letter to his friends Emil and Franciszka Haecker wished that they might all "live to see the quick demise of this sad regime of 'moral recovery'."[72] Limanowski also confided to these friends that he read *Forward* regularly and carefully and that this refreshed him in spirit,"though physically I am only half alive."[73] A stringent censorship of the press had been instituted under Pilsudski's rule. The central organ of the PPS, the *Worker*, suffered over 300 confiscations from May 1926 to August 1933. The PPS commemorated the 300th confiscation by issuing a proclamation signed by several veterans of the party, among whom were Limanowski and Daszynski; in it the PPS stressed the importance of the socialist press in Poland and appealed to the public for financial assistance. The proclamation was promptly confiscated by the authorities.[74]

Meanwhile, from 1930 to 1935 the "Great Depression" affected Poland; besides causing impoverishment of both peasants and workers it inspired political extremism of both left and right. After the XII PPS congress held in May 1931, which called for "a coalition of all democratic forces in the country," the PPS collaborated with the now united Peasant Party (SL). The two parties demanded a "government of workers and peasants" and "immediate nationalization of big estates, without compensation." (The Center-Left soon disintegrated, due in part to the arrest of its leaders in the fall of 1930 and ideological differences as well.) Pilsudski's colonels dealt with the general crisis by authoritarian means. The "Enabling Act" of March 23, 1933, which empowered the President to rule by decree, and Moscicki's smooth re-election in May of that year—both reflected the evolution toward a purely dictatorial regime in Poland.[75]

Limanowski availed himself of the rights accruing to him in virtue of his great age, and being a veteran of the PPS, to oppose the "colonels" at meetings, in interviews and in articles. His last years were overshadowed by this struggle. Though he continued to fight for a democratic Poland, all his writings from 1929 on were confiscated by the authorities. Some of the ruling elite still considered themselves his "disciples"; yet, in general, he was now extremely unpopular with Pilsudski's followers. The authorities took petty revenge against him by awarding him the Medal of Independence on March 10, 1931, which was usually given—in contrast to the Independence Crosses—to merely third-rate public figures. They justified this insult to Limanowski by arguing that he never took part in an armed struggle. When Limanowski learned of their intention to humiliate him in this manner, he wrote a vehement letter to the *Worker* (which this paper, however, did not dare to print). In it Limanowski not only refused to accept the Medal, but stressed that he wished to publicize his refusal, because "a mere quiet protest would have constituted an act of unforgivable betrayal on his part toward his political friends." Thus, the regime was finally cutting itself off from its nationalist socialist traditions, the mainspring of the Polish movement for independence in the period from 1880 to 1918. The tragic conflict between his patriotism and social conscience, which Limanowski had experienced in his youth, returned, as it were, to embitter the last few years of his life.[76]

In 1924, Limanowski had received an Honorary Doctorate from his Alma Mater, the University of Lvov. And in 1927 he was made an honorary member of the Polish Historical Association based in that city. Finally, when he had reached his one-hundredth birthday, the University of Warsaw, too, decided to grant him an Honorary Doctorate. On this occasion, the historian Marceli Handelsman (who was to perish in a Nazi concentration camp assassinated by the native fascist NSZ) spoke of Limanowski as "an original scholar." According to Handelsman, "in many areas Limanowski was truly a pioneer. . . . As a writer and scholar he . . . expressed the longing of the last three generations of [Poles] who . . . sought social justice and the independence of the fatherland [striving] . . . to reconcile historical truth with championship of the Polish cause. . . ."[77] An anonymous writer in a journal entitled *Man in Poland (Czlowiek w Polsce)*[78] deplored the all too modest ceremony at the University of Warsaw. In his view, it was the University which was being honored in granting Limanowski a doctorate *Honoris Causa*, and not vice versa.

Every public figure of considerable moral stature produces, as it were, a myth. At times his "legend" becomes more important for contemporaries than the real man. This was the case with Limanowski.[79] When he died on February 1, 1935 (in the words of the Chairman in the Senate, Wladyslaw Raczkiewicz), "the news of his death reverberated throughout the country, which now in parting with the late Senator parted, as it were, with a symbol of the struggle for liberation and reconstruction of the fatherland that had lasted a hundred years. . . . Boleslaw Limanowski was dear to us, for he linked [us] with the past. His personality brought us near to the great Poles of a by-gone era of heroic struggles for freedom, because he was truly one of them surviving amongst us."[80]

Almost every periodical in Poland included an obituary of Lima-nowski. Even the right-wing *National Thought (Mysl Narodowa)* paid him a generous tribute, having described him as a distinguished historian, an idealist who led a blameless personal life, and a "prominent historical personality."[81] Moreover, numerous condolences came from private individuals as well as social and cultural organizations in Poland, and from abroad. Almost all socialist parties of Europe and North America were represented at Limanowski's funeral, which took place on February 5. The funeral procession of 50,000, described by Limanowski's Polish biographer as "an attempt to create the appearance of national unity," included the representatives of minority nationalities, workers, intelligentsia, cabinet ministers, generals, and the Premier himself.[82]

A symbol of Polish struggles for independence in the past, Lima-nowski in a sense also symbolized the Polish commitment to parliamentary rule. His passing almost coincided with the demise of the last vestiges of political democracy in Poland. Shortly after his death, the government promulgated the new Constitution of April 23, 1935; on January 26, 1935, its final draft had been adopted by the now docile Sejm. Though the principle of cabinet responsibility to parliament was formally preserved under the new Constitution, in all important matters the President was no longer required to have his decrees countersigned by the Prime Minister. The legislative power of the Sejm was also restricted by the creation of an influential Senate, in part appointed by the government, and in part elected by the political and intellectual elite in the country. After Pilsudski's death in May 1935, a new system of elections to the Sejm became law on July 8. This enabled the government, in spite of the democratic suffrage, both to control the choice of candidates and to manipulate voting. The entrenchment of the "colonels' regime" in Poland was now complete.[83]

Limanowski's political treatises, published in interwar Poland, fall within two broad categories. First, in a number of articles and pamphlets he discussed various topics related to current politics: the origins and significance of the First World War; European diplomacy and the Treaty of Versailles; the aims and significance of the League of Nations; federalism; and national minority problems in both his native Latvia and in the Polish *irredenta* in the west (i.e., Silesia, Marienwerder, Ermeland and Prussian Mazuria). Second came his writings on a more abstract level. In his *Sociology* (1919 and 1921) and in several articles and pamphlets, he produced the final version of his political theory, commented on the idea of progress and on socialism in general and discussed the nineteenth century concept of a "Polish mission."

I

In 1899, when Tsar Nicholas II proposed his Peace Conference—which was held at the Hague—Limanowski had objected to this démarche as *inter alia* aimed at perpetuating an unjust status quo in Poland. He believed that the First World War could not have been avoided. In their struggle against war, socialists in Europe had disposed of but one—and that probably a futile weapon, that is, the general strike; Nevertheless, he admitted a possibility of their succeeding; but this, in his opinion, would have benefited only Russia. He was convinced that, in a sense, the war had been justified for it resulted in the demise of the Tsarist regime which he hated.[1]

Having put down the revolution in its domains, by 1908 the Tsarist government, in Limanowski's opinion, had become the champion of Slavdom in order to ensure social stability at home as well as to enhance Russia's prestige abroad. Allegedly, Russia resented both Austria's "evolution toward federalism" as well as the latter's toleration of both the Polish and Ukrainian independence movements, and was bound to incite Serbia to provoke Austria. Yet, he believed, Germany could not allow the disintegration of her ally and the potential subjugation by Russia of the Austrian Slavs; hence, Germany had no choice but to declare war in defense of Austria against Russia.[2]

Limanowski held the latter responsible for the outbreak of hostilities in 1914 for yet another reason. Russia, allegedly, provoked the war in order to avert a potential revolution at home. (Rather than averting the revolution, in fact it was this war which accelerated the upheaval in Russia, as conservative Russian statesmen like Durnovo had feared.) Undoubtedly, Limanowski's far-fetched theorizing on the origins of the war, in so far as it influenced the public, only intensified Russophobe feelings in Poland.[3]

Limanowski was ambivalent toward the First World War. He admitted that, unexpectedly, it proved disastrous for Europe. As a result, European hegemony in world affairs had almost ended. But he sometimes referred to the war as a "holy war"; for instance, in 1919 he argued that it was a "holy crusade" and another "battle of Marathon," as it were, in the twin liberal causes of democracy and national self-determination.[4] He had hailed President Wilson's famous address on war aims in the American Senate on January 22, 1917—in which the latter attempted to reconcile the idea of a collectively guaranteed peace with the principle of national self-determination— "as a harbinger of the liberation of nationalities from oppression by [alien] states."[5] Limanowski argued that the war had resulted in the birth of new European states, including the reconstruction of Poland; this had come about as a result of the demise of both "Russian Tsardom and Prussian Caesarism"—hitherto the mainstays of European reaction—whose defeat represented a victory of the principle of representative government over the monarchical one in Europe.[6]

Limanowski considered the early part of the twentieth century— before the war and just after the war—as a transitional period in the history of mankind, when "state politics" clashed with "nationality politics," that is, "old diplomacy" clashed with "new diplomacy." He felt that the latter had scored but a theoretical victory; in practice, the old ruling classes of Europe still held their sway, while the masses— just emerging from centuries of subjection—remained ignorant and superstitious. Yet, in comparing the Paris Peace Conference with the erstwhile Congress of Vienna, he detected a measure of progress in international relations. In 1815, it was monarchs who debated the idea of the balance of power, paying but scant attention to the wishes of their subjects. But in 1919, it was peoples (and not monarchs), as represented by their delegates, who decided for themselves their respective fates. However, these delegates—he felt—were not truly

representative of their peoples; they were spokesmen for the economically privileged classes. Therefore, he saw a certain ambiguity in
the proceedings of the Conference. He asserted that, when considering problems which affected their respective states, the delegates to
the Paris Peace Conference proved themselves to be egoistic "etatists"
and, conversely, when they discussed affairs of other nations, which
did not touch their self-interest, they might act as disinterested "nationalists." (But, we may point out, such behavior was in fact quite
natural.)

Limanowski went on to argue that if Wilson had spoken directly
to the "peoples," they might have sent truly "national" delegations
to the Peace Conference; these would have been in a position to remove the many obstacles placed by the conservative leaders of the
old Europe in the path of the American President. (Wilson, in fact,
in the case of Fiume did appeal directly to the Italian people, bypassing their official delegation. Thereupon, the members of the Italian delegation walked out of the Conference. And Wilson's gesture
came to nothing, having alienated the Italian populace.)

Influenced as he had been by the romantic ideology of Mazzini
and Mickiewicz, Limanowski assumed that abstract "peoples" were
bound to be more reasonable than were their "bourgeois" governments, though he also claimed that the European masses were ignorant and superstitious. He unrealistically attributed the lack of full
consensus at the Peace Conference to the unequal development of
the countries represented, rather than to clashing national interests.[7]

Limanowski deplored that, in drafting the Treaty of Versailles of
June 28, 1919, European statesmen were only partly guided by the
idea of national self-determination. He felt that they had proved incapable of freeing themselves from "that greatest of all superstitions,"
the concept of state sovereignty. His distinction between a "natural"
association, like a "people," and the coercive entity which he considered the state to be, blinded him to the realities of politics, whether
internal or external. It should be stressed that Limanowski confused
the concept of sovereignty of a state vis-à-vis its subjects and that of
sovereignty of a state in its external relations with other states. Moreover, he assumed that in democracies the people were sovereign rather
than the state. Hence he argued thus: "Why should the state 'pretend'
to be sovereign, in order to commit injustices? We see numerous examples of this. Governments, aided by the idea of state sovereignty,
which is 'fictititious,' and having at their disposal both police and
army, proceed to oppress both religious and national minorities in

the state and, in an internal class struggle usually they even support that class which is stronger economically. The Treaty of Versailles did not abolish this 'fiction' [of state sovereignty], though by creating the League of Nations it went a long way toward doing so."[8]

Believing that the underlying cause of most wars was struggle for national liberation, he argued that the League must not become a mere instrument for perpetuation of the status quo. To remove the causes of such wars, it should foster creation of independent states which would unite on the basis of strategic, cultural and economic affinities, so as to form a wider federation to ensure lasting peace. But Limanowski's proposal was surely tantamount to creation by the League of new sources of international conflict rather than their attenuation, at least in the immediate future.[9]

Dominated as it was by two powers, England and France, some socialists mistrusted the League. Just as Limanowski had always favored the idea of socialist participation in the existing parliaments in Europe (this, he believed, had resulted in ameliorating workers' lives, had facilitated dissemination of socialist propaganda as well as socialist organization, and had given socialists a sense of confidence in themselves and in their cause), he now advocated that socialists support the League in its present form. Yet the analogy which he drew between democratic parliaments and the League was not quite appropriate; the latter, in fact, was not really a sovereign body,[10] and Limanowski was well aware of this. While deploring the preponderance of the two powers in the League, he considered this institution to be an important step toward the realization of the idealism of the French Revolution: "Liberty, Equality and Fraternity." He hoped that the member nations would eventually allow it to become a supra-national organization. Even the present hegemony of France and England in Europe he considered less harmful to the cause of national self-determination than the erstwhile rule of "Russian Tsardom and Prussian Caesarism" had been.[11]

Linked to Limanowski's concern for a new international order was his interest, long nurtured, in the nationality problems of Poland's western and eastern borderlands. Between 1919 and 1925, he published several articles and pamphlets dealing with the problem of the western borderlands disputed by Poland with Germany, that is, Silesia, Marienwerder, Ermeland, and Prussia Mazuria. In addition, Limanowski, having in 1921 visited his family estate Podgorz after an absence of forty-three years, commented on the situation in his Livonian homeland in the east—now part of Latvia—as it affected the Poles still resident there.[12]

Comparing Latvia and Poland, as regards general politics, educational policy and agrarian reform, Limanowski gave his verdict in favor of Latvia. However, he found the educated Polish element there very dispirited. Back in the 1860s, it had not yet given up its hopes; and it survived repression by the Tsarist regime. Now, Polish landlords in Latvia, expropriated as a result of a recent agrarian reform, and unaccustomed to physical labor, had no choice but to emigrate to Poland. On the other hand, the petty Polish noblemen, who like peasants tilled the soil themselves, had benefited from this land reform. The mainstay of the Polish element in Latvia, they were well liked by the native Latvian population and intended to remain. They did not fear the Latvian authorities who were in general mistrustful of the Latvian Poles.

Both Latvia and Poland were new republics and both had come into being as a result of the collapse of the Tsarist regime. Limanowski felt that new countries like Latvia, which included minorities with a "higher" culture and with memories of having been the ruling class (as was the case with the Latvian Poles), tended in the interest of self-preservation to become rather suspicious of these minorities. (But, back in September 1860, Limanowski had been impressed by the high level of Latvian technology and culture!) Limanowski argued that states which were "clinging tenaciously to the idea of sovereignty" were usually newly created republics like Latvia; that is, states devoid of a long tradition of independence; they remembered the humiliations of their bondage and feared to lose their freedom again. However, he seems to have forgotten that Poland, which had a long tradition of nationhood, still clung as tenaciously to its sovereignty as did Latvia. In Limanowski's opinion, however, independence was only necessary for the purpose of consolidating a reborn—or a newly created—nation-state and of securing its recognition *de jure* by the international community. He reiterated his belief that "egoistic nationalism" was but a temporary phenomenon in Europe. Eventually, history itself, i.e., an evolutionary process, was bound to lead European nations toward federalism.[13]

Back in 1916 in his *One Hundred and Twenty Years of Struggle by the Polish People for Independence* Limanowski wrote: "The democratic movement of our times, which has enhanced the importance of the lower classes . . . has aided our national cause . . . . in Silesia as well as in both West and East Prussia."[14] Like Lelewel, he believed that "it was aristocracy which had lost Silesia and Prussia; but democracy was bound to restore these lands to Poland."[15] As

Limanowski pointed out, it was the *émigré* Democratic Society which, during the 1840s, had been the first to publicize the fact that there existed a native peasant element in these formerly Polish lands.[16]

In 1919, while the fate of Upper Silesia still hung in the balance, prior to the plebiscite which took place there in 1921, Limanowski wrote a pamphlet entitled *The Anti-Polish Policy of the Prussian Government*, which was published by the Polish Committee to Defend Silesia.[17] In it he outlined the recent history of the attempted Germanization of Poles inhabiting Poznania, Silesia, West Prussia and Prussian Mazuria, against the background of the centuries-old Prusso-Polish rivalry. Limanowski argued that this policy of Germanization had proved self-defeating; in addition, contrary as it had been to the democratic principle of people's sovereignty, it had even corrupted German liberalism. He claimed that chauvinism, having permeated the program of progressive German parties, engendered a spirit of resistance among Poles in the lands occupied by Germany, and that this had intensified their national consciousness.[18]

In the final version of his pamphlet on Silesia, which he entitled *The Rebirth and Development of Polish Nationality in Silesia* and published shortly before the 1921 plebiscite took place in that area,[19] Limanowski pointed out that systematic Germanization in that part of Silesia, which Prussia had acquired in 1763, had begun only under Prussian rule. In the middle of the nineteenth century, however, the region (Upper Silesia) experienced a national revival which was due in part to the influence of resurgent patriotism among the neighboring Czechs and Lusatian Serbs, and in part a result of Polish propaganda emanating from the Lower Silesia's University of Breslau (Wroclaw). Many Polish students attended this university from 1831, after the Russian authorities had closed the University of Warsaw. Polish struggles for independence affected the rise in national consciousness among the indigenous Silesians as well.[20]

In addition, German clergy and intellectuals, who were influenced by the folk cult of Romanticism, began to create Polish schools in Silesia, thereby contributing, too, to the rise of a Silesian conscious Polish patriotism. In its struggle with Protestantism, the German Roman Catholic clergy wooed the Polish-speaking peasantry.[21] Yet, when Bismarck abandoned his Kulturkampf, the German clergy ceased to support the cause of national-cultural autonomy for Polish-speaking Silesians.

Limanowski believed—correctly—that during the 1890s in Prussian Silesia the PPS had exerted little influence on the indigenous popula-

ation. In 1901, however, when the German Social Democrats concluded an alliance with Rosa Luxemburg's SDKPiL, the PPS had had
to remove its organ for Prussian Poland—the *Workers' Gazette* (which
had begun appearing in 1893 and was at first subsidized by the German Social Democrats)—from Berlin to Katowice in Upper Silesia.
Katowice then became the center of Polish socialist agitation in the
Prussian-ruled regions inhabited by Poles. Nevertheless, by 1907 it
was the Polish Christian Democratic movement which gained the
upper hand in Polish politics in Upper Silesia. Yet in 1921 we find
Limanowski, rather surprisingly, still contending—as in the 1911
original edition of his pamphlet—that nowhere else was there as great
opportunity to combine socialist agitation with patriotic activity as
in this part of Silesia; for here, he argued, the proletariat was Polish
while the capitalists were German, both in industry and agriculture.[22]

Limanowski was convinced that in Upper Silesia the industrial
workers considered themselves as Poles, but he feared that the peasants, who were conscious merely of their Catholicism, might not
vote for Poland in the forthcoming plebiscite (as actually happened).
And his fear proved indeed justified in view of the actual outcome
of the plebiscite which represented only a partial victory for Poland.
Also, he felt that the Allied Commission, which had been created to
supervise the conduct of the plebiscite, by leaving intact the old
Prussian organs of administration had prejudiced the result in favor
of the Germans. Limanowski appealed to both the Polish people and
its government to do their utmost in order to reunite Upper Silesia
with Poland. He endorsed a promise of broad autonomy (in fact to
remain unfulfilled), which the Polish government extended to the
Upper Silesians in the event of a pro-Polish vote. And he assumed
that this promise was bound to create a good impression among the
Roman Catholic clergy of the area, which considered Silesians as an
ethnically mixed group, neither German nor Polish but, rather, simply
Roman Catholic.[23]

Limanowski argued that, in contrast to the clericalist Prussian Silesia, the socialist movement in the Austrian part of this region (i.e.,
the Duchy of Teschen) had indeed made some headway. After Daszynski had suffered electoral defeat in Galicia in 1907, he stood again
for election in the Duchy of Teschen and won. Limanowski considered
Daszynski's victory a great triumph for both the socialist and Polish
causes in Austrian Silesia. He deplored, however, the dominance by
the German element of the local administration and diet; he cited
evidence that this had proved economically detrimental to both the
Poles and Czechs who inhabited Teschen.[24] After the war, when in

1920 the powers had awarded the Czechs the major part of the Duchy of Teschen, despite its predominantly Polish-speaking population, Limanowski charged that the Czech administration in the former Duchy of Teschen was attempting to abolish Polish schools. Nevertheless, he was confident that this policy would fail in the long run, as had been the case with repressive nationality policies in the past, whether pursued by Metternich, Bismarck, Muraviev or Berg. He believed optimistically that a peaceful process of historical evolution was bound—in the long run—to restore the Czech-ruled parts of Teschen to Poland.[25]

In 1920. while the Russo-Polish war was in progress, a plebiscite conducted in the northern districts disputed with Germany, that is, Warmia (Ermeland), Olsztyn (Allenstein) and Powisle (Marienwerder), resulted in an overwhelming defeat for Poland. This prompted Limanowski to devote several articles and one pamphlet to the problem of Polish-speaking populations in these regions, concentrating on the "Mazurians" inhabiting the southern parts of East Prussia (who numbered around 350,000). Limanowski based his articles and pamphlet on research which he carried out on the spot during brief excursions to the areas subjected to the plebiscite of 1920 (in 1923, 1924 and 1925). His pamphlet, entitled *The Prussian Mazuria*, represented a fairly comprehensive study of East Prussia in general, and Prussian Mazuria in particular. In it he outlined the history of this land; described its topography and climate and the character of its population; and analyzed contemporary East Prussian economic and educational policies, against the background of the social conditions of the country.[26]

Limanowski pointed out in his pamphlet that originally East Prussia was not a Teuton land. Its early inhabitants had been Lithuanians and the non-German "Prussians" (who had been either exterminated or assimilated as a result of their conquest by the Teutonic knights in the thirteenth century). Both Polish and German colonists arrived in several stages. The first Poles to settle in East Prussia were Mazurian peasants, who came in the late Middle Ages. Subsequently, in the course of the seventeenth century the country sheltered Polish-Lithuanian Protestants, some of whom were to play an important role in the founding and development of the University of Koenigsberg (Krolewiec). By the First World War, East Prussia had become an ethnically mixed region, with a German-speaking majority including the assimilated Mazurians.[27]

Limanowski had been keenly interested in the history of East Prussia ever since, in his childhood, he had read Mickiewicz's epic

Konrad Wallenrod (where there is mention of the early inhabitants of East Prussia). He himself had come in contact with a group of Prussian Mazurians in Koenigsberg in the fall of 1860, on his way to Paris. Without clear-cut national allegiance, these Mazurians spoke in broken Polish; yet their German was inadequate as well. Dressed in rags, they apparently subsisted on bread and vodka. An illiterate and beggarly people, the Prussian Mazurians—as Limanowski reminisced in his pamphlet—were treated by Germans with contempt, and they seemingly lacked self-esteem. Yet the Mazurian workers, whom Limanowski had met in Koenigsberg in 1860, were conscious of being exploited; they had complained to him bitterly of having just been cheated on pay-day. Shortly afterwards, he again met a number of Mazurian workers, this time while riding in a fourth-class carriage of a train bound for Berlin, and once in Dirschau (Tczew) he had encountered a number of them at a working-class hostel, in which he stopped for one night. Later on, in Lvov in 1870 Limanowski met the historian Wojciech Ketrzynski, who had been educated first in Mazuria and then at the University of Koenigsberg; he was the author of the first thorough study of Prussian Mazuria in Polish. Limanowski learned a great deal about this region from Ketrzynski.

Though Limanowski was most disappointed when in July 1920, the Prussian Mazurians voted overwhelmingly in favor of the status quo in East Prussia, he still hoped for a change in their feelings toward Poland. It was to help bring this about that he had written his brief history of Mazuria.[28] Limanowski explained the outcome of the plebiscite thus. First, the Russian invasion of East Prussia in 1914 had proved disastrous for the local population. Hence, Mazurians were grateful to the German armies for expelling the Russians. Then, after the war, Germany aided East Prussia generously in rebuilding its towns and in feeding its hungry population. Limanowski felt that this solicitude on the part of Germany for the population of East Prussia was bound to have made a good impression on the Mazurians. At any rate, he speculated that they were usually too ignorant and too economically dependent on their German employers—whether landlords or industrialist—to have voted for a union with Poland. Moreover, he knew that the Mazurians, as Protestants, had been subjected to the influence of their pro-German pastors and feared discrimination in a predominantly Catholic Poland. Finally, Limanowski argued that the pro-German agitation, the existing Prussian organs of administration, the feeling of terror inspired by the local paramilitary formations, and the Bolshevik invasion of Poland in July 1920, all contributed to the setback for Poland in Mazuria.[29]

Concerned with both moral and material welfare of the Prussian Mazurians, Limanowski also felt that the East Prussian "irredenta" was vital to Poland, both strategically and economically. Thus, his argument on behalf of Prussian Mazurians was somewhat tainted with chauvinism. He asserted that in possessing the Mazurian land, Poland would advance by some 300 kilometers toward the Baltic Sea. And the greater a nationality—both qualitatively and in terms of numbers—the more it was likely, in his opinion, to rise to a "high level of spitual existence!" On September 18 and November 20, 1920, in the Warsaw *Tribune* Wladyslaw Gumplowicz had argued that a union between Germany and Austria be permitted in exchange for creating an independent East Prussia. The idea seemed reasonable to Limanowski, who felt that this Prussian "colony" had never been organically united with other German lands. Geographically, it formed a single entity with Poland, Belorussia and Lithuania. In addition, he believed that—economically—East Prussia was of greater importance to these countries than it ever had been for Germany. He assumed that the port of Danzig (Gdansk) provided a natural outlet for Poland's trade; and he believed that acquisition of another access to the Baltic Sea at Koenigsberg, in addition to that at Danzig, would lead to improvement of Polish relations with the Free City.[30]

Limanowski argued that Poles so far had been half-hearted in their attempt to attract Mazurians toward Poland. In their anti-Polish propaganda, directed at the Mazurians, the authorities in East Prussia compared the German efforts in East Prussia, in regard to assuring the welfare of the people at large, with conditions in Poland. This comparison, even Limanowski was compelled to admit, actually went in favor of the German regime in East Prussia. Moreover, in spite of rampant inflation, Berlin had expended much money for agitation in Mazuria. In contrast, Poland did almost nothing for the Prussian Mazurians, whether materially or in terms of spreading pro-Polish propaganda. He noted that in the border zone there was almost no information available on Mazuria; therefore, he advocated that bookstores in this area carry appropriate political literature for the edification of Mazurians on either side of the frontier. Limanowski even charged that, rather than aiding patriotic agitation among Mazurians, the Polish government actually impeded it by making border crossings difficult.

Limanowski believed that the greatest obstacle in attracting Mazurians to Poland was their fear of her Roman Catholicism; the educated Mazurians, moreover, felt culturally drawn to Germany. He assumed that the best means of gaining the confidence of the Mazurians was by spreading Polish socialist propaganda among them, despite the fact that they were almost entirely a rural people, albeit a proletarianized one. He noted that socialist activity had significantly contributed to a national revival in "Red" (Austrian) Silesia. Limanowski argued that Polish socialist propaganda ought to make Mazurians aware of their "Prussian yoke," as well as guaranteeing them their enjoyment of civil equality in Poland. He was convinced that they would not fear a Polish commonwealth, for such would be based on the Christian principle of brotherhood. Yet he omitted to take into consideration the fact that in the early 1920s socialism was far from being the dominant ideology in Poland.[31]

Limanowski went on to argue that the PPS ought to aid Mazurians as the German Social Democrats had been doing; the latter had become champions of the idea of a cultural-national autonomy for Mazurians. He felt that, in addition to agitating for civil and national rights for these people, the PPS should render them financial aid; this would enhance the party's prestige in Poland at large. Moreover, in order to improve their status in East Prussia, the Mazurians—he urged—should form their own political organizations. He felt that in this way they would win allies among liberal Germans, and indirectly help in spreading Polish national consciousness among the non-Mazurian Poles, who were scattered throughout East Prussia.[32] While agitating on behalf of the Prussian Mazurians, Limanowski referred to them, perhaps exaggerating somewhat, as "the most unhappy people in Europe . . . a people of the same Mazurian origin as the Mazurians whose capital is Warsaw."[33]

After the plebiscite of 1920 they continued to live in poverty and fear. During the early 1920s, while travelling by train in East Prussia Limanowski noted that "at the sight of him—a stranger—a Polish-speaking Mazurian, gripped with fear, would cease talking or whispered instead."[34] He pointed out that 98% of Prussian Mazurians were peasants, most of whom were landless. The local authorities were contemptuous of this exploited proletariat. Suffering frequently from unemployment, and without means to redress their grievances, these people—in Limanowski's opinion—were ripe for "Bolshevism."[35]

In 1923, after the local authorities in East Prussia went back on their earlier promise to give national-cultural autonomy to the Mazurians, Limanowski noted a resurgence of patriotism among these people. But, on the whole this patriotism was "Mazurian" rather than Polish. Emerging during the 1840s when Prussian Poland exhibited increased patriotic activity—and influenced by the uprising of 1863 and later by the revolution of 1905—this Mazurian patriotism had never been separatist. Unlike the more nationally conscious Roman Catholic Poles in Marienwerder and especially in Ermeland, the Protestant Mazurians were politically quite passive. The former had taken advantage of the new Weimar Constitution of November 1920, which provided for freedom of speech, assembly and association, to create a Polish Union in order to struggle for Polish nationality rights in Marienwerder and Ermeland by legal means.[36]

In his pamphlet on Mazuria Limanowski wrote: "It would be a noble task to arouse these Mazurian people, so that they would realize their shame, attempt to shake off their yoke, and obtain civil equality as well as the right to autonomous development."[37] We should note that not all nationality rights were equally important to Limanowski, whose special concern was cultural autonomy.[38] Moreover, at this time he was not advocating an armed struggle for the sake of incorporating Mazuria into Poland. Mindful of the horrible consequences of the First World War, Limanowski could not now conceive another war in the foreseeable future; he had become a pacifist. At the same time, he believed that—as had always happened in the past—a forceful policy of assimilation by Germany was doomed to failure, especially under the Weimar Republic. Instead, he expected a peaceful evolutionary process to lead to the national liberation of the Mazurians (as in the case of the Poles under Czech rule in Teschen). He pointed to "economic separatism," allegedly existing among German businessmen in East Prussia, as a factor which might assist this development.[39]

II

It is interesting that Limanowski now professed a kind of economic determinism. He argued that "in our present-day democracies, economic interest is of more immediate concern to the masses than politics and it exerts an ever increasing influence on the latter."[40] Limanowski, we have seen in Chapter IX, believed that pre-historic man was moulded by his natural environment; his political life was determined by the economic factors of his existence. However, with the rise of the state in antiquity, politics came to shape the economic relations of man; and ancient states were all based on the institution of slavery, due to conquest. Slavery and, subsequently, serfdom represented a direct exploitation of one man by another. The development of an economy based on money gave rise to the indirect exploitation of labor which became a commodity.[41] However, in modern times not only in the *Rechtstaat* did the economic factor again become dominant, but exchange of goods between countries led to an exchange of beliefs, of scientific data, and even of political institutions. Limanowski was convinced that economic affinities between constitutionally ruled peoples tended to result in ideological affinities between them as well.[42]

From 1918 to 1929, in his *Sociology* as well as in a number of articles and in several pamphlets Limanowski developed the final (and, on the whole, the most plausible) version of his political theory. In the first of these works, a treatise titled *The Nationalized State and Nationalization of Land* (1918), he partly reverted to his earlier idea of the origins of the state (put forward in 1876), that is, that a state might well come into being by means of an evolutionary process rather than by a sudden conquest, as Krauz had maintained. He explained that, after man had settled down to till the soil, this settled life soon resulted in inter-tribal conflicts due to the eventual shortage of good arable land. And, as a consequence of the division of labor and differentiation of the primitive society, a new military class formed for the purpose of defense or aggression. It was this class, argued Limanowski, which had founded the state. He asserted that some states were the result of federation of kindred tribes, aimed at waging a *defensive* war. However, he nevertheless continued to assume that a majority of states had formed as a result of armed *conquest*. It is worth noting that in both cases he implied an *external* stimulus as the prerequisite for the rise of a state. When arguing in favor of the first alternative, Limanowski based himself on the evidence presented by certain Polish historians of the mid-nineteenth and early twentieth centuries like Lelewel, Smolka and Bobrzynski; whereas, when espousing the second theory of the origins of the state, he showed the influence of a number of Polish as well as French scholars writing in that same period, notably Ludwik Gumplowicz,[43] Szajnocha and Augustin Thierry.[44]

Limanowski viewed history in terms of a struggle between the principles of war and peaceful labor. The "original" state, he argued, owed both its origins and aims to the pursuit of warfare. The closer it approximated the "original" type, the more a state was militaristic. Conversely, the further it evolved, the more it epitomized peaceful labor which was characteristic of a subjugated people. In powerful states victorious wars intensified militarism and enhanced the status of the ruling classes. In contrast, weaker states aimed at maintaining their integrity by taking into account the needs of the masses; this facilitated a process of democratization in these states. (We might point as an example to Sweden.) Limanowski identified democratiza-tion with the process of formation of a modern nationality; that is, pas-sive subjects becoming full-fledged citizens of a state. He reiterated that he considered the middle class—whose leader he believed was the *déclassée* nobility—to be the instrument of this progress; in the West, allied with the king against the feudal aristocracy, the middle class had bridged the gulf between the (originally) ethnically divergent gentry and people. And it was the rise of this middle class which, in his opinion, had transformed the state from being a private *dominium* of the ruler into a *res publica.*[45]

Believing that the French Revolution had initiated a process of democratization, henceforward he detected a trend toward "social engineering" (i.e., socio-political progress) in Europe. He argued that the "metaphysical" principle put forward by the French Physiocrats of the eighteenth century—"Laisser faire, laisser passer"—had now been superceded in European social theory by the Positivistic for-mula of Comte: "Savoir pour prévoir, afin de pourvoir."[46]

Yet, he considered progress as a continuation of the natural evo-lution of man, believing that primitive man had been only vaguely conscious of the notion of change, but that eventually man began to strive deliberately toward improvement of his existence. In stressing that progress became a conscious process, Limanowski came closer to the ideology of Comte than to the theory of Spencer. In accord-ance with the triple slogan of the French Revolution—"Liberty, Equality and Fraternity"—he distinguished three aspects of progress: political, material and moral. And he defined political progress in terms of an attempt at reconciliation of individual freedom with social solidarity, and of nationalism with internationalism.[47]

According to his Lassallian faith in the proletariat as the modern instrument of progress, he argued that the proletariat—on becoming the most numerous and hence the ruling class in the state—would transform the principle epitomizing itself, i.e., labor, into a universal principle. And the liberated proletariat, no longer a commodity, would become a means of liberating mankind as a whole. Limanowski out-lined the economic advantages of socialism thus: "First, socialization of the economy would produce an increase in productivity by per-mitting the full utilization of the existing labor force. This today is as yet not possible, because production follows the demand, while the majority are forced to remain at the level of bare subsistence. Secondly, in terms of employing the most modern means of produc-tion, both in industry and agriculture the large enterprise is bound to prove more profitable than a small factory or farm. . . . Socialization of the means of production would lead to a more efficient, than hitherto, deployment of labor."[48]

Notwithstanding his utilitarian argument, however, Limanowski continued to view socialism in terms of an ethical creed: "Socialism is a new religion . . . because, in spite of its scientific base, for an overwhelming majority it represents the twin causes of . . . justice and brotherhood. . . . [Modern] socialism has reawakened among the masses a misty and wistful memory of the erstwhile communal rule. . . ."[49]

Yet Limanowski also underlined the intellectual appeal of socialism, arguing that the respective realms of emotions and intellect were not strictly delimited:

> On the contrary, both are bound together with many threads to form a single spiritual whole. With the rise in education, there was improvement in morals; hence, the early manifestations of socialism like fraternal organizations, cooperatives and trade unions. Socialism, by illuminating religion with knowledge, will lead mankind to fulfillment—here, on earth, rather than toward a salvation in an unknown Heaven. The traditional religions, in concentrating on the egoistic idea of salvation of the individual soul and on the idea of reward in the afterlife, had actually impeded the perfectibility of man. . . . The aim of religion ought not to be salvation of the individual man, but the collective salvation of mankind as a whole. . . .[50]

According to Limanowski, Comte, the creator of the new science of Positivistic sociology, and Marx the pioneering economist, had both laid a "scientific" basis for socialism. He was impressed with the logical coherence of Marxism and the Marxian analysis of man's economic development, deploring that Marxists in stressing class struggle had almost obliterated the Christian idea of brotherhood, which Comte had defined in secular terms as "altruism." Limanowski pointed out that it was brotherhood which Saint-Simon—the mentor of both Comte and Marx—had considered as Christianity's most creative principle. Brotherhood, reiterated Limanowski, constituted the essence of both Christianity and socialism. He concluded that socialism was the result of a perfect synthesis of religion and science; it was Christianity transformed into a social rather than individual creed.[51]

Limanowski argued that socialism applied was "bourgeois" democracy made universal. Thus, a sudden transition from despotism to people's rule was impossible: *"Natura non facit saltum."* Social transformation was a gradual process; sudden and violent changes caused chaos and the regression of a society for an extended period of time.

Putting aside his usual idealization of the French Revolution, Limanowski pointed to this upheaval, as well as to the Bolshevik revolution, in order to back up this thesis. He defined a successful revolution as an accelerated evolution, citing Bauer who had paraphrased Marx's famous definition of force as a midwife assisting the birth of a new order: "Force cannot deliver a new society, if the latter has not yet matured in the womb of the old society."[52]

Limanowski attempted to explain why, in general, so few of the intelligentsia in Europe supported socialism. He attributed this to its egoism, i.e., narrowly conceived self-interest. But he was optimistic that the intelligentsia might change its attitude. He argued that, although even enlightened intellectuals of the pagan world like Aristotle considered slavery as an indispensable socio-economic institution, nevertheless slavery and later its less oppressive form, serfdom, had both almost ceased to exist. He believed that hired labor, labor as a commodity, which he defined as the last vestige of slavery, was bound to disappear in the long run, too. And he still considered science as the mainspring of both material and moral progress.[53]

In an attempt to reconcile his idealism with the "scientific" aspect of his "integral" socialism, that is, socialism which was grounded in sociology conceived as a natural science, Limanowski rejected the political and social implications of Darwinism. He believed that existing society placed certain individuals in a privileged position, whereas this was not the case in the world of nature with respect to members of the same species. Therefore, he appealed for equality of opportunity as in nature. He believed that man's struggle for existence would be alleviated both by birth control and by planning human and material resources.[54]

In attacking social Darwinism as a means of justifying aristocracy of birth, he admitted that members of the nobility usually were physically attractive. This, he felt, was due to their comfortable life, rather than to biological superiority over the masses.[55] In addition, Limanowski rejected the biological argument in favor of subjection of women (i.e., that women were supposedly the weaker and the less intelligent sex). He argued that in general, their status in society had actually deteriorated with the advance of civilization. Though he was uncertain whether artificially induced mental inferiority in women might be inherited by their male descendants, he was convinced that it was bound to adversely affect men as husbands or sons. He argued that great men were sons of enlightened mothers. Realizing that women's inferior role in society made them prone to accept myths and superstitions, and that they were subjected to sex-stereotyping, he felt that it was very necessary to give them a sound education. He insisted that modern feminism was in the interest of society as a whole. Only those nations might prosper materially and morally where all were full-fledged citizens. And, though history recorded many instances of sheer brutality, Limanowski was convinced that in the long run social justice and true morality (in which he included the liberation of women) were bound to prevail.[56]

He believed that communes and states became more or less "organic," depending on the feeling of solidarity binding their citizens. The state, he argued, constituted an organic entity when it was based on a "social contract" rather than on brute force. Nevertheless, he realized that the idea of a "social contract" in a political entity was, at least in part, fictitious. He agreed with Fourier and Saint-Simon that both commune and association epitomized the same principle of voluntary participation, but in the case of commune (or community) this voluntarism was less real. Limanowski now rejected Spencer's artificial classification of states into military and industrial organisms; instead, he opposed the concept of an absolutist rule to the idea of a democratic republic.[57] It would seem that Limanowski subordinated racial factors to psychological ones in socio-political life, although—inconsistently—in his *Sociology* he postulated social organicism "not in a metaphorical and somewhat metaphysical sense, but in an actual sense. What was social life," he argued there, "but a tendency of the social organism to preserve its existence and develop?"[58]

Limanowski rejected Gobineau's concept of race.[59] He was convinced that biological and anthropological research had shown this theory to be a myth—and one which he considered to be politically pernicious. Though he agreed with Gobineau that the Aryan race had been the most active and intelligent, he believed that mankind had become a mixture of races; there were no pure Aryans left anywhere. The common traits of a nationality, according to Limanowski, were psychological rather than racial ones. And in view of the modern trend toward equality that he detected in Europe, he strongly favored all types of mixed marriages, believing that this would help the process of democratization by attenuating religious, ethnic and social strife.[60]

Pointing to Switzerland as an example, Limanowski argued that the most highly civilized nations were those which were most mixed ethnically, religiously and socially. And, realistically, he no longer made any distinction on racial grounds between tribe and nationality. He defined the latter as "that social organism in which the consciousness of tribal separatism is subordinated to the realization that there exists a higher common interest in preserving and maintaining the nation as a whole."[61]

He assumed that modern nationalities resulted from an ethnic blending between "active" and "passive" stocks, distinguishing the early, prehistoric conquests from later ones. The former, he believed, had activated the "passive" peoples, aiding their gradual development. The smaller the aggressor tribe, or the closer it was ethnically to the passive one, the sooner the two assimilated. Eventually, from this assimilation arose the middle class, which became a vital factor of consolidating nationality. We even find Limanowski arguing like Dmowski (whose amoral national egoism was indeed the opposite of Limanowski's idealistic social theory) that the influx into Poland of the Jews, "very different racially and in religion [from the Poles] . . . had impeded the development of a strong Polish bourgeoisie . . . . This has been the principle cause of Poland's weakness vis-à-vis her neighbors."[62]

Though Limanowski's romantic messianism was not based on the idea of any racial superiority of the Poles, but—rather—on their alleged special "moral" qualities, it nevertheless could have been misunderstood after Poland regained her independence. It was significant, Limanowski argued in his last treatise titled *The Development of Polish Social Thought* (1929), that (excepting Bulgaria) the Slavs had been the principal ethnic group to benefit from the First World War. He felt that history, as it were, now presented them with the opportunity to lead mankind toward socialism.[63] In his attempt to decide which of the Slav nations was the potential "leader," Limanowski ignored the southern Slavs altogether, while his argument to "disqualify" the Czechs has already been discussed in the previous chapter. A disciple, in a sense, of such nineteenth century writers as Duchinski, Limanowski—in contradiction to his general social theory—dismissed Russia as unfit to lead Slavdom and Europe, in part because (as he alleges was admitted by Lenin) the Russians had very little Slav blood in them, being mixed with Finns and Tartars! Hence, modern Russia was more "Asiatic" than European. In addition, Limanowski argued, her "All-Muscovite Communism'" which was heir to the expansionist Tsarist tradition, was both oppressive and undemocratic.[64]

The Bolshevik experiment, he thought, had shown that the erstwhile concept of the SDKPiL, *"ex-oriente lux,"* was an illusion. Only Poland he judged fit to lead the European nations toward socialism, due to her long tradition of libertarian radicalism. He noted that as early as the Renaissance, the Polish "Arians," having been liberated from the tutelage of the Church, originated Unitarianism—the most theologically radical creed of its time; while in social theory they leaned toward a type of communism. And he held that late eighteenth century Polish reformers, like Kollataj and Staszic, became heirs—like himself—to the humanistic tradition of these "Arians."[65]

In the nineteenth century Poles were the first Slavic people to preach socialism. He pointed by way of example to Worcell who had combined the idea of Poland as an embodiment of brotherhood among nations with the Saint-Simonian theory of progress; to Libelt as the co-author of the Prague Slav Congress Manifesto of 1848; and to Mickiewicz who, in his *La Tribune des peuples* in 1849, "stood under the banner of international socialism."[66]

Limanowski asserted that the "French ideology" had stressed the principle of liberty and the "German ideology" the principle of equality. And now it was the turn of "brotherhood," the slogan of the noblest among Polish poets, publicists, and scholars. He suggested that perhaps it was to be the "Polish mission" to lead Slavdom and Europe toward the implementation of the principles of brotherhood among both individuals and nations. Yet he was realistic enough to know that there was much that was undesirable in the life of interwar Poland.[67]

Limanowski's final treatise of 1929 was still symbolic of the optimism of the 1920s. We do not know exactly how the "troubled 1930s"—the impact of the Great Depression and the victory of authoritarianism in Poland, as well as the rise of Hitler abroad—affected his thinking. Nevertheless, his letters to his friends[68] indicate that he never relinquished his belief in an eventual triumph of democracy in Poland as well as in other lands which had succumbed to dictatorship.

CHAPTER XII

CONCLUSION

Limanowski's political friends exaggerated his merits just as, conversely, his enemies tended to underrate him as a public figure. Almost no one, however, has questioned his moral stature; the nobility of his character was universally acknowledged. Limanowski's most remarkable characteristic was moral strength. A time came when he had to limit his political activity and scholarship, and when his creativity diminished. But his moral strength was inexhaustible. He remained extremely sensitive to social injustice and political oppression until the end of his life. In moments when nobody counted on him, because—having reached a great age—he had well earned his rest, Limanowski, a voice of collective conscience, defended democracy, becoming a symbol as it were of protest against Pilsudski's authoritarianism in interwar Poland.

Yet, in his lifetime, Limanowski's moral qualities in a sense obscured his concrete activity. Moreover, the very length of his life makes it difficult to produce a balanced appraisal of his career. If Limanowski had died before the turn of the century, he might well have been considered, primarily, as one of the most prominent Polish historians of his time. When he wrote his first sociological treatises, for the handful who studied them in Poland they appeared almost as a revelation, despite their basic lack of originality. It would have been impossible for a centenarian like Limanowski to remain intellectually influential during the entire span of his adult life; sooner or later he would have to surrender his leadership to younger people (even though, it is true, Limanowski remained a moral influence even beyond his death in 1935).

Limanowski became the first modern Polish socialist to combine in his theory the struggle for independence and the pursuit of socialist aims, without ideologically subordinating one to the other. A contemporary of Polish Utopian socialists of the post-1863 emigration, he stood on a border line between Utopian and "scientific" socialism. Unlike the radical Russian intellectuals of the 1860s (notably their eminent representative, Chernyshevsky), Limanowski rejected metaphysics without abandoning—at least in theory—sentimental romanticism for hard-headed realism; he attempted to synthesize elements of Romanticism with the Positivist creed of Auguste Comte. He harked back to the Enlightenment as Chernyshevsky had done; indeed, he was as much a disciple of Condorcet as of Darwin. Influenced by Buckle and the English Utilitarians through John Stuart Mill, he believed that the pursuit by the individual of an enlightened self-interest was a motive force of progress. In addition, Limanowski had a naive faith in the sheer idealism of the intelligentsia as well as of the masses. Both Limanowski and Chernyshevsky assumed that ignorance rather than vested interest was the main obstacle to the realization of socialism. Their faith in progress was simpler, more direct, and more naive than was the optimism of Marx.[1]

As a socialist, Limanowski did not belong wholeheartedly to any of the current schools of thought. He attempted to produce a unique "Polish" socialist ideology, deriving his socialism from keen patriotism as well as democratic and populist convictions. His socialist creed was a product of direct acquaintance with social injustice in his childhood and of the resultant humanitarianism which this generated; of compassion combined with logical deduction and intellectual conviction—as inspired by traditions of Polish democracy and of the nineteenth century struggle for Polish independence.

Limanowski stressed the ethical aspect of socialism. Thus, he became ideologically akin to the "revisionist" Western socialists, Eduard Bernstein and Jean Jaurès. Both Bernstein and Limanowski felt that what made socialism almost inevitable was not capitalism as such, but man's quest for social justice, freedom and brotherhood. Neither economic conditions, nor self-interest, in themselves, were sufficient to explain the hardships which socialists endured in their struggle for a better society. A socialist humanist like Jaurès, Limanowski stressed the idea of social consensus rather than class struggle. Though acknowledging class struggle as a historical fact, he argued for class collaboration as a means of achieving social reform.[2]

True, prior to 1918 Limanowski's advocacy of class solidarity was mainly due to his emphasis on national unity in the struggle for Polish independence. Actually, Limanowski was never particularly interested in the day-to-day struggle of the industrial workers in Poland, perhaps because he had gone abroad before industrialization took place there, and in emigration had had no contacts with the masses. He never understood the nature of the "historical mission of the proletariat" in the sense that orthodox Marxists claim to understand this. A liberal democrat, he rejected the idea of the dictatorship of the proletariat: he interpreted the *Communist Manifesto* in his own way, assuming that Marx was a protagonist of a bloodless transition to socialism, wherever possible. Faithful to the idealism of his youth, Limanowski argued that Polish socialists in partitioned Poland should first of all aim at national liberation, both as an end in itself and as a means of achieving socialism. Unrealistically, he envisaged—after Poland had regained her independence—the creation by peaceful means of a "neutral" Polish socialist state which would be based on the principles of the mythical Slav commune of antiquity. Hence, according to Polish Marxists, he was a romantic Utopian as well as an "opportunist."[3]

Polish Marxists point to a basic incoherence in Limanowski's socio-political theory. In it he combined a naturalistic idea, organicism, with a class concept of society (as derived, for instance, from Augustine Thierry's historical writing and from the ideology of Saint-Simon). Limanowski tended to employ the organicist conception of society in his sociological works and that of class society in his historical and political writing. Moreover, he usually referred to nationality as an organism, whereas he used the concept of class society in discussing the origins of the state. He contrasted the "organic" nationality with the oppressive state. Hence his distinction between "nationalism" and "etatism" (i.e., patriotism and chauvinism).

On the one hand, Limanowski's socio-political theory was influenced by the contemporary preoccupation in Europe with the biological sciences and, on the other hand, it stemmed from his awareness of the European peoples' alienation from their oppressive, and sometimes foreign, governments. Although this theory was syncretic, it nevertheless contained several elements worthy of note.

Limanowski believed in a multiplicity of environmental factors reacting on man in society, but he considered social life to be an extension of the biological existence of the individual. At the same time, he espoused an idea of progress as a conscious process involving man's intellect as well as his emotional and moral faculties. Notwithstanding his adoption of an idealistic teleological concept of history aimed at the realization of social justice and brotherhood, Limanowski assumed that an ever-increasing conscious participation of ordinary people in the making of history was a prerequisite for progress. He had a rationalistic faith in progress, that is, in the victory of socialism, but he rejected the Marxian—somewhat metaphysical—impersonal concept of "historical necessity"; instead, he believed that mass education and effective propaganda were bound to lead to a universal commitment to socialism. In respect to the idea of "federalism" which usually he viewed in terms of an innate, "natural" tendency,

Limanowski nevertheless defined federations, realistically, as permanent alliances formed in order to counteract a threat from without. This concept of federalism was rooted in a rationalistic faith which he had derived from Polish as well as foreign thinkers of the Enlightenment, and in the Polish Romantic historical writing of Lelewel.

As an "international socialist," i.e., socialist and prophet of a united Europe, Limanowski was philosophically rooted in the Enlightenment. He usually favored moral force (i.e., education and peaceful persuasion); and he was a believer in the idea of progress, whether moral or material, which he assumed to be due to the perfectibility of human nature. On the other hand, as a "nationalistic socialist," he was rooted in Romanticism, and he was guided by a naive faith in the public spirit of his compatriots. Two contradictory philosophies of history—a regressive movement to a golden age in the past and a progressive movement to a golden age in the future—coexist within his somewhat anarchistic political creed.

Limanowski belonged to the "optimistic" school of Polish historiography which comprised on the one hand the idealistic Romantic school founded by the great Polish historian Joachim Lelewel and, on the other, the Positivistic Warsaw school which crystallized after the uprising of 1863, and laid claim to historical objectivity. Whereas Lelewel tended to idealize all Polish history prior to the partitions of 1772-1795 that was truly "Polish" and free of foreign "corruption," the Warsaw school instead pointed to the era of Polish Enlightenment, which preceded the final partittion of 1795, as proof of the essential vitality of the old Polish-Lithuanian Commonwealth. The "optimists"—both "slavophiles" and "westernizers"—stressed the external rather than internal causes of Poland's fall. They believed that Poland fell due to the greed of her powerful neighbors—Austria, Prussia and Russia. On the other hand, the conservative Cracow school was "pessimistic"; it stressed the internal political and moral decay as the determinant factor in the demise of an independent Poland. As a historian of the Polish reform, Limanowski belonged to the liberal Positivistic Warsaw school; however, as a spokesman for the Polish "revolution" (i.e., insurrection) he was usually a disciple of the democratic Romantic historian Lelewel.

He became the first historian to analyze systematically Polish insurrections as well as democratic and socialist groupings in the Polish emigration in the West. Patiently assembling historical source materials available both at home and abroad, he attempted to bring out the vitality and modernity of those progressive elements in the Polish nation, which were struggling for independence and social justice. His pioneering works differed from those of the conservative historians of the Cracow school both as to theme and as to the political and social radicalism of the author. Although, it is true, Limanowski tended to exaggerate the import of progressive Polish opinion in the past, he earned for himself a unique place in Polish historiography. Only after 1945 has an attempt been made in Poland to produce works on the history of modern Polish democratic and libertarian tradition as comprehensive as those which had been undertaken by Limanowski. In addition to pioneering an entirely new trend in Polish historical writing, that is, the study of the period of partitions, he initiated historical research in several aspects of modern Polish history: he was the first historian of the uprisings of 1846 and of 1863, as well as of the Great Emigration—especially of Polish Utopian socialism.

Himself a participant in the patriotic activity of the 1860s, Limanowski wrote with an empathy for the events and the individuals he described. Yet he was not uncritical in evaluating sources as well as persons in Polish history. He distinguished himself both as a popular historian who aimed at informing the Polish masses of the struggles for independence and peasant emancipation in the late eighteenth and in the nineteenth centuries, and as a severe critic of the Polish gentry. His historical writing appealed especially to young people.

Limanowski had been a capable and thorough researcher, who usually based his works on primary sources. Professional historians of high caliber, like Handelsman and Prochnik, appreciated his historical scholarship; even present-day writers remain impressed by a number of his works. Some of Limanowski's histories are still being published in Poland today. His reputation as a historian, it should be pointed out, was due not only to his earlier writing, but also to works which he produced in his old age.

His prestige in Polish politics, it is true, was due in large measure to his achievement as a historian who in his books had formulated a political program, appealing to non-socialist democrats, too. Yet it would be unfair to attribute his reputation as a political leader solely to the popularity of his historical writing (or, conversely, to consider his historical—and even his sociological works—as merely an instrument of his politics).[4]

Nevertheless, Limanowski considered sociology as a means of implementing his socio-political program, rather than as a "purely" scientific discipline. Although other pioneer sociologists were to a greater or lesser extent ideologists—for instance, Comte, Marx, Spencer or Durkheim—and Limanowski in this respect was not unique, what distinguished him from these thinkers was the somewhat superficial grasp of the ideas of his mentors and his imperfect construction. It is difficult to determine to what extent this was due to his lack of first-rate ability in the field, or to his defective and unsystematic philosophical training—his too brief and interrupted study and his distrust of "metaphysics." Already in the 1880s European sociology was becoming "modern," but Limanowski failed to keep up with this development and stopped mid-way between social philosophy, political ideology and social science. Notwithstanding his theoretical assertion that sociology was a quasi-concrete or empirical discipline (which he made in his doctoral dissertation on Comte), in fact, like Comte Limanowski tended to consider sociology as a speculative philosophy generalizing on the basis of empirical data obtained from history and other social sciences.

It is interesting to note that in his lifetime Limanowski's compatriots either viewed his sociological works as political writing, or alternatively, ignored them. Prior to 1918, Limanowski received almost no recognition as a sociologist (though as a popularizer of Western social theories—that is, positivism, organicism, social Darwinism, and evolutionism—he had played an important and pioneering role in Poland). Aside from the question of the merit of his sociology, this neglect was mainly due to the fact that, in contrast to the positivism of the Warsaw school, Limanowski's treatises proved to be too erudite for the average Pole, whereas Polish scholars were in the main hostile toward socialist ideology and, hence, refused to read Limanowski's works. In addition, Polish Marxists—as mentioned above—were contemptuous of his eclectic social philosophy. Limanowski's major sociological work, *Sociology*, remained in manuscript form for nineteen years before it was finally published in interwar Poland. No doubt, his prestige as the "Nestor of Polish socialism" then contributed to the demand for his writings. But again it was the political aspect of his social theory which, as in the past, made him popular with his readers.[5]

In attempting to assess Limanowski's significance as a political activist rather than a "pure" scholar or social theorist, we should stress the lack of ideological continuity which existed in Poland in the three decades following the defeat of the uprising of 1863. The political trend known as Warsaw Positivism of the 1870s and 1880s represented a sharp break with the intellectual climate of the first half of the century, i.e., Polish Romanticism (though in some respects it harked back to the "Reform Era" associated with the Polish Enlightenment). In a sense, Limanowski attempted to bridge the gap thus created. Ideologically, he linked Polish Romanticism with the "neo-Romantic" era in Polish politics and intellectual history, which had begun shortly before the turn of the century. Secondly, he became an intermediary, as it were, between the last generation of old-style Polish democrats and the new generation of Polish socialists at the end of the century. Thirdly, in linking the ideology of Polish Utopian socialists with Polish "scientific" socialism, he occupied a position between a socialist ideology appealing primarily to the emotions (e.g., Mickiewicz's) and one based on rationalist presuppositions. This bridging of ideological gaps became perhaps the most significant of Limanowski's contributions; he represented continuity in nineteenth century Polish intellectual history. The post-1863 Polish Positivists and international socialists both negated the immediate past. The latter underlined their conflict with the tradition of insurrections for the sake of preserving their ideological purity. Not so Limanowski; he derived socialism both logically and historically from Polish democratic ideology which was rooted in the era preceding the last insurrection.

Limanowski's basic ideology, though elaborated upon in the course of time, remained essentially as he had formulated it about 1875. Confirmed in his evolutionary socialist creed during his exile in the West, which lasted from 1878 to 1907, throughout his long life he remained faithful to this principle: Independence first and social reconstruction on gaining independence. He changed his tactic from social revolution to national insurrection very early in his career. His apparent ideological and tactical waverings reflected, on the one hand, the international diplomatic situation and, on the other, the state of the Polish or international revolutionary movement, as in the case of many earlier or contemporary to him Polish émigré radicals.

The steadfastness of his political convictions, which his political opponents—with some justification—liked to call inflexibility, prevented Limanowski from joining the anti-patriotic Polish socialist movement, i.e., the Proletariat and, subsequently, its successors—the SDKP and SDKPiL. (The latter was allied with the Russian Social Democratic Labor Party and eventually became the Polish Communist Party.) Had Limanowski died prior to the foundation of the PPS in 1892, it might well have been said of him that his political inflexibility condemned him to obscurity. But, in fact, Limanowski's ideology triumphed in being adopted by the PPS; and the PPS became the strongest Polish leftist party, remaining such as long as Limanow-ski was alive.

However, Limanowski's critics were justified in asserting that he was perhaps too naive to become a true political leader. Like Marx, he was above all an ideologist, thinker and scholar, and not a fighter in the day-to-day political struggle. Yet, on the other hand, this was due not merely to his own predilections, or even to his lack of talent as an organizer; it resulted in large measure from the circumstances in which he lived, including the struggle to survive in emigration. For instance, exiled to north Russia, he had been able to play but an in-significant role in the Polish patriotic struggles of the 1860s, whereas during crucial events like the Revolution of 1905 and the First World War, Limanowski was already an old man. Though not a "charismatic" leader, he made an indelible impression on almost all who came into contact with him, due to his sense of justice, kindness, modesty and serenity. The modern Polish socialist movement remains greatly in-debted to his pioneering activities.[6]

NOTES

NOTES TO PREFACE

1. A. Prochnik, *Boleslaw Limanowski* (Warsaw, 1934), pp. 3-4.

2. W. Studnicki-Gizbert, "Boleslaw Limanowski," in *Z okolic Dzwiny* (Wilno, 1912), p. 93.

3. J. P. Nettl, *Rosa Luxemburg* (London, 1966), II, 848. This passage was translated by Mr. Nettl.

4. This phenomenon was attributed also to John Stuart Mill by his biographer in the *Encycopaedia Britannica* (166), XV, 463.

5. Z. Zechowski, *Socjologia Boleslaw Limanowskiego* (Poznan, 1964), pp. 180-181.

6. Limanowski began writing his *Memoirs* in 1912, at the age of seventy-six. In 1918 he completed the so-called Cracow draft, covering the period from 1835 to 1907. In 1919, in Warsaw, he commenced another record of his activity from 1835 to 1919, the so-called Warsaw draft, which he finished in 1928. Volume I (1835-1870) is based essentially on the Cracow version but, because a systematic account there ended in 1861, this volume also contains material taken from the Warsaw draft. Subsequent volumes—II (1870-1907) and III (1907-1919)—represent the Warsaw version. Also, between March 12, 1914 and November 23, 1928, Limanowski wrote a diary in the form of valuable brief comments inserted in the margins of his memoirs, or on loose sheets placed between the pages of his notebooks.

Limanowski's diary, describing his political life in independent Poland, has much merit; it is not a record of a senile mind. However, Volume III of his *Memoirs* produced in that same period (1919-1928) contains a great deal of insignificant detail and is lacking in literary merit. In contrast, Volume II not only constitutes an important source for the history of the Polish socialist emigration of the 1870s and 1880s in Switzerland as well as of the 1890s and 1900s in Paris, but it also comments on its contacts with foreign socialists in the West. It provides many biographical details concerning Russian, Ukrainian and Polish radicals abroad who were active subsequently to the peasant emancipation in Russia.

Volume I became my principal source for Chapter II, due to its objective value and the scarcity of other data for this period in Limanowski's life (1835-1870). Subsequently, I supplement his *Memoirs* with an ever growing variety of other sources. This volume centers around the Polish uprising against Russia in 1863, and it might well serve as a source for Russian attitudes vis-à-vis Polish exiles in Russia, in the 1860s. Here Limanowski offers insights into his early intellectual development, against the background of several different milieus.

He describes rural life in the borderlands separating Poland from Russia, prior to the emancipation of the local serfs. It is a cruel world of severe landlords and starving peasants who unsuccessfully rebel in 1847. Presenting a vivid picture of these borderlands, he comments on the local social relations and customs. He describes the lesser Polish gentry in Vitebsk *guberniia*; Belorussian and Latvian peasants; and Jewish drivers and innkeepers. He goes on to recreate the atmosphere of fear pervading the Russian gymnasia in Moscow, which he calls "correctional institutions." He describes his teachers and professors, as well as high school and university fellow students, both in Moscow and Dorpat; clashes between students and officers over dancing girls that took place in Moscow; an excursion to a nearby town of Zagorsk noted for a famous monastery inhabited by dissolute monks; and student riots in Moscow as well as Polish conspiratorial organizations. Yet another milieu awaits him in Paris, where he attends a Polish military school in the suburb of Batignolles (which was eventually transferred to Italy). In Paris Limanowski meets several members of both the old and the new Polish emigrations. Back in the borderlands, he notes some opinions on the agrarian question, held by the Lithuanian gentry on the eve of the uprising of 1863. Later on, exiled to north Russia, he encounters various interesting individuals, both conspirators and loyal citizens: judges, gendarmes, and soldiers, *guberniia* officials and their wives as well as fellow exiles who work in the Governor's chancellery—Russians, Ukrainians and Poles. He travels by stages from jail to jail between Archangel and his destination—Voronezh *guberniia* in south Russia. Lasting four and a half months, this trip provides him with an opportunity to meet ordinary Russian people—peasants, convicts and soldiers. Finally, having travelled by peasant and Jewish carts, river boats and stagecoach, Limanowski arrives in Warsaw to confront yet another alien milieu. Here he meets radicals as well as conservatives, newspaper editors and repatriated fellow exiles. Some of the latter are wealthy and others are very poor—unemployed and homeless—and still others are clerks in the newly Russified local administration.

7. Limanowski, *Pamietniki*, ed. A. Prochnik (1937; rpt. Warsaw, 1957), I, 17-18.

8. *Biblioteka Narodowa*, MS 6418/I, fols. 1-2.

NOTES TO CHAPTER I

1. H. Jablonski, *Miedzynarodowe znaczenie polskich walk wyzwolenczych XVIII i XIX w.*, 2nd ed. (Warsaw, 1966), pp. 12-16.

2. B. Baczko, "Niektore wezlowe problemy rozwoju polskiej mysli spoleczno-politycznej i filozoficznej XIX w. (do lat siedemdziesitych)" in Baczko, ed., *Z dziejow polskiej mysli filozoficznej i spolecznej. Wiek XIX.* (Warsaw, 1957). III. 11-13.

3. T. Korzon, *Wewnetrzne dzieje Polski za Stanislawa Augusta, 1764-1794*, 2nd ed. (Cracow-Warsaw, 1897), I, 320; R. F. Leslie, *Polish Politics and the Revolution of November 1830* (London, 1956) pp. 8-9.

4. Leslie, *Polish Politics*, pp. 25, 27-28, 67.

5. Ibid., pp. 24-25. 67-69. Cf. H. Mogilska, *Wspolna wlasnosc ziemi w polskiej publicystyce lat 1835-1860* (Warsaw, 1949), p. 6.

6. Leslie, *Polish Politics*, pp. 30-31; M. Turski, "Spor od pokolen," *Polityka*, XII (June 15, 1968), 3. In the first partition (1772) Prussia took Ermeland, Royal Prussia, the Netze District and parts of Poznania; Austria acquired Galicia; whereas Russia took the regions north of the Western Dvina and large areas in Belorussia up to the Dnieper and the valley of the Druc (Polish Livonia, the regions of Vitebsk as well as Mohilev). In 1793, by the second partition, Poland lost to Prussia almost all of Poznania, and to Russia all her provinces east of a line running in the north from a point on the Dvina near Dünaburg southward to the Austrian frontier on the river Zbrucz, a tributary of the Dniestr (Central Belorussia with Minsk, Eastern Volhynia, Right-Bank Ukraine and Podolia). See Leslie, *Polish Politics*, p. 17.

7. Leslie, *Polish Politics*, pp. 33-34.

8. Ibid., pp. 28-29; Turski, *loc. cit.*.

9. B. Limanowski, *Rozwoj przekonan demokratycznych w narodzie polskim* (Cracow, 1906), pp. 13-17; Baczko, *op. cit.*, pp. 23, 25, 47-51.

10. Leslie, *Polish Politics*, p. 50.

11. Baczko, *op. cit.*, pp. 22-23; Mogilska, *op. cit.*, p. 6.

12. Prince Adam Czartoryski to Prince Leon Sapieha, letter dated March 24, 1848, cited in M. Kukiel, *Czartoryski and European Unity* (Princeton, 1955), p. 263; Baczko, *op. cit.*, pp. 23-24.

13. P. Brock, "The Socialists of the Polish 'Great Emigration'," in Asa Briggs and John Saville, eds., *Essays in Labour History* (London, 1960), p. 140.

14. Brock, *loc. cit.;* Mogilska, *op. cit.*, pp. 5-7; Baczko, *op. cit.*, pp. 23-25; B. Limanowski, "Pierwsze przejawy mysli socjalistycznej na emigracji polskiej po powstaniu 1831 r.," *Pobudka*, No. 1 (1889), pp. 12-13.

15. B. Suchodolski, *Polskie tradycje demokratyczne* (Wroclaw, 1946), pp. 7-11.

16. Brock, "The Socialists," p. 140. Cf. Baczko, *op. cit.*, pp. 25-27; F. Romaniukowa, "Zagadnienie wlasnosci gminnej i spoldzielczo-zrzeszeniowej w pogladach radykalnych demokratow polskich," *Zeszyty naukowe SGPiS*, No. 19 (1960), pp. 123-24; *Manifest Towarzystwa Demokratycznego Polskiego* (Paris, 1836), pp. 16-17.

17. The Norse theory was developed in Russia by the German historian G. S. Bayer nearly a century prior to its appearance in Poland. The early Russian rulers were undoubtedly of Scandinavian origin. Not so in Poland; in prehistoric times there were undoubtedly Germanic and Sarmatian peoples, but it is presumed that they were not numerous and became completely assimilated with the native population. See W. Dzieciol, *The Origins of Poland* (London, 1966), pp. 32, 46-47, 96, 204-7; Leslie, *Polish Politics*, pp. 63-64, also 24; W. Smolenski, *Szkoly historyczne w Polsce* (Warsaw, 1898), pp. 30-32, 99; Limanowski, *Historia demokracji polskiej w okresie porozbiorowym*, 4th ed. (Warsaw, 1957), I, 11-12.

18. J. Lelewel, "Stracone obywatelstwo stanu kmiecego w Polsce," in I. Chrzanowski, ed., *Joachim Lelewel, czlowiek i pisarz* (Warsaw, 1946), pp. 181-97; M. Serejski, *Joachim Lelewel* (Warsaw, 1953), pp. 86-98; Brock, "The Socialists," pp. 142-44; Mogilska, *op. cit.*, pp. 41-42, 73; Lidia i Adam Ciolkosz, *Zarys dziejow socjalizmu polskiego* (London, 1966), I, 99-102.

19. Limanowski, "Pierwsze przejawy," pp. 13-14; Chrzanowski, ed., *op. cit.*, p. 182; Siermiega, "Historia poddanstwa w Polsce, Roz. II," *Pobudka*, No. 2 (1889), p. 12n.

20. Ciolkosz, *op. cit.*, I, 98; Nina Assorodobraj, "Ksztaltowanie sie zalozen teoretycznych historiografii Joachima Lelewela," in Baczko, ed., *op. cit.*, pp. 190-191.

21. Brock, "The Socialists," pp. 141-42, 144-45; Limanowski, "Pierwsze przejawy," pp. 15-16, 21; A. Prochnik, *Ku Polsce socjalistycznej* (Warsaw, 1936), pp. 6-7.

22. Jablonski, "U kolebki polskiego socjalizmu," *Wiedza i zycie*, XV (1946), 373-74.

23. Brock, "The Socialists," pp. 147-49; Romaniukowa, "Zagadnienie," p. 123.

24. Brock, "The Socialists," p. 150; Limanowski, "Stanislaw Worcell," *Przeglad Socjalistyczny*, Nos. 2/3, 1893, pp. 68-71. The appeal of Saint-Simonism to Latins and Western Slavs was due to the Catholic roots of this ideology. For the Poles it represented a religion of consolation, embodying the essence of religious faith without the dogma and rites. See Limanowski, *Stanislaw Worcell* (1910; rpt. Warsaw, 1948), p. 201.

25. Brock, "The Socialists," p. 150; Limanowski, "Stanislaw Worcell," *Przeglad Socjalistyczny*, Nos. 2/3, pp. 69-71; Baczko, *op. cit.*, pp. 55-56; Jablonski, "U kolebki," p. 373; W. Lukaszewicz, ed., *Postepowa publicystyka emigracyjna, 1831-46* (Warsaw, Cracow, Wroclaw, 1961, pp. 243-44.

26. Baczko, *op. cit.*, p. 56; Ciolkosz, *op. cit.*, I, 118-20.

27. Romaniukowa, "Zagadnienie," p. 124.

28. Romaniukowa, "Dalsze dokumenty do historii Gramady Rewolucynej Londyn," *PH*, LI (1960), 548; Jablonski, "U kolebki," pp. 375-76; S. Mikos, "W sprawie skladu spolecznego i genezy ideologii Gromad Ludu Polskiego w Anglii, 1835-1846," *PH*, LI (1960), 680-81.

29. Brock, "The Socialists," pp. 169-172; Romaniukowa, "Dalsze dokumenty," pp. 555-56; Limanowski, "Stanislaw Worcell," pp. 63-66.

30. Romaniukowa, "Dalsze dokumenty," p. 556. On Swietoslawski see Limanowski, *Historia ruchu spolecznego w XIX stuleciu* (Lvov, 1890), pp. 457-480; Brock, "Zeno Swietoslawski (1811-75): a Forerunner of the Russian Narodniki," *ASEER*, XIII (1954), 566-587.

31. Brock, "The Socialists," pp. 160, 163-68; W. Weintraub, "Adam Mickiewicz, the Mystic Politician," *Harvard Slavic Studies*, I (1953), 137-78. Cf. Prochnik, *Ku Polsce*, p. 7; A. Mickiewicz, "Sklad Zasad," in S. Pigon, *Studia literackie* (Cracow, 1957), pp. 181-83; Limanowski, "Pierwsze przejawy," pp. 14-15.

32. K. Marx and F. Engels, *Manifesto of the Communist Party*, trans. S. Moore (1888; rpt. Moscow, n.d.), pp. 107-8.

33. Ciolkosz, *op. cit.*, I, 86-87; Prochnik, *Ku Polsce*, pp. 5-6; Mogilska, *op. cit.*, p. 73.

34. S. Kieniewicz, *Historia Polski, 1795-1918* (Warsaw, 1968), p. 212; W. Tatarkiewicz, *Historia Filozofii* (Warsaw, 1948), II, 315-323. H. Struve, *Historia Filozofii w Polsce* (Warsaw, 1900), pp. 40-55.

35. Baczko, *op. cit.*, pp. 38-41, 56-62. Some democrats, however, included all working people in the category of the "people," and potentially gentry as well.

36. Brock, "The Socialists," pp. 156-57; Kieniewicz, *Historia*, pp. 155-71; *Ruch chlopski w Galiciji w 1846 r.* (Wroclaw, 1951), p. 183; Limanowski, "Rozwoj mysli ludowo-socjalistycznej w granicach dawnej Rzeczpospolitej," *Pobudka*, No. 3 (1889), pp. 9-10. See also Thomas W. Simons, Jr., "The Peasant Revolt of 1846 in Galicia: Recent Polish Historiography," *Slavic Review*, XXX (December, 1971), 795-817.

37. Limanowski, *Historia demokracji*, II, 44-45; "Rozwoj mysli," p. 10. For further details of the conspiracy see M. Tyrowicz, *Sprawa Ksiedza Piotra Sciegiennego* (Warsaw, 1948).

38. Prochnik, *Ku Polsce*, p. 8; Baczko, *op. cit.*, pp. 26-27. Cf. Limanowski, "Charakterystyka H. Kamienskiego," in *Filozofia ekonomii materialnej ludzkiego spoleczenstwa* (Warsaw, 1911), pp. 14-16 (an abridged version of Kamienski's work in two volumes, published in Poznan in 1844-45); Limanowski, "Henryk Kamienski jako mysliciel spoleczny," *Pobudka*, No. 4 (1889), pp. 21-27. Kamienski (1813-1866) wrote his treatise on mass guerilla war under the pseudonym Filaret Prawdowski. His *Prawdy zywotne narodu polskiego* (Brussels, 1844), reprinted under the pseudonum XYZ as *Wojna Ludowa* (Bendlikon, 1866), was eventually translated by J. Tepicht for the use of the French underground during the Second World War, under the title *L'insurrection est un art* (1943). See Kieniewicz, *PSB*, XI, 534-36; *Historia*, pp. 156-57.

39. E. Dembowski, "Mysl o przyszlosci filozofii," *Pisma*, IV, 368, cited in Baczko, *op. cit.*, pp. 64-65; Kieniewicz, *Historia*, pp. 157-58; Ciolkosz, *op. cit.*, I, 254-55; Serejski, *Lelewel*, p. 87.

40. Ciolkosz, *op. cit.*, I, 254, 239-40; Brock, "Socialism and Nationalism in Poland, 1840-46," *Canadian Slavonic Papers*, IV (1959), 145-46; Baczko, *op. cit.*, pp. 17-18; 27-28.

41. Kieniewicz, *Historia*, p. 272; J. Borejsza, "Oblicze polityczne polskiej prasy emigracyjnej na Zachodzie Europy, 1864-1870," Diss., University of Warsaw, 1962, pp. 19-20.

42. Romaniukowa, *Radykalni democraci polscy, 1863-1875* (Warsaw, 1960), pp. vii-ix; S. Walczak, "Rola genewskiego osrodka emigracyjnego polskich socjalistow w ksztaltowaniu sie swiadomosci socjalistycznej w kraju," *Ze Skarbca Kultury*, No. 1/7 (1955), pp. 27-28.

43. Romaniukowa, "Zagadnienie," pp. 120-123.

44. Ibid., pp. 130-31; Kieniewicz, *Historia*, p. 273; Borejsza, "Oblicze," pp. 240, 261-276; Limanowski, *Patriotyzm i socjalizm*, 2nd. ed. (Paris, 1888), p. 20.

45. Romaniukowa, "Zagadnienie," pp. 124-27; Borejsza, "Oblicze," pp. 281-82, 394. For the text of the Manifesto and decree based on it see Limanowski, *Historia ruchu narodowego od 1861 do 1864 r.* (Lvov, 1882), II, 9-11, 242.

46. Romaniukowa, "Zagadnienie," p. 128. Mroczkowski eventually joined Limanowski's organization, the Association of Polish People, in 1881.

47. Borejsza, "Oblicze," pp. 121-132, 391, 402-7; K. Szajnocha, "Lechicki poczatek Polski," in his *Dziela* (Warsaw, 1876), IV, 85-291.

48. Limanowski, *Patriotyzm*, pp. 20-21; Borejsza, "Oblicze," pp. 206-9.

49. Borejsza, "Oblicze," pp. 400-1.

50. Ibid., pp. 455-460. For the program of the Social Democratic Association see Limanowski, *Patriotyzm*, pp. 23-24; ZHP, APPS, AM 877/I, fol. 9.

51. Borejsza, "Oblicze," pp. 551-52, 364-66.

52. Ibid, pp. 366-67.

53. Ibid, pp. 501-2, 412-14, 585e.

54. Romaniukowa, "Zagadnienie," p. 126; Borejsza, "Oblicze," pp. 582-85; Kieniewicz, *Historia*, pp. 276-78; Baczko, *op. cit.*, pp. 43-44.

55. Kieniewicz, *Historia*, p. 278; Borejsza, *W kregu wielkich wygnancow, 1848-1895* (Warsaw, 1963), pp. 102-113; Walczak, *op. cit.*, pp. 28-31. Cf. M. Zlotorzycka, "Dzialalnosc Zwiazku Ludu Polskiego w Anglii, 1872-1877," *Niepodleglosc*, XIII (1936), 165-197. Lavrov was the most "western" of theorizers of Russian Populism and closest of all to Marxism. See F. Venturi, *Roots of Revolution*, trans. Francis Haskell (London, 1964), p. 465.

56. Kieniewicz, *Historia*, pp. 127-28, 317-22; Baczko, *op. cit.*, p. 65; Romaniukowa, "Zagadnienie," p. 119. The *Weekly Review* was the chief mouthpiece of Warsaw Positivism in the 1870s.

57. Borejsza, *W kregu*, p. 113; Walczak, *op. cit.*, pp. 58-59; Borejsza, "Oblicze," p. 585d. Cf. Zlotorzycka, *op. cit.*, pp. 196-97.

NOTES TO CHAPTER II

1. H. Jablonski, "Wstep," in Limanowski, *Pamietniki*, ed. J. Durko (Warsaw, 1958), II, vii; Limanowski, *Pamietniki*, ed. A. Prochnik (1937; rpt. Warsaw, 1957), I, 21 (both volumes referred to below as *Pam.*); P. Brock, "Boleslaw Wyslouch, tworca ideowy ruchu ludowego w Galicji," Diss., Jagiellonian University of Cracow, 1950, p. 37; W. Studnicki-Gizbert, "Boleslaw Limanowski," in *Z okolic Dzwiny* (Wilno, 1912), pp. 93-94.

2. *Pam.*, I, 21. Cf. Brock, "Boleslaw Wyslouch," pp. 39-40.

3. The coat of arms of the Limanowskis in Livonia was a maiden sitting on a bear. A standard armorial of Polish gentry lists another Limanowski family in Kovno *guberniia* in 1807, with crest "Alemanni" (of German origin?). Limanowski was unaware of this. See Count S. Uruski, ed., *Rodzina: Herbarz szlachty polskiej* (Warsaw, 1912), IX, 51. Cf. "Pamietniki," BN, MS 6417/I, fols. 5-7 and BN, MS 6418/I fols. 2-3.

4. *Pam.*, I, 19-21.

5. *Konrad Wallenrod* (1828) was a tale of a captive Lithuanian boy, who became a Grand Master of the Order of the Teutonic Knights to avenge his fatherland. "Wallenrodism" in Polish is tantamount to justification of evil tactics by a noble end. See *Pam.*, I, 37-39.

6. Limanowski read this book in the Polish translation by S. Porajski (1845). See *Pam.*, I, 34-39; Limanowski, "Kartka z mojego zycia," *Przedswit*, XXX, No. 11 (1910), 706-9; "Pamietniki," BN, MS 6417/I, fols. 47-48.

7. Limanowski shared his experience of serfdom with other gentry children growing up in the Russian Empire, as described in Marc Raeff's *Origins of the Russian Intelligentsia* (New York, 1966). In the Ukraine, Podolia and Volhynia serfdom was less onerous, but a greater gulf separated peasant and his lord than in the north-eastern regions, due to the wealth of the gentry and different religion. See Limanowski, "Udzial narodu polskiego w rewolucji 1848 i 1849r.,"

in Karol Marks, *Rewolucja i kontrrewolucja w Niemczech*, trans. by Limanowski from the German (London, 1897), pp. 80-81; *Pam.*, I, 37, 46-48. (The term "White Tsar" might have been a product of Tartar influence, the Muscovite Tsar being considered the successor to the Tsardom of the White—or Golden—Horde.)

8. *Pam.*, I, 49, 51-52.

9. Ibid., I, 74-78. At this confirmation, which took place around this time, Limanowski chose another name, Tadeusz, thus honoring the memory of Kosciuszko. See ibid., I, 77.

10. The letter, which Limanowski wrote to his sister Kamila in the fifth year of gymnasium, indicates that his Polish was considerably Russified. See *Pam.*, I, 85-89; *Pam.*, II, 673.

11. Ibid., I, 89-91, 516. Cf. J. B. Bury, *The Idea of Progress* (1932, rpt. New York, 1955), pp. 275-76.

12. *Pam.*, I, 99.

13. Ibid., I, 99-100.

14. Ibid., I, 102-4; Limanowski, *Socjologia* (1919, rpt. Cracow, 1921), I, 5-6 (referred to below as *Socjologia*).

15. *Pam.*, I, 108.

16. *Socjologia*, I, 6-7.

17. *Pam.*, I, 133-35; Limanowski, *Stanislaw Worcell* (1910; rpt. Warsaw, 1948), p. 6; Limanowski, "Jak stalem sie socjalista," in *Socjalizm-Demokracja-Patriotyzm* (Cracow, 1902), p. 84 (referred to below as *SDP*). It was the first student demonstration in the Russian Empire, initiating a student movement which culminated in the student riots of 1861. See William L. Mathes, "The Origins of Confrontation Politics in Russian Universities: Student Activism, 1855-1861," *Canadian Slavic Studies*, II (Spring 1968), 28-34.

18. *Pam.*, I, 135-36; Limanowski, *Worcell*, pp. 5-6; *SDP*, pp. 83-84.

19. *Pam.*, I, 136-37, 152-53; *SDP*, pp. 84-85; Limanowski, *Rozwoj przekonan demokratycznych w narodzie polskim* (Cracow, 1906), pp. 23-24. The Poles in Moscow and St. Petersburg at this time—both students and military men—were subjected to the influence of Herzen and Chernyshevsky, after having been previously exposed to Polish democratic literature. An interesting interaction of ideas took place: the Polish *émigrés* in the West in the 1850s acquainted Herzen with the idea of the ancient Slav commune, the latter influenced Chernyshevsky, and both Russians subsequently influenced other Poles. See

Limanowski, *Historia ruchu spolecznego w XIX stuleciu* (Lvov, 1890), pp. 489-90.

20. *Pam.*, I, 153-54; *Socjologia*, I, 7-8.

21. *Pam.*, I, 156-57; Limanowski, *Worcell*, pp. 6-7.

22. Limanowski, *O powstaniu polskim 1863-4 roku* (Cracow, 1913), pp. 9-12; S. Kieniewicz, *Historia Polski, 1795-1918* (Warsaw, 1968), pp. 227-28.

23. Kieniewicz, *Historia*, p. 228. Cf. N. N. Ulashchik, *Predposylki krestianskoi reformy 1861 g. v Litve i na Zapadnoi Belorusi* (Moscow, 1965).

24. *Pam.*, I, 155-58.

25. Ibid., I, 157-58, 168-171, 184; *Socjologia*, I, 8-9.

26. *Pam.*, I, 192.

27. Ibid., I, 202-4, 214-15; *SDP*, p. 87.

28. *SDP*, pp. 86-89.

29. *Pam.*, I, 209-210, 223-24.

30. Ibid., I, 225-26; George Woodcock, *Anarchism* (1962; rpt. New York and Cleveland, 1967), p. 29.

31. *Pam.*, I, 241-43, 247-49; "W Podgorzu po wojnie po 43 latach," *Papiery Wyslouchow,* vol. XXIX, BN Mfm 30099, p. 367.

32. *SDP*, pp. 91-92; Limanowski, "Ludwik Mieroslawski," *Swiatlo,* III (1900), 49-51; *Pam.*, I, 250-51, 256-57, 263-64.

33. *Pam..*, I, 261-63; Limanowski, "Ludwik Mieroslawski," p. 52. Patriotic demonstrations in Russian Poland began in June 1860, lasting until the outbreak of the uprising in January 1863. In February 1861, the Agricultural Society met to consider the imminent agrarian reform. Originally, its ultimate aim was conversion to rents, but under pressure from the radicals it adopted conversion to rents as a basis for later emancipation with land. See Kieniewicz, *Historia*, pp. 234-35.

34. Limanowski, "Pierwsza manifestacja w Wilnie 1861 r.," in *W czterdziesta rocznice powstania styczniowego* (Lvov, 1903), pp. 5-13, BN, Mfm 29470; *Pam.*, I, 280-98. Cf. J. Gieysztor, *Pamietniki z lat 1857-1865* (Wilno, 1913), I, 41-44, 103-7. "Boze cos Polske" originally was a counterpart of "God Save the Tsar *(Bozhe Tsaria khrani),*" the official hymn of the Congress Kingdom, which by 1861 had become anti-Russian.

35. *Pam.*, I, 299-308; Janko Plakan [B. Limanowski] , "Wybrzeza morza Bialego," *Klosy*, X (March 1870), 208; *SDP*, p. 96.

36. *Pam.*, I, 309-310, 315-16.

37. Ibid., I, 322, 318; *SDP*, pp. 97-98. Cf. E. H. Carr, *What is History?* (1961; rpt. London, 1967), p. 36.

38. *Socjologia*, I, 10; *SDP*, pp. 98-99.

39. Ibid., pp. 99-100.

40. Ibid., pp. 101-3; "Urywek z opowiadania mego przyjaciela," in *Upominek, Ksiazka zbiorowa na czesc Elizy Orzeszkowej, 1866-1891* (Cracow, 1893), pp. 364-67; *Pam.*, II, 423.

41. *Pam.*, I, 331-32; *SDP*, pp. 103-4.

42. *Pam.*, I, 335-39.

43. Ibid., I, 342; Limanowski, *Powstanie narodowe 1863 i 1864 r.*, 2nd ed. (Lvov, 1900), pp. 55-57, 65-66; Kieniewicz, *Historia*, pp. 230-48. The "Reds" aimed at forcing a solution of the agrarian question, and to awaken patriotism among the peasants, thus creating a basis for a mass people's war in the future. See Limanowski, *Historia ruchu narodowego od 1861 do 1864 roku* (Lvov, 1882), II, 241 (Document No. 18).

44. *Pam.*, I, 360-378. In Limanowski's native Livonia peasants burned manors and delivered tied insurgents to the authorities; such was the fate of his brothers, Lucjan and Jozef. In Belorussia fighting soon ceased, due to peasant reluctance to support it; whereas in the Ukraine the peasant masses exhibited an even greater hostility toward the insurgents than was the case in Polish Livonia. However, in Samogitia (Zmudz) the struggle proved popular. Conducted under the slogans of "land" and "liberty," it threatened to spread into Courland and the remainder of Livonia, where the Baltic Germans constituted the oppressive ruling class descended from the German Knights of the Sword. See Limanowski, *Historia ruchu*, II, 71, 102-8. Cf. M. Zielenczyk, Rev. of Limanowski's *Pamietniki*, I, in *Polityka*, I (August 14, 1957), 5.

45. Chubinsky's circle included the woman teacher Elizabeta Pavlova, the Ukrainian doctor Lipnitsky, the primary school teacher Chizhniakov and the high school history teacher Mak. The local Director of Excise, Neronov, who was a very wealthy but extremely radical local official, was also associated with the circle. (See *Pam.*, I,

352-54, 379-82; *SDP*, p. 109.) It seems that in his youth Chubinsky was a Ukrainian separatist. See *Pam.*, I, 353. Cf. A. N. Pypin, *Istoriia russkoi etnografii*, Vol. III (St. Petersburg, 1890-92), pp. 356-58, 347-56; H. K-a, *Entsiklopedicheskii slovar* (1903), XXXIII[a], 932-33.

46. *Pam.*, I, 384-85; Bury, *op. cit.*, pp. 309-10. The manuscript of Limanowski's article was lost.

47. *Socjologia*, I, 13; *SDP*, pp. 110-11.

48. *SDP*, pp. 114-15, 109-10; *Socjologia*, I, 12-13. Mikhail Muraviev, "The Hangman," was the ruthless Governor-General of Lithuania, whereas Frederick Berg was the Governor-General in the former Congress Poland. Both were appointed to pacify the Poles after the defeat of the insurrection.

49. *SDP*, pp. 110-14.

50. *Pam.*, I, 353, 375, 400-4.

51. *SDP*, p. 115; *Pam.*, I, 405-20.

52. *SDP*, p. 116; *Pam.*, I, 389-390.

53. *SDP*, pp. 116-19; *Pam.*, I, 437-38; Limanowski, "Wzajemna pomoc," *Opiekun Domowy*, III (July 24, 1867), 238-39.

54. *Pam.*, I, 434-35, 437-38; *SDP*, pp. 119-20. Like Mill, Limanowski rejected Comte's religious system with its priestly hierarchy. See *Pam.*, II, 287.

55. *Pam.*, I, 441-42. Before his departure, Limanowski received almost 100 rubles from the Governor of Voronezh for the period which he spent in Pavlovsk. This transaction is perhaps indicative of the fair treatment accorded to political exiles of gentry origin in Russia at this time. Boleslaw Limanowski to Lucjan Limanowski, letter dated September 10, 1867, BN, MS 2900, fols. 19-20.

56. *Pam.*, I, 448-57.

57. Ibid., I, 460-62, 469-71; Boleslaw Limanowski to Lucjan and Jozef Limanowski, letter dated July 21, 1868, BN, MS 2900, fol. 28; Bury, *op. cit.*, pp. 207-9.

58. *Pam*, I, 52, 484-85.

59. Ibid., I, 481, 485-86. Limanowski secured permission to go home, in August, 1869, due to the efforts made on his behalf by the aide-de-camp of the Governor of Vilna, the Latvian Kori, who wished to buy Podgorz. The Limanowskis apparently had powerful friends

in Vitebsk who prevented the compulsory sale of the estate. See *Pam.*, I, 477, 486.

60. *Pam.*, I, 471-72, 493-501.

61. Limanowski, Rev. of *Auguste Comte and Positivism* (1865) by John Stuart Mill, *Przeglad Tygodniowy*, IV (1869), 128-29, 149-51, 159-60. Cf. Limanowski, Rev. of *George Washington*, 5th ed. (Paris, 1868), by Cornelis de Witt with Introduction by M. Guizot, *Przeglad Tygodniowy*, IV (1869), 279-81. Limanowski differed from the Polish People which had regarded Washington as an enemy of the masses. See W. Lukaszewicz, ed., *Postepowa publicystyka emigracyjna, 1831-1846* (Warsaw, Cracow, Wroclaw, 1961), p. 239.

62. Limanowski, "Tegoczesna daznosc spoleczna," *Przeglad Tygodniowy*, IV (1869), 261-62; Limanowski, "Przeglad zycia spolecznego, I," *Przeglad Tygodniowy*, IV (1869), 345-47.

63. Limanowski, "Bibliografia Polska" [Rev. of *Banki rolnicze powiatowe, czyli zaliczkowe stowarzyszenia* (1869), by T. Romanowicz], *Przeglad Tygodniowy*, IV (1869), 363; Limanowski, "Przeglad pismiennictwa polskiego" [Rev. of *Pogadanki popularne streszczone podlug Juliusza Martinelli* (1869), by H. Elzenberg], *Przeglad Tygodniowy*, IV (1869), 406-8. Cf. Woodcock, *op. cit.*, pp. 11-12, 20, 26, 113-15.

64. Limanowski, "Koniecznosc ograniczenia liczby godzin pracy w fabrykach in rekodzielniach" [Rev. of *Praca dzieci po rekodzielniach* (1869), by L. Wolowski; "Zmniejszenie liczby godzin pracy w fabrykach i pracowniach rzemieslniczych" from *Gazeta Polska*, by A. Makowiecki], *Przeglad Tygodniowy*, IV (1869), 373-75; Limanowski, "Przemysl i prawodawstwo przemyslowe w Szwajcarii," *Ateneum* (1890), IV, 79-92.

65. Limanowski, "Przeglad pismiennictwa polskiego" [Rev. of *Szkice ekonimiczne* (1869), by K. Wzdulski], *Przeglad Tygodniowy*, IV (1869), 399-400.

66. In *Two Paths* Limanowski intended to tell the story of two friends who decided to serve the Polish cause each in his own way. The first married a peasant girl, identified himself with the people and converted many of them to the idea of struggle for independence. The second, like another Wallenrod, decided to fight the enemy from within by means of gaining great wealth. An employee of a bank, to achieve his purpose he married a daughter of his boss and became a very successful businessman. However, in the process of acquiring financial power, he became reconciled to the political

status quo. See *Pam.*, I, 409, 499-500.

67. Janko Plakan, *Dziewczyna nowego swiata* (Lvov, 1872), p. 33. The reception of this novelette, when published in 1872, was mixed, in accordance with the political philosophy of the reviewer— Romantic or Positivist. Favorable reviews praised it for simplicity, naturalness, charm and a realistic description of exile life, whereas unfavorable reviews attributed to its author morbid sentimentality, as well as lack of common sense and literary talent. See Limanowski to Katarzyna Sawicka (mother), letter dated September 24, 1869, BN, MS 2901, fols. 84-85; Limanowski, "Korespondencja z zagranicy," *Bluszcz*, No. 15 (1872), p. 118; *Pam.*, II, 52-53. Cf. *Tygodnik Wielkopolski*, No. 14 (1872), pp. 192-93.

68. *Pam.*, I, 508-511; Kieniewicz, *Sprawa wloscianska w powstaniu styczniowym* (Wroclaw, 1953), pp. 389-90; Limanowski, "Wspomnienia z pobytu w Galicji," *Przedswit*, XXI, No. 11 (1901), 401; J. Trabczynski [L. Kulczycki], "Boleslaw Limanowski," *Krytyka*, No. 3 (1900), p. 179.

NOTES TO CHAPTER III

1. Limanowski, *Pamietniki*, ed. J. Durko (Warsaw, 1958), II, 5-8 (referred to below as *Pam.*); Limanowski, "Wspomnienia z Galicji," *Przedswit*, No. 11, XXI (1901), pp. 401-2.

2. P. Brock, "Boleslaw Wyslouch, tworca ideowy ruchu ludowego w Galicji," Diss., Jagiellonian University of Cracow, 1950, pp. 11-16; Limanowski, *Galicja przedstawiona slowem i olowkiem* (Warsaw, 1892), pp. 42-43.

3. S. Kieniewicz, *Historia Polski, 1795-1918* (Warsaw, 1968), pp. 296-97, also 194. Cf. Brock, "Boleslaw Wyslouch," pp. 8-29; Brock, *W zaraniu ruchu ludowego* (London, 1956), pp. 11-15. Limanowski attributed the backwardness of Galicia to the lesser role which this province played in the armed struggle for independence, when compared to Prussian and Russian Poland. See his review of *Ostatnie chwile powstania styczniowego*, by ZLS, *Przeglad Spoleczny*, II (1887), 398.

4. Kieniewicz, *Historia*, p. 297; Limanowski, *Galicja*, pp. 44-45; 113-16; Brock, "Boleslaw Wyslouch," p. 30; Brock, *W zaraniu*, p.11; I. Daszynski, *Pamietniki* (Cracow, 1925-26), I, 190-92.

5. Limanowski, *Galicja*, p. 43; *Pam.*, II, 19-20.

6. W. Feldman, *Stronnictwa i programy polityczne w Galicji, 1846-1906* (Cracow, 1907), II, 5-7, 15; *Pam.*, II, 11-17.

7. Feldman, *Stronnictwa*, II, 18-21.

8. For instance this was the position of T. Romanowicz, the editor of the *Lvov Daily*. See *Pam.*, II, 18-19.

9. Feldman, *Stronnictwa*, II, 6-7.

10. Ibid., II, 18-20; Z. Zechowski, *Socjologia Boleslawa Limanowskiego* (Poznan, 1964), p. 43. Cf. J. Plakan [B. Limanowski], "Kilka slow o pracy organicznej," *Przyjaciel Domowy*, XXVI, No. 13 (1875), pp. 99-101.

11. J. Borejsza, *Emigracja polska po powstaniu styczniowym* (Warsaw, 1966), p. 421. Cf. Limanowski, *Stanislaw Worcell* (1910; rpt. Warsaw, 1948), pp. 7, 414, 425.

12. The socialist movement in Galicia lagged behind most other areas of Central Europe. The Austrian Social Democratic Party came into being in 1868. The German Marxist party was formed a year later in Eisenach. In 1875 it merged with the General Association of German Workers (Allgemeine Deutsche Arbeiterverein), founded by Lassalle in 1863, to form the Sozialistische Arbeiterpartei Deutschlands, the forerunner of the modern Social Democratic party. See E. Haecker, *Historia socjalizmu w Galicji i na Slasku*, Vol. I: 1846-1882 (Cracow, 1933), pp. 1, 93-97; Z. Zygmuntowicz, "Boleslaw Limanowski w swietle akt austriackich," *Niepodleglosc*, XI (1935), 130; W. Najdus, "Poczatki socjalistycznego ruchu robotniczego w Galicji," *Z Pola Walki* III, No. 1/9 (1960), pp. 4-5; Limanowski, "Wspomnienia," pp. 402-3.

13. Haecker, *op.cit.*, pp. 110, 120-128; *Pam.*, II, 87-88. Cf. A. Prochnik, *Ku Polsce Socjalistycznej* (Warsaw, 1936), p. 10; K. Marx and F. Engels, *The Manifesto of the Communist Party*, trans. S. Moore (1888; rpt. Moscow, n.d.), pp. 19n and 33n.

14. Limanowski, *O kwestii robotniczej* (Lvov, 1871), pp. 18-25.

15. Ibid., pp. 3-18, 26-29; Haecker, *op. cit.*, p. 112; Res [Feliks Perl], "Boleslaw Limanowski," *Przedswit*, XXX, No. 11 (1910), pp. 701-2; "Boleslaw Limanowski," *Swiatlo*, III (1900), 3. Limanowski recovered less than half the cost of the printing. See *Pam.*, II, 30-31.

16. Limanowski, "W sprawie Zwiazku Towarzystw," *Rekodzielnik*, No. 15 (November 19, 1871), p. 1; Haecker, *op. cit.*, pp. 97-101; *Pam.*, II, 28; Z. Gross, "Poczatki ruchu zawodowego w zaborze austriackim," *Robotniczy Przeglad Gospodarczy*, XVIII (1947), 19.

17. *Pam*, II, 31.

18. Ibid., II, 21-23; Limanowski, "Korespondencja z Zagranicy," *Bluszcz*, No. 11 (1873), p. 86; Limanowski, "W sprawie *Przegladu Tygodniowego,*" *Gazeta Narodowa,* May 18, 1872, p. 1.

19. Limanowski, "Marzyciele," *Gazeta Literacka*, Nos. 6, 7, 8, 9 (1871); Boleslaw Limanowski to Lucjan and Jozef Limanowski, letters dated December 13, 1870 and March 15, 1871, BN, MS 2900, fols. 59-60, 61-62; *Pam.*, II, 25-26; Limanowski to Katarzyna Sawicka, (mother), letter dated April 12, 1871, BN, MS 2901, fols. 108-9.

20. *Pam.*, II, 32-35, 43-45; Limanowski to Katarzyna Sawicka (mother), letter dated October 8, 1871, BN, MS 2901, fols. 112-13; Boleslaw Limanowski to Lucjan Limanowski, letters dated November 8, and December 8, 1871, BN, MS 2898, fols. 2,4.

21. *Pam.*, II, 89-90, also 56-62; Limanowski, "Plato i jego Rzecz-pospolita," *Na Dzis*, II (1872), 87-117; Limanowski, *Dwaj znakomici komunisci: Tomasz Morus i Tomasz Campanella i ich systematy: Utopia i Panstwo Sloneczne* (Lvov, 1873); Limanowski to Katarzyna Sawicka (mother), letter dated February 10, 1872, BN, MS 2901, fols. 116-17; Boleslaw Limanowski to Lucjan Limanowski, letter dated March 18, 1872, BN, MS 2898, fols. 5-6.

22. *Pam.*, II, 64-66. The *National Gazette* had a circulation of at least 4000, which was most unusual at a time when the leading dailies in Poland appeared in not more than 2000 copies. This paper derived much of its income from advertising. It was the most popular daily in Galicia, which skillfully guided public opinion and aspired to represent a "national" ideology. See *Zarys historii prasy polskiej*, ed. by T. Butkiewicz and Z. Mlynarski, Part II, No. 1 (Warsaw, 1959), pp. 93-95.

23. During the war of 1877-78 more than 100 Poles fought on the side of the Turks against Bulgaria. Among those who perished was one of the most prominent members of the post-1863 emigration, Benedykt Rahoza, a friend of Limanowski and Wroblewski. See Borejsza, *Emigracja*, pp. 322-3; *Pam.*, II, 129-31, 134-35, 140; Limanowski, "Rewolucja w Petersburgu," *Gazeta Narodowa*, December 23, 1876, p. 1. Cf. Limanowski, rev. of *Pisma wojskowo-polityczne Generala W. Chrzanowskiego*, Vol. I (1871) ed. by L. Chrzanowski, *Gazeta Literacka*, No. 9 (1871), pp. 10-11.

24. *Pam.*, II, 90-92; Limanowski, "W zwiazku ze sprawa zydow-ska," *Gazeta Narodowa*, November 18, 1873, pp. 1-2; November 22, 1873, p. 2; November 30, 1873, pp. 1-2; December 7, 1873, pp. 1-2; December 11, 1873, p. 1. However, nothing was done for the peasants

in Galicia prior to the rural agitation by the priest S. Stojalowski, beginning in 1877. He utilized the idea of village assemblies and edited special periodicals for the peasants. See Brock, *W zaraniu*, pp. 15-17.

25. *Pam.*, II, 50-51; Limanowski, *Galicja*, p. 44; Limanowski, "Korespondencja ze Lwowa," *Przeglad Tygodniowy*, VII (1872), 111.

26. Limanowski, "Korespondencja ze Lwowa," *Przeglad Tygodniowy*, VIII (1873), 80; *Pam.*, II, 88-89; Limanowski, "Towarzystwo Naukowej Pomocy dla Ksiestwa Cieszynskiego," *Tydzien*, No. 1 (1877), p. 1. These lectures were published under the title: *Losy narodowosci polskiej na Slasku* (1874). A revised edition appeared in 1911 and was reprinted in 1921; both were entitled: *Odrodzenie i rozwoj narodowosci polskiej na Slasku*.

27. (Lvov, 1875); *Pam.*, II, 74-76; Limanowski to mother, letter dated November 24, 1872, BN, MS 2901, fols. 120-21; Boleslaw Limanowski to Lucjan Limanowski, letter of the same date, BN, MS 2898, fol. 9.

28. *Pam.*, II, 96-97; Limanowski to Wincentyna Szarska, letter dated February 10, 1874, BN, MS 2903, fols. 40-41; Limanowski to M. Dubiecki, letter dated June 19, 1874, BJ, Korespondencja M. Dubieckiego, MS Add./10/67; Limanowski to mother, letter dated July 25, 1874, BN, MS 2901, fols. 130-31.

29. Lucjan bribed officials in Vitebsk to ensure that the family estates Podgorz and Jassy remain in the hands of the Limanowskis. See *Pam.*, II, 139, 143-44, 146, 148-49, 159-60, 167.

30. This was the contention of an anonymous author, writing in 1891 in *Przedswit*, No. 16—probably Stanislaw Mendelson, who was a student activist in Warsaw in the late 1870s. See K. Orthwein, "Teoria i praktyka pierwszych socjalistow polskich," *Kultura i spoleczenstwo*, II (1958), 128, 120-127.

31. *Pam.*, II, 128-29; Haecker, *op. cit.*, p. 134; Limanowski, "Wspomnienia z Galicji," pp. 404-5.

32. Haecker, *op. cit.*, pp. 130-32; *Pam.*, II, 184; Limanowski, rev. of *Die Quintessenz des Sozialismus* (1877), by Dr. A. Schäffle, *Tydzien*, No. 14 (1877), pp. 221-22; Limanowski, "Wspomnienia z Galicji," p. 405.

33. Limanowski, "Kto poprowadzi nasz lud," *Tydzien*, No. 5 (1877), p. 65; Limanowski, rev. of *Jak spelnic mozna cud pozadany calkiem naturalnymi srodkami* (1876), by Julia Goczalkowska, *Tydzien*, No. 6 (1877), p. 91; Limanowski, rev. of *Dzieje Polski. Czesc III. Krolestwo Polskie po rewolucji listopadowej* (1877), by F.

Skarbek, *Tydzien*, No. 8 (1877), p. 123; *Pam.*, II, 184-85; Limanowski, "Tadeusz Kosciuszko," *Towarzysz Pilnych Dzieci*, III, Nos. 3, 4, 5, 6, 7, 8, 9, 10 (1877).

34. *Pam.*, II, 140-41.

35. On Drahomanov see *The Annals of the Ukrainian Academy of Arts and Sciences in the United States*, Vol. II, No. 1/3, Spring 1952; Wincentyna Limanowska to K. Jesipowicz, letter dated July 30, 1877, BN, MS 2903, fols. 1-2; Wincentyna Limanowska to Lucjan Limanowski, letter dated July 13, 1877, BN, MS 2899, fols. 251-52; Boleslaw Limanowski to Lucjan Limanowski, letters dated August 25 and September 9, 1877, BN, MS 2898, fols. 58-59; Zygmuntowicz, *op. cit.*, pp. 131-33; *Pam.*, II, 169-70, 176-79; Haecker, *op. cit.*, pp. 134, 144-46.

36. Boleslaw Limanowski to Lucjan Limanowski, letter dated November 17, 1877, BN, MS 2898, fols. 62-63; Limanowski to Katarzyna Sawicka (mother) letter dated November 19, 1877, BN, MS 2901, fols. 146-47; Zygmuntowicz, *op. cit.*, p. 134.

37. Limanowski, "Wspomnienia z Galicji," p. 407; *Pam.*, II, 182-83.

38. Limanowski, rev. of *Socjalizm jako objaw choroby spolecznej* (1878), by Dr. Juliusz Au, *Tydzien*, No. 56 (1878), pp. 316-17.

39. Limanowski, *Socjalizm jako konieczny objaw dziejowego rozwoju* (Lvov, 1879), p. 12. He relegated his polemics with Au to the footnotes. Au felt that a socialist regime might well result in despotism, nepotism, corruption and economic chaos; and Limanowski failed to answer Au's charges in a convincing manner. However, Limanowski's approach to socialism was pragmatic; he believed that the exact features of the ideal society could only be determined on the basis of actual experience, the criterion being welfare of individuals, rather than dogmatic "social engineering." See for instance, ibid., pp. 9n-10n, 33n-35n.

40. Limanowski, *Socjologia Augusta Comte'a*, pp. 152-53.

41. Limanowski, *Socjalizm*, pp. 12, 29-31.

42. Ibid., pp. 10-11.

43. Ibid., pp. 1-2.

44. The word "socialism" was invented independently in England and France around 1830. See J. B. Bury, *The Idea of Progress* (1932; rpt. New York, 1955), p. 234. Cf. *International Encyclopaedia of the Social Sciences*, ed. David L. Sills (1968), XIV, 506; Zechowski, *op. cit.*, p. 116.

45. Limanowski, *Socjalizm*, pp. 59-61.

46. Ibid., pp. 63-66.

47. Ibid., pp. 83-84.

48. Ibid., pp. 68-98.

49. Ibid., pp. 91-95.

50. Ibid., p. 93.

51. Ibid., p. 98.

52. Ibid., pp. 85-86.

53. "Kilka slow o udziale Jenerala W. Wroblewskiego w obronie Komuny Paryskiej," in W. Rozalowski, *Zywot Jenerala Jaroslawa Dabrowskiego*, trans. from German by Jozef Plawinski (Lvov, 1878). This sketch was a translation from P. O. Lissagaray, *Histoire de la Commune de 1871* (1876); *Pam.*, II, 188-89.

54. Boleslaw Limanowski to Lucjan Limanowski, letter dated April 23, 1878, BN, MS 2898, fol. 67; Dziennik Podawczy Akademii Umiejetnosci w Krakowie, 1873-1891, April 26, 1878, Archiwum BPAN, MS PAU I, 206.

55. Boleslaw Limanowski to Lucjan Limanowski, letter dated May 29, 1878, BN, MS 2898, fol. 68; Haecker, *op. cit.*, pp. 151-52; *Pam.*, II, 190-93.

56. Ibid., II, 198-99; Haecker, *op. cit.*, pp. 156-57. Cf. Najdus, "Poczatki," pp. 6-7.

57. Ibid., pp. 7-8. Cf. Haecker, *op. cit.*, pp. 157-58, 164-70.

58. Polski Komitet Narodowy we Lwowie, AGAD, Kancelaria Gubernatora Warszawskiego, Referat I Tajny, 3/1874. Cf. *Pam.*, II, 194-96.

59. Chief of Police Gustanowski to Limanowski, letter dated August 21, 1878, BJ, MS Add./245/62; *Pam.*, II, 196-98. The authorities in Vienna alleged that Kobylanski was sent to Galicia from Switzerland to form a socially revolutionary party there. In Galicia, the ruling circles feared all rumors of conspiracies—whether nationalist or socially revolutionary—which Vienna lumped together indiscriminately. (See Intelligence Bureau of Foreign Ministry in Vienna to Potocki, letter dated August 25, 1877, No. 2995/1B ex 1877, ZHP, AM 877/2, pp. 659-60.) It seems that in Galicia the authorities particularly feared solidarity in subversive activities on the part of the various ethnic groups; for instance, Limanowski's alleged contacts with Ukrainian socialists in 1877, and with Jewish socialists like Mendelson in Warsaw and Adolf Inlaender in Lvov. See Potocki to Foreign Minister Taafe, letter dated December 19, 1879, No. 1004/pr 2603/1B, ZHP, AM 877/2, pp. 850-56.

60. *Pam.*, II, 201-2.

NOTES TO CHAPTER IV

1. Limanowski, "Garsc wspomnien z pobytu w Genewie," *Kalendarz Robotniczy na rok 1922*, Warsaw, p. 2.

2. Limanowski to Felicja Petraszewska (aunt), letter dated April 20, 1881, BN, MS 2903, fol. 42; Limanowski, *Pamietniki*, ed. J. Durko (Warsaw, 1958), II, 203-4 (referred to below as *Pam.*); Limanowski, "Korespondencja z Genewy," *Tydzien*, No. 68 (December 15, 1878), p. 508.

3. *Pam.*, II, 218-20, 223-25; Limanowski, "Garsc wspomnien," pp. 84-86.

4. *Pam.*, II, 215-17; Limanowski, "Garsc wspomnien," pp. 82-83.

5. *Pam.*, II, 216-17, 220, 223, 228, 713, 310-11; Limanowski, "Garsc wspomnien," p. 86. Such fraternal societies existed in all major Polish *émigré* centers in Europe.

6. *Pam.*, II, 225-26.

7. Ibid., II, 226-28.

8. Cut off from the earlier emigration, the new Polish exiles in Geneva were in a difficult position financially, and had to depend on help from their Russian friends in securing the necessities of life. Eventually, most of these Poles earned their living by drawing labels for drug containers. See *Pam.*, II, 229-30; Limanowski, "Korespondencja z Genewy," *Tydzien*, No. 1 (January 5, 1879), p. 13.

9. Wislicki to Limanowski, letter dated April 23, 1879, BJ, MS 6871, fol. 163a; Limanowski, "Listy o wspolczesnej literaturze francuzkiej," *Przeglad Tygodniowy*, XV (1880), 301; *Pam.*, II, 227, 230-31.

10. Limanowski's writings about Switzerland, including his later contributions in the Warsaw *Ateneum* and the *Voice (Glos)*, would make up a thick book, if ever collected in an anthology.

11. Limanowski, "Korrespondencja z Szwajcarii,"*Przeglad Tygodniowy*, XIII (1878), 222, 277-78, 316-17, 376-77, 442-44, 501, 514; Ibid., XIV (1879), 258-59, 294-95, 390-91, 464-65, 537-38, 610-11, 620-21; *Pam.*, II, 230-31.

12. Limanowski, "Korrespondencja z Szwajcarii," *Przeglad Tygodniowy*, XVI (1881), 247-48.

13. *Pam.*, II, 233, 826, 829.

14. Limanowski, "Korespondencja z Genewy," *Tydzien Polski*, No. 46 (1879), p. 728. Cf. ibid., Nos. 4 and 5 (1879), pp. 60, 76-77.

On the basis of developments in the socialist movement in Western countries, as reported in Eduard Bernstein's *Jahrbuch für Sozialwissenschaft und Sozialpolitic*, published in Zurich, Limanowski became convinced that he lived in an era which was characterized by the gradual demise of the "hero in history." For the role of the leader, in the opinion of Limanowski, diminished commensurately to the rise of national and class consciousness among the masses. See ibid., No. 38, p. 600.

15. Limanowski, "Korespondencja z Genewy," *Tydzien Polski*, No. 46 (1879), p. 728.

16. *Pam.*, II, 234-35; Boleslaw Limanowski to Lucjan Limanowski, letter dated October 28, 1880, BN, MS 2898, fols. 97-98; Drysdale to Limanowski, letters dated February 5, 1881 and January 11, 1882, BJ, MS 6871, fols. 38-41; Ludwik Straszewicz to Limanowski, letter written in 1887 or 1888, BJ, MS 6871, fols. 154-55.

17. Limanowski, *Socjologia*, (1919; rpt. Cracow, 1921), II, 140; *Zasady nauki spolecznej przez Doktora Medycyny*, trans. B. Limanowski (Geneva, 1880), pp. 368-76; Boleslaw Limanowski to Lucjan Limanowski, letter dated April 9, 1881, BN, MS 2898, fols. 101-2.

18. *Pam.*, II, 235.

19. Limanowski, "Z powodu spodziewanego glodu w Galicji," *Rownosc*, No. 3 (1879), pp. 1-5. Limanowski derived his statistics from W. Rapacki, *Ludnosc Galicji* (Lvov, 1879).

20. Limanowski, "Z powodu," pp. 6-7. Because of his vagueness in this article, Limanowski appeared more radical to the conservatives in Galicia than he really was; henceforth his writings would be promptly confiscated in Cracow and Lvov. See, for instance, letters of the Viceroy Potocki to the Sub-prefects in Galicia, January 30, 1880, APKr., St. GKr. 809, No. 98 and of Potocki to Directorate of Police in Cracow, February 21, 1880, APKr., St. GKr. 809, No. 150.

21. *Pam.*, II, 213, 712. Of the *Equality* group, Mendelson and Dluski were most affluent and financed its activity. The future brother-in-law of Madame Marie Curie-Sklodowska and a distinguished medical doctor in later years, Dluski eventually became a fellow member of Limanowski in the nationalistic Polish Socialist Party (PPS). As to Mendelson, he became the author of this party's first program, published in 1893, but shortly after the formation of the PPS, he abandoned the Polish socialist movement. (As an exile in London, Mendelson was befriended by Engels.)

22. "Program Socjalistow Polskich," in A. Molska, ed., *Pierwsze pokolenie marksistow polskich, 1878-1886* (Warsaw, 1962), I, 5-9; K. Orthwein, "Teoria i praktyka pierwszych socjalistow polskich,"

Kultura i Spoleczenstwo, II (1958), 132-35; K. Marx, *The Critique of the Gotha Program* (Moscow, 1947), pp. 21-29, 50-51. Cf. G. Woodcock, *Anarchism* (1962; rpt. Cleveland and New York, 1967), pp. 9-34.

23. Molska, ed., *op. cit.*, I, lxii-lxiii; Ibid., I, 338; Ibid., II, 687-91; Orthwein, *op. cit.*, pp. 135-38; *Pam.*, II, 213.

24. Molska, ed., *op. cit.*, I, 103ff.

25. *Pam.*, II, 242-43; Boleslaw Limanowski to Lucjan Limanowski, letters dated January 4, and February 17, 1880. BN, MS 2898, fols. 85-86.

26. *Pam.*, II, 243-45, 255-56; E. Haecker, *Historia socjalizmu w Galicji i na Slasku*, Vol. I: 1846-1882 (Cracow, 1933), pp. 224, 234; "Sprawa Krakowska," in Molska, ed., *op. cit.*, I, 10ff; Ibid, II, 694ff.

27. Devoted to Limanowski, Kobylanski was to solicit financial support for Limanowski's writings. It was due to Kobylanski's efforts that a group of Polish students in Kiev eventually published his works on European social movements. See Limanowski, *Historia ruchu spolecznego w drugiej polowie XVIII stulecia* (Lvov, 1888); *Historia ruchu spolecznego w XIX stuleciu* (Lvov, 1890); *Pam.*, II, 208-12.

28. *Pam.*, II, 256-57; J. Borejsza, *W kregu wielkich wygnancow, 1848-1895* (Warsaw, 1963), pp. 136-37, 140-46. Limanowski appended the London letter to his *Patriotyzm i Socjalizm* (Geneva, 1881 and Paris, 1888).

29. Molska, *op. cit.*, I, 402-5.

30. Borejsza, *W kregu*, p. 148.

31. S. Walczak, "Rola genewskiego osrodka emigracyjnego polskich socjalistow w ksztaltowaniu sie swiadomosci socjalistycznej w kraju," *Ze Skarbca Kultury*, No. 1/7 (1955), pp. 34, 52-59.

32. Haecker, *op. cit.*, pp. 200-3; W. Pobog-Malinowski, *Najnowsza historia polityczna Polski*, Vol. I: 1864-1914 (London, 1963), pp. 167-68; J. Buszko, "Polskie czasopisma socjalistyczne w Galicji," *Zeszyty Prasoznawcze*, II, No. 1-2 (1961), pp. 30-32.

33. W. Najdus, "Poczatki socjalistycznego ruchu robotniczego w Galicji," *Z Pola Walki*, III, No. 1/9 (1960) 28-29. Cf. *Pam.*, II, 265-66; Haecker, *op. cit.*, pp. 261-62. For text of the program see Materialy pierwszych polskich organizacji socjalistycznych i niepodleglosciowych, Socjalisci Galicyjscy, 1881-4, ZHP, AM 877/1.

34. Haecker, *op. cit.*, pp. 265-67. Cf. Najdus, "Poczatki," p. 12; S. Kieniewicz, ed., *Galicja w dobie autonomicznej, 1850-1914* (Wroclaw, 1953), p. xxxiv.

35. Limanowski's letter appeared in *Hromada*, No. 5 (1882), pp. 236-39 and was reprinted in Res [Feliks Perl], *Dzieje ruchu socjalistycznego w zaborze rosyjskim do powstania PPS* (1910; rpt. War-

saw, 1958), pp. 505-9. Cf. M. Drahomanov, "Ukrainski hromadivtsi pered pol'skim socializmom i pol'skim patriotizmom," *Hromada*, No. 5 (1882), pp. 231-35.

36. Drahomanov, "Ukrainski hromadivtsi," pp. 240-41. Drahomanov objected to the vagueness of the phrase "we respect the independence of Lithuania and Ukraine" employed in the "Program of Polish Socialists" of 1878. He pointed out that Lithuania, like Poland and Russia, was a geographic entity and not just a nationality; it was inhabited by Belorussians and Latvians as well as by the Lithuanians. Limanowski defended this program and the *Equality* group in an article published in Becker's organ. See Limanowski, "Quelques mots à propos de l'article de M. Drahomanov sur les révolutionnaires et les socialistes polonais," *Le Précurseur* (1881), cited in *Pam.*, II, 252-53. Cf. Res, *Dzieje*, pp. 504-5.

37. *Pam.*, II, 252-53; Limanowski, "Garsc wspomnien z Genewy," pp. 86-87; Limanowski to Wislicki, letter dated December 27, 1879, BP, Korespondencja Adama Wislickiego; Limanowski, *Historia ruchu narodowego od 1861 do 1864 roku* (Lvov, 1882), I, 9-10.

38. For Drahomanov's arguments see his *Hromada*, No. 4 (1879) devoted to the great Ukrainian poet Taras Schevchenko, cited in Limanowski, "Korespondencja z Genewy," *Tydzien Polski*, No. 21 (1879), pp. 331-32; Agaton Giller to Limanowski, letter dated October 31, 1880, cited in *Pam.*, II, 257-59. Notwithstanding their controversy, Limanowski defended Drahomanov publicly when the latter conflicted with the Russians. The authorities in Vienna even suspected Limanowski to have been Drahomanov's secret agent in Lvov, and felt that Drahomanov was the foremost leader of Russian "nihilism"; they were convinced that Drahomanov continued to be the alleged mentor of Limanowski, after the latter moved to Geneva and resided in proximity to Drahomanov. This self-deception on the part of the Austrian authorities helps to explain why Limanowski remained a *persona non grata* in Galicia for almost thirty years. See Limanowski, "Garsc wspomnien z Genewy," p. 87. Cf. Memorandum of the Intelligence Bureau of the Foreign Affairs Ministry in Vienna, dated May 19, 1880, ZHP, AM 877/2, No. 1173/1B.

39. Limanowski, *Historia ruchu narodowego od 1861 do 1864 roku*, 2 vols. (Lvov, 1882). A subsequent, also anonymous edition bore the title *Historia powstania narodu polskiego 1863 i 1864 r.*, 2 vols. (Lvov, 1894). A revised edition in one volume appeared in Lvov in 1909, under the auspices of Pilsudski's PPS Revolutionary Fraction; in it Limanowski's general outlook and interpretation remained unchanged.

40. Limanowski, *Historia powstania narodu polskiego 1863 i 1864 r.,* 2nd revised ed. (Lvov, 1909), iii-iv.

41. Ibid.; A. Giller, *Historia powstania narodu polskiego w 1861-4 r.*, 4 vols. (Paris, 1867-1871); J. Borejsza, *Emigracja polska po powstaniu styczniowym* (Warsaw, 1966), pp. 173-74.

42. J. Trabczynski, [L. Kulczycki], "Boleslaw Limanowski," *Krytyka*, No. 3 (1900), pp. 179-80; *Pam.*, II, 253; K., "Boleslaw Limanowski," *Tydzien, dodatek do Kurjera Lwowskiego*, VIII, No. 35 (1900), p. 273; K. Czachowski, "Szermierz wolnosci," *Niepodleglosc*, II (1930), 209; Limanowski, *Historia ruchu narodowego*, Introduction.

43. S. Kieniewicz, *Sprawa wloscianska w powstaniu styczniowym* (Wroclaw, 1953), v; Kieniewicz, "Historiografia polska wobec powstania styczniowego," *PH*, XLIV (1953), 13-14. Limanowski reduced every movement for independence in Poland to the struggle between "Mountain" and "Gironde." In all his books on Polish insurrections he maintained that it was moderation and hesitation on the part of the insurgent authority, mindful of vested interest and foreign diplomacy, which brought the defeat. He maintained that it was necessary to arouse the dormant peasant mass and to struggle until victory was secured, or perish. He did not have at his disposal the sources available to later scholars and he lacked the necessary detachment of an academic historian; yet, he was so close psychologically to the spirit of events and personages about whom he wrote that his works are still relevant to the student of Polish politics in the nineteenth century.

44. Kieniewicz, "Historiografia," p. 14. Cf. J. Sokulski, fragment of an article on Limanowski's relationships with Giller, BPAN, MS 6683, pp. 3-4; *Pam.*, II, 237-40, 257-60, 273.

45. Giller to Limanowski, letter dated October 3, 1882, BJ, MS 6871, fols. 73-74.

46. *Pam.*, II, 251, 681-83. Around this time his socialist rivals in Geneva translated Marx's *Capital* and *The Communist Manifesto*. On the other hand, Limanowski apparently ignored such immediate problems for socialism as the rise of the Polish proletariat, and was unconcerned with the strike movement in Russian Poland, the result of a severe depression in 1881-82, which facilitated the formation of the first Polish socially revolutionary party, the Proletariat, by Warynski. See Z. Zechowski, *Socjologia Boleslawa Limanowskiego* (Poznan, 1964), p. 27. Cf. Limanowski, *Patriotyzm i Socjalizm*, 2nd ed. (Paris, 1888), p. 28.

47. See W. Liebknecht, *W obronie prawdy* (Geneva, 1882); J. F. Becker, *Manifest do ludnosci rolniczej*, trans. K. Sosnowski (Geneva,

1883). Liebknecht's speech was originally translated into Polish by students in Warsaw, who had it published in Cracow; it was the printer of this pamphlet who denounced Warynski and his comrades to the authorities, causing their arrest in Cracow in 1879. (See Liebknecht, *op. cit.*, p. iii.) An avowed enemy of Russia and a former participant in Mieroslawski's campaign in Baden in 1848, Liebknecht was the author of the "Gotha Program" of the German Social Democratic Party, which came into being in 1875 and was led by Liebknecht together with August Bebel.

48. See Limanowski, *Ferdynand Lassalle i jego agitacyjne pisma* (Geneva, 1882); *Patriotyzm i Socjalizm* (Geneva, 1881), reprinted in a revised edition (Paris, 1888); *Polityczna a spoleczna rewolucja* (Geneva, 1883), reprinted in his *Socjalizm-Demokracja-Patriotyzm* (Cracow, 1902) as "Kwestia narodowa a spoleczna." Also, see note 71.

49. Limanowski, *Patriotyzm i Socjalizm*, 2nd ed. (Paris, 1888), pp. 4-5.

50. Ibid., pp. 5-6.

51. Ibid., pp. 7-10. Cf. Limanowski, *Socjalizm jako konieczny objaw dziejowego rozwoju* (Lvov, 1879), p. 65.

52. Limanowski, *Patriotyzm*, pp. 6-7.

53. Following the suppression of the *Narodnaia Volia*, due to the assassination of Tsar Alexander II in March 1881, Limanowski again became disillusioned with the Russian revolutionary movement. Yet back in the spring of 1880 he had hoped for a Constitution soon to be granted by the Tsar—as a "step forward"—anticipating, however, that such Constitution might well be inferior to that operative in Austria. See Boleslaw Limanowski to Lucjan Limanowski, letter dated April 6, 1880, BN, MS 2898, fol. 92.

54. For text of Limanowski's program see his *Pam.*, II, 684-87; 267-69.

55. Molska, ed., *op. cit.*, I, 555-58. The board of *Equality* had split in this same summer of 1881 on the issue of anarchism, when Warynski offered his allegiance to the anarchist "International" scheduled to meet in London in July 1881. Thereupon, Mendelson withdrew his financial support from *Equality* and founded a new paper *L'Aurore* on August 15, 1881. However, by this time Warynski rejected his purely "economic" approach to socialism, accepted the validity of a temporary program of immediate demands, and joined the editorial board of *L'Aurore* in solidarity against Limanowski and his new organization, the Polish People. See Leon Baumgarten, *Dzieje Wielkiego Proletariatu* (Warsaw, 1966), pp. 32-35.

56. *Pam.*, II, 648-87; Molska, ed., *op. cit.*, I, 566n, 539-40.

57. Haecker, *op. cit.*, pp. 200-3, 261-64; Pobog-Malinowski, *op. cit.*, pp. 167-68; Walczak, *op. cit.*, p. 57; Najdus, "Poczatki," pp. 13-15. For text of the program of January 1881, see Materialy pierwszych polskich organizacji socjalistycznych i niepodlegloscio-wych, Socjalisci Galicyjscy, 1881-4, ZHP, AM 877/1, especially pp. 5, 8, 11-15.

58. Haecker, *op. cit.*, pp. 265-67. Cf. Najdus, "Poczatki," pp. 15-17. For text of the program of May 1881, seee Kieniewicz, ed., *Galicja*, pp. 182-191.

59. *Pam.*, II, 269-71; Limanowski, "Kongres w Chur," *Przeglad Tygodniowy*, XVI (1881), 533. Cf. Najdus, "Poczatki," pp. 28-29; G. D. H. Cole, *The Second International, 1889-1914*, Part I (London, 1963), p. 2.

60. Between 1881 and 1883 Limanowski prepared weekly reports on the Polish and Russian revolutionary movements for Schneeberger, to augment his income. See *Pam.*, II, 263-65, 269, 272-73.

61. Boleslaw Limanowski to Lucjan Limanowski, letter dated October 24, 1881, BN, MS 2898, fol. 108; Molska, ed., *op. cit.*, I, lxi-lxiii, 540-43, 560-62, 570-72; Limanowski to the editorial board of *Light (Swiatlo)*, letter dated December 3, 1899, ZHP, Korespond-encja i rachunki redakcji *Swiatla*, AM 1050/11, No. 198.

62. Molska, ed., *op. cit.*, I, 552-57; Limanowski, *Patriotyzm i Socjalizm* (Geneva, 1881 and Paris, 1888), p. 3 and Limanowski's introduction to the appended letter from Marx and associates; Li-manowski, "Socjologia Wojciecha Schäfflego" [Rev. of *Bau und Leben des socialen Körpers (1875-78)*, by A. Schäffle], *Ateneum* (1882), III, 573-78; Limanowski, *Historia ruchu spolecznego w XIX stuleciu*, pp. 251-55, 264-65; Limanowski, "Monizm w historii," *Krytyka*, No. 11 (1908), 334-43; Limanowski, "Z dziedziny socjo-logicznej," *Przeglad Tygodniowy*, XXII (1887), 173.

63. *Pam.*, II, 271-72. Cf. Molska, ed., *op. cit.*, II, 718. J. Borejsza, *W kregu*, pp. 160-61.

64. Limanowski, "Korespondencja z Genewy," *Tydzien Polski*, No. 2 (1879), 24; Limanowski, "Socjologia Schäfflego," p. 587. Cf. J. P. Nettl, *Rosa Luxemburg* (London, 1966), I, 46-47. It would seem that this author exaggerates the importance of a "personality cult," as opposed to tactics and theory, in causing the rift between Limanowski's followers and Warynski's group.

65. Limanowski, "Socjologia Schäfflego," pp. 583-93; Limanow-ski, "Z dziedziny socjologicznej," *Przeglad Tygodniowy*, XXII,

(1887), 173; Limanowski, "Kwestia narodowa a spoleczna," in *Socjalism-Demokracja-Patriotyzm* (Cracow, 1902), pp. 152-53 (referred to below as *SDP*).

66. Ibid., pp. 142-48, 153-154, 156, 161.

67. *SDP*, pp. 156-58. "It has been said that revolutions are made in spite of the revolutionaries in the sense that the coincidence of economic and political crises produces the mass discontent and the uprising of the workers at the seat of power, which the revolutionary class, though surprised, exploits. . . . Insurrections are necessarily acts of free will, decisions to take up arms made by the insurgents themselves, who make rather than are overtaken by events, but free will is not distributed evenly within the insurgent class. It is exercised most liberally by the younger and least experienced who terrify the propertied classes. For this reason there appears the paradox of the insurrection—the appearance of the counter-revolution before the insurrection and the revolution implied by it have occurred." (See R. F. Leslie, *Reform and Insurrection in Russian Poland, 1856-1865*, London, 1963, pp. vii-viii.) It is perhaps the realization on the part of Limanowski of this "voluntarism" characteristic of an insurrection, as opposed to the "determinism" of a revolution, which convinced him that insurrections were "easier"—more likely to be successful—than revolutions. See also *SDP*, p. 141.

68. *SDP*, p. 153. Like Mieroslawski, Limanowski realized that the foreign regimes in Poland were allied with the "counter-revolution" at home; in order to succeed the insurgents had to ultimately struggle against the combined forces of the foreign regime and of the "counter-revolution." See ibid., p. 141.

69. Ibid., pp. 156-57, 162-66.

70. Ibid., pp. 155-56.

71. L. Wasilewski, *Boleslaw Limanowski w setnym roku zycia* (Warsaw, 1934), p. 12; J. Myslinski, *Grupy polityczne Krolestwa Polskiego w Zachodniej Galicji, 1895-1904* (Warsaw, 1967), p. 34n; Haecker, *op. cit.*, 280. Cf. J. Buszko, *Ruch socjalistyczny w Krakowie, 1890-1914* (Cracow, 1961), pp. 27-28, 43; Police Directorate in Cracow to Viceroy in Lvov, letter dated March 20, 1889, APKr, St. GKr. 510, No. 125.

72. Memorandum of Police Directorate in Vienna, dated February 16, 1882, ZHP, AM 877/2, No. Z 692/Pr.

73. M. Mazowiecki [L. Kulczycki], *Historia ruchu socjalistycznego w zaborze rosyjskim* (Cracow, 1903), p. 40; Molska, ed., *op. cit.*, II, 720; W. Feldman, *Stronnictwa i programy polityczne w Galicji,*

1846-1906 (Cracow, 1907), II, 89; Feldman, *Dzieje polskiej mysli politycznej, 1864-1914*, 2nd ed. (Warsaw, 1933), pp. 231-34, 238; Czachowski, *op. cit.*, p. 208; Wasilewski, *op. cit.*, pp. 11-12. See also *Pam.*, II, 277-78.

74. *Pam.*, II, 290-91. See ibid., II, 688-91 for text of Limanowski's Open Letter. The article by Wscieklica, entitled "Rojenia socjalistow polskich wobec nauki ich mistrza," appeared in *Ognisko: Ksiega Jubileuszowa ku czci T. T. Jeza* (Warsaw, 1882), which also included Swietochowski's credo at the time: "Wskazania polityczne." Wsieklica's article was reprinted in Molska, *op. cit.*, II, 765-72. For reply by K. Dluski and W. Piekarski, *Mistrz Wscieklica i Spolka* (Geneva, 1883), see ibid., I, 583-683, praising Limanowski for his Open Letter to Wscieklica on the national issue (ibid., p. 599). Finally, for a comparison of Limanowski's ideology with that of Swietochowski, see B. Suchodolski, *Polskie tradycje demokratyczne* (Wroclaw, 1946).

75. *Pam.*, II, 318-19.

NOTES TO CHAPTER V

1. Wincentyna Limanowska to mother-in-law, undated letter, BN, MS 2901, fols. 21-22; Limanowski, *Pamietniki*, ed. by J. Durko (Warsaw, 1958), II, 286-89 (referred to below as *Pam.*); Boleslaw Limanowski to Lucjan Limanowski, letters dated May 21, 1880, July 10, 1882 and May 7, 1883, BN, MS 2898, fols. 94, 114-15, 124-25.

2. Limanowski to Felicja Petraszewska (aunt), letter dated April 20, 1881, BN, MS 2903, fols. 42-43; *Pam.*, II, 280-81, 284, 302-8.

3. *Pam.*, II, 309-10, 321-334; Boleslaw Limanowski to Lucjan Limanowski, letters dated May 6, August 17, November, 1885 and December 1, 1885, BN, MS 2898, fols. 137-38, 139, 144-46; Limanowski to Korzon, letter dated December 30, 1885, BN, Korespondencja Tadeusza Korzona, MS 5937, fols. 103-4; J. Sokulski, MS article about Limanowski's relationship with Giller, BPAN, MS 6683, pp. 3-4.

4. Boleslaw Limanowski to Lucjan Limanowski, letters dated July 23 and August 3, 1886, BN, MS 2898, fols. 151-54; *Pam.*, II, 335-38.

5. The *Social Review* provided a forum for leftist writers in Poland, ranging from Marxists to various types of nationalistic populists, and was subject to the influence of Russian populism, too. It ceased publication in mid-1887. See P. Brock, "Boleslaw Wyslouch, tworca ideowy ruchu ludowego w Galicji," Diss., Jagiellonian University of Cracow, 1950, pp. 57, 60-61, 278-80; *Zarys historii prasy polskiej*, ed. T. Butkiewicz and Z. Mlynarski, Part II, No. 1 (Warsaw, 1959), p. 105; K. Dunin-Wasowicz, ed., *Przeglad Spoleczyn, 1886-7* (Wroclaw,

1955), pp. 12-14, 34-35. In 1887 Wyslouch purchased the only progressive daily then published in Lvov, the *Lvov Messenger (Kurjer Lwowski)*; Limanowski was to contribute to the weekly supplement to this paper initiated by Wyslouch in 1893. The circulation of the *Lvov Messenger* was about 3000 copies in 1886 and 8000 in 1900. See *Zarys*, p. 115.

6. B. Wyslouch, "Szkice Programowe," in Dunin-Wasowicz, ed., *op. cit.*, pp. 53-64.

7. Limanowski, "W kwestii Szkicow Programowych," *Przeglad Spoleczny*, II (July 2, 1886), 50-54. Wyslouch hastened to assure Limanowski that he was not contemplating a compulsory emigration of Jews from Poland; this would have been, in Wyslouch's opinion, contrary to the democratic principle of national self-determination and to the old tradition of toleration of the Jews in Poland. (See Brock, "Boleslaw Wyslouch," p. 86.) Like Wyslouch, Limanowski distinguished himself by his enlightened attitude toward the Polish Jews. See for instance *Pam.*, III, 107-8, 627. Cf. Dr. Filip Eisenberg, "Wilhelm Feldman: Szkic biograficzny," in *Pamieci Wilhelma Feldmana* (Cracow, 1922), pp. 21-24.

8. See "Od czego zalezy przyszlosc Rosji" [Fragment from unpublished memoirs of V. Debogorii-Mokrievich], *Przeglad Spoleczny* III (1887), 49-61; Limanowski, "Dyskusje" [Rev. of "Od czego zalezy przyszlosc Rosji," by V. Debogorii-Mokrievich], *Przeglad Spoleczny*, III (1887), 159.

9. During his exile in north Russia Limanowski had been impressed by exaggerated rumors of a peasant uprising in the Volga regions, fomented by the so-called "Kazan conspiracy," referred to above. Limanowski considered this "conspiracy" the beginning of an active collaboration between Polish patriots and the Russian revolutionaries. See W. Studnicki-Gizbert, "Boleslaw Limanowski," in *Z okolic Dzwiny* (Wilno, 1912), pp. 94-95; Limanowski, *Powstanie narodowe 1863 i 1864 r.*, 2nd ed. (Lvov, 1900), p. 137; Limanowski, "Dyskusje," pp. 159-62; V. Debogorii-Mokrievich, *Vospominaniia* (Paris, 1894-98), pp. 490-521.

10. Limanowski, "Powszechne glosowanie," *Przeglad Spoleczny*, II (1886), 421-24; Godin to Limanowski, letters dated October 2, 1886 and January 22, 1887, BJ, MS 6871, fols. 76-78; *Pam.*, II, 336-37; Limanowski, "Andrzej Godin" [Obituary], *Glos*, II (1888), 103. Limanowski's article was reprinted in Godin's *Le Devoir* of February 6, 1887.

11. Limanowski, "Powszechne glosowanie," pp. 420-23; G. Sabine, *A History of Political Theory*, 3rd ed. (New York), 1962), p. 696. Cf. Limanowski, *Bolszewickie panstwo w swietle nauki* (Warsaw,

1921), pp. 13-15. Under the influence of Lambert Quetelet, the Belgian astronomer who originated the statistical approach to social science, Limanowski concluded that universal suffrange, though not a perfect system, would better reflect the general will than rule by a few, because if suffrage were universal, the prevailing point of view would tend toward the average opinion. Limanowski was introduced to some of the works by Quetelet while in exile in north Russia and also in Lvov. See *Pam.*, I, 384; Limanowski, *Panstwo unarodowione i unarodowienie ziemi* (Warsaw, 1918), pp. 7-9.

12. Limanowski, "Powszechne glosowanie," pp. 423-25.

13. Around the time of Limanowski's correspondence with Lukaszewski, the Viceroy in Lvov informed the Police Directorate in Cracow that the *émigrés* had failed in their attempt to unite, due to the issue of socialism. He alleged that a nationalistic socialist organization was about to be created in Paris, the Swiss branch of which was to be organized by Limanowski. See Viceroy to Police Directorate in Cracow, letter dated November 28, 1886, APKr., St.GKr. 534, No. 446. Cf. Limanowski to Lukaszewski, letters dated March 18, September 20, December 21, 1886 and February 1887, BN, Listy do Juliana Lukaszewskiego, 1886-7, Mfm. 28062, Nos. 255-260. Cf. Lukaszewski to Limanowski, letter dated July 1, 1886, BJ, MS 6871, fols. 133-34; *Pam.*, II, 293, 375, 717, 719; S. Kieniewicz, *Historia Polski, 1795-1918* (Warsaw, 1968), pp. 333-34; Jan Kancewicz, "Powstanie Socjaldemokracji Krolestwa Polskiego i Litwy," p. 11 (Unpublished monograph at the University of Warsaw, 1959).

14. *Pam.*, II, 375, 386-87.

15. Kieniewicz, *Historia*, p. 334; Limanowski to Lukaszewski, letter dated February 1887, BN, Listy do Juliana Lukaszewskiego, Mfm. 28062, No. 260.

16. Kieniewicz, *Historia*, pp. 332-33. Like the *Social Review*, the *Voice* considered the peasantry a decisive force in national affairs. The *Review*, however, was more critical of Russian populism than the *Voice* and less inclined toward Slavophilism. The *Voice* stressed that there was a cultural gulf between the gentry and the people and rejected all Polish traditions as being gentry traditions, while the *Review* aimed at making these traditions the heritage of the entire nation and was convinced that democratization in Poland would assure everyone his rightful place in society. Unlike Milkowski's League, both papers reflected the historic boundaries of 1772. See Dunin-Wasowicz, *op. cit.*, pp. 35-37; Brock, "Boleslaw Wyslouch," p. 90.

17. Poplawski to Limanowski, letter dated December 15, 1886, BJ, MS 6872, fol. 20.

18. *Koalicja kapitalu i pracy* (Poznan, 1868). See Limanowski, "Obecne polozenie wloscian w Szwajcarii," *Glos*, II (1887), 504, 519; Limanowski, "Z dziedziny socjologicznej," *Przeglad Tygodniowy*, XXII (1887), 452; Limanowski, "Unarodowienie ziemi," *Pobudka*, Nos. 10/11/12 (1890), pp. 4-5. A study of Dluski convinced Limanowski that the situation of the peasant in Russian Poland was deteriorating, too. See also Michael Tracy, "Agriculture in Western Europe: The Great Depression, 1880-1900," in Charles K. Warner, ed., *Agrarian Conditions in Modern European History* (New York-Toronto, 1966), pp. 98-112.

19. Limanowski, Rev. of *La question agraire, étude sur l'histoire politique de la petite propriété*, by R. Meyer and G. Ardant, *Glos*, II (1887), 539.

20. Limanowski, "Wlasnosc ziemska kolektywna" [Rev. of *La propriété collective du sol en differents pays* (1886), by Emile Laveleye], *Przeglad Tygodniowy*, XXI (1886), 374-75.

21. Limanowski, "Z dziedziny socjologicznej," p. 452; Limanowski, "Unarodowienie ziemi," pp. 3-4.

22. Limanowski, "Wlasnosc ziemska," pp. 373-74.

23. See Adam Krzyztopor [Tomas Potocki] *O urzadzeniu stosunkow rolniczych w Polsce* (Poznan, 1859), pp. 461-62, cited in Limanowski, "Z dziedziny socjologicznej," p. 452. Cf. Limanowski, "Obecne polozenie wloscian w Szwajcarii," p. 519; Limanowski, *Historia demokracji polskiej w epoce porozbiorowej*, 4th ed. (Warsaw, 1957), II, 45. 1st ed., 1901.

24. *Pam.*, II, 339-42, 364-65, 367, 388, 403. Limanowski to Katarzyna Sawicka (mother), letter written in 1886 and one dated June 24, 1887, BN, MS 2901, fols. 188-191; Limanowski to Wislicki, letter dated August 31, 1887, BP, Korespondencja Adama Wislickiego; Boleslaw Limanowski to Lucjan Limanowski, letters dated January 4, March 12, April 22 and December 5, 1887, BN, MS 2892, fols. 157-162, 167-68.

25. *Pam.*, II, 380-81, 719; Limanowski, "Galicja jako deska ratunku z powodu ucisku zaborcy w dwoch innych zaborach: Rzecz o Galicji ze stanowiska przyszlosciowego niepodleglosci polskiej," *Pobudka*, No. 6 (1889), p. 19.

26. Slaz [Feliks Daszynski] *Pod pregierz: Szopka Wigilii Bozego Narodzenia* (Geneva, 1888). According to the sociologist Ludwik

Krzywicki, the incident and the pamphlet might well have frustrated a potential reconciliation between Limanowski and the party Proletariat II, the successor to Warynski's Proletariat in Russian Poland. Apparently, the Proletariat II had a delegate at the congress in Zurich; he negotiated with Limanowski. See L. Krzywicki, *Wspomnienia* (Warsaw, 1959), III, 82. Cf. Limanowski, "List do Redakcji," *Pobudka*, No. 6 (1889), pp. 18-19; *Pam.*, II, 375, 380-81; I. Daszynski, *Pamietniki* (Cracow, 1925-6), I, 42-43.

27. Limanowski, "Galicja jako deska ratunku," p. 19; *Pam.*, II, 388; Daszynski, *Pamietniki*, I, 43.

28. W. Rupniewski, ed., "Program niepodleglosciowy *Pobudki*," *Niepodleglosc*, II, (1930), 353-54. Cf. Limanowski program in *Pam.*, II, 684-87.

29. Rupniewski, ed., *op. cit.*, p. 352; "Credo," *Pobudka*, No. 1 (1891), p. 1; W. Feldman, *Dzieje polskiej mysli politycznej, 1864-1914*, 2nd ed. (Warsaw, 1933), pp. 234-35. Cf. Kancewicz, *op. cit.*, p. 12; J. Myslinski, *Grupy polityczne Krolestwa Polskiego w Zachodniej Galicji, 1895-1904* (Warsaw, 1967), p. 18.

30. Myslinski, *op. cit.*, p. 18; W. Pobog-Malinowski, "Umowa Ligi Narodowej [*sic*] z Paryska Gmina Narodowo-Socjalistyczna, 1889," *Niepodleglosc*, VII (1933), 432-34.

31. Some officials in Vienna grossly exaggerated the numerical strength, the influence and the radicalism of the Commune, and were convinced that it was an internationalist party allied with the Russian revolutionary *Narodnaia Volia* (now almost extinct). However, in Cracow the police felt that Limanowski's nationalistic socialism had so far only a limited appeal; he had been unsuccessful, to date, in his efforts to create a political party of his own. See Memorandum for Minister Taafe by Intelligence Bureau of the Foreign Affairs Ministry in Vienna, dated February 2, 1889, ZHP, Odpisy dokumentow CK MSZ w Wiedniu i Dyrekcji Policji w Krakowie, 1877-1890, AM 877/2, No. Ad. 356/4. Cf. Viceroy in Lvov to Directorate of Police in Cracow, letter dated March 20, 1889, APKr., St. GKr. 510, No. 125.

32. Limanowski, "Unarodowienie Ziemi," *Pobudka*, Nos. 10/11/12 (1890), pp. 3-13; *Pam.*, II, 409.

33. Limanowski, "Unarodowienie ziemi," pp. 5-8.

34. Ibid., pp. 8-10.

35. Ibid., p. 11.

36. Ibid., pp. 10-11.

37. Ibid., p. 12.

38. Limanowski, "Unarodowienie przemyslu," *Pobudka*, No. 1 (1891), pp. 2-3. In this article Limanowski was not fully committed to evolutionary social change; it would seem that his stress on evolutionism in his review of Schäffle's *Bau und Leben des socialen Körpers* (1875-78), published in the *Ateneum* in 1882, was dictated by the stringent censorship to which this periodical was subjected in Warsaw. See *Ateneum* (1882), III, 564-596.

39. Limanowski, "Unarodowienie przemyslu," p. 3.

40. Limanowski, "Unarodowienie podzialu," *Pobudka*, No. 3 (1891), pp. 2-3.

41. Ibid., p. 3.

42. For instance, around this time Limanowski became popular with a new student group in Cracow, the "Campfire *(Ognisko)*," which came into being in 1888 and was influenced by Limanowski's nationalistic socialism as well as by the ideology of "Zet." (See Myslinski, *op. cit.*, pp. 13-17; J. Gorzycki to Limanowski, letter dated February 3, 1890, BJ, MS 6871, fols. 80-81.) The police in Cracow now felt that the "Zet," the Campfire, Limanowski and *La Diane* group—all espoused the same ideology and were closely allied with Drahomanov! See Report by the Directorate of Police in Cracow to the Prime Minister, dated March 18, 1890, ZHP, Odpisy ... AM 877/2, No. Ad. 953/4-1136. Cf. W. Najdus, "Poczatki socjalistycznego ruchu robotniczego w Galicji," *Z Pola Walki*, No. 1/9 (1960), p. 32.

43. Limanowski, *Historia ruchu spolecznego w drugiej polowie XVIII stulecia* (Lvov, 1888), and *Historia ruchu spolecznego w XIX stuleciu* (Lvov, 1890); Najdus, "Poczatki," pp. 31-32. A number of chapters from Limanowski's first book appeared in Polish periodicals. See the *Week (Tydzien)*, 1877; *Equality (Rownosc)*, 1879 and 1880, and the *Social Review (Przeglad Spoleczny)*, 1886 and 1887. For bibliographical details regarding above see *Pam.*, II, 825-32. Parts of Limanowski's first book also appeared in foreign periodicals. See Limanowski, "Morelly," *Die Zukunft*, No. 16 (May 15, 1878), pp. 488-95; "Morelly," *New Generation (Uj Nemzedék)*, 1880 (in Hungarian); and "Mably, Rousseau, Morelly," the *Hammer (Molot)*, 1879 (in Ukrainian). Also see the *Ateneum* (1886-89) for Limanowski's discussion of Owen, Saint-Simon, Fourier, Proudhon, Blanc and Jan Colins.

44. Boleslaw Limanowski to Lucjan Limanowski, letter dated September 5, 1887, BN, MS 2898, fols. 167-68; Limanowski, *Socjologia* (1919, rpt. Cracow, 1921), I, 15-16.

45. Limanowski, *Historia . . . XVIII stulecia*, pp. 5-6, 8-9, 17-18, 32.

46. *Pam.*, II, 368-69. This book was well received by a Russian historian Karaiev; and a favorable review of it appeared in a Swiss journal, written by Liske, a professor at the University of Lvov. It was also discussed in Sybel's *Historische Zeitschrift*, XXVIII (1890) and in Malon's *Revue Socialiste*, IX (Jan. to June 1889), 190-208. See W. M. Kozlowski, "Zycie i praca Boleslawa Limanowskiego," in *Socjalizm-Demokracja-Patriotyzm* (Cracow, 1902), pp. 66-67.

47. Limanowski, *Historia . . . w XIX stuleciu*, pp. 1-2.

48. *La Revue Socialiste*, XIV (July to Dec. 1891), 199, 254; Boleslaw Limanowski to Lucjan Limanowski, letter dated July 15, 1891, BN, MS 2898, fol. 199. A Czech edition *Dejiny socialniho hnuti XIX stol.*, trans. Ant. Hojn and Ab. Hojn (Praha, 1891) did not include the chapters on Poland for the sake of economy. Limanowski derived no material benefit from the Czech edition. See Casopis ceskeho studentsva v Praze, letter to Limanowski dated December 13, 1890, BJ, MS 6872, fols. 15-16; *Pam.*, II, 427.

49. While considering these books dated, the contemporary Polish socialist historians Adam and Lidia Ciolkosz feel that Limanowski's chapters on Poland are still of value, because the earlier history of the Polish socialist movement was neglected or omitted in the works of foreign authors dealing with socialism in general. See their *Zarys dziejow socjalizmu polskiego* (London, 1966), I, 498. Cf. Myslinski, *op. cit.*, p. 34n; Z. Zechowski, *Socjologia Boleslawa Limanowskiego* (Poznan, 1964), pp. 224-25; S. Glabinski, Rev. of *Historia . . . XVIII stulecia*, by B. Limanowski, *KH*, III (1889), 351-56; his Rev. of *Historia . . . w XIX stuleciu*, by B. Limanowski, *KH*, V (1891), 434-42; J. Trabczynski [L. Kulczycki], "Boleslaw Limanowski," *Krytyka*, No. 3 (1900), pp. 182-84; K., "Boleslaw Limanowski," *Tydzien*, *dodatek do Kurjera Lwowskiego*, VIII, No. 35 (1900), 273; *Swiatlo*, III (1900), 4; W. F. [Wilhelm Feldman], "Boleslaw Limanowski," *Krytyka*, No. 11 (1910), p. 188; L. Krzywicki, "Artykuly i rozprawy," *Dziela* (Warsaw, 1959), III, 601-4; W. M. Kozlowski, *op. cit.*, pp. 66-68; K. Czachowski, "Szermierz wolnosci," *Niepodleglosc*, II (1930), 207. Aleksander Swietochowski, in his *Utopie w rozwoju historycznym* (Warsaw, 1910), does not mention the works of Limanowski. According to Marceli Handelsman, the Polish historian writing in 1935, "This positivistic or idealist socialist writing the history of . . . French revolutions prior to Aulard, Jaures and Stern, at a time when Taine was publishing his three volumes [*Les origines de la France contemporaine*], (Paris, 1876-1894), independently . . . reached similar

conclusions to those arrived at by French scholars," See M. Handelsman, "Boleslaw Limanowski jako badacz historii powszechnej," *PH*, XXXIII (1936), 329-30.

50. I. Tymolski and I. Kupiewski to Limanowski, letter dated March 25, 1889, BJ, MS 6872, fol. 81. Cf. S. Kobukowski to Limanowski, letter dated December 6, 1889, BJ, MS 6871, fols. 94-95; Polskie Akademickie Stowarzyszenie "Ognisko" in Vienna to Limanowski, letter dated July 15, 1892, BJ, MS 6872, fol. 38.

51. Boleslaw Limanowski to Lucjan Limanowski, letters dated February 19 and March 29, 1889, BJ, MS 2898, fols. 180-83; *Pam.*, II, 389-90; S. Laguna to Limanowski, letter dated February 12, 1889, BJ, MS 6872, fols. 9-10.

52. Limanowski, "Rok uchwalenia swieta robotniczego," *Jedniodniowka Majowa* (Warsaw, 1920), pp. 3-4; *Pam.*, II, 390, 393-94. There were 612 delegates at the Possibilists' Congress as opposed to 357 present at the Marxists' meeting. The congress agreed on broad autonomy in tactics, in accordance with a motion by Hyndman of the British Social Democratic Federation, and defined universal suffrage as the primary aim for socialists in all countries. See P. L., "Dwa kongresy socjalistyczne w Paryzu," *Pobudka*, Nos. 7/8/9 (1889), pp. 46-49; G. D. H. Cole, *The Second International, 1889-1914*, Part I (London, 1963), pp. 5-9.

53. *Pam.*, II, 395-96; Limanowski, "Rok," p. 4. The entire speech of Limanowski was cited in P. L., "Dwa kongresy," pp. 49-51. Limanowski explained in it that there existed an economic rivalry between Russian industrialists and landlords and their counterparts in Russian Poland, which resulted in discriminatory railway tariffs being imposed by the Russian authorities on Polish goods imported to Russia. Also, he stressed that national oppression in Poland directly affected workers' movement abroad: in Prussian Poland expropriation of Polish peasants caused mass emigration, especially to Westphalia and the United States. These *émigré* Polish workers were ignorant politically and depressed wages of local labor. See ibid., p. 50.

54. P. L., "Dwa kongresy," pp. 51-52, also 48; Limanowski, "Rok," pp. 4-5; *Pam.*, II, 396-97.

55. P. L., "Dwa kongresy," p. 49; *Pam.*, II, 397-400.

56. *Pam.*, II, 403-6; Boleslaw Limanowski to Lucjan Limanowski, letter dated February 14, 1890, BN, MS 2898, fols. 188-89.

NOTES TO CHAPTER VI

1. For Limanowski's speech see his *Pamietniki*, ed. J. Durko (Warsaw, 1958), II, 692-94; 407, 409-10, 412, 425-27 (referred to below as *Pam.*); Boleslaw Limanowski to Lucjan Limanowski, letter dated February 14, 1890, BN, MS 2898, fol. 188. Limanowski considered Mickiewicz a socialist, due to the poet's contributing to the Paris *Tribune des Peuples* in 1849 and the program drafted by him for the Polish legion which he had formed in Italy in 1848. (See Limanowski, "Przedmowa," in *Artykuly polityczne Adama Mickiewicza*, Cracow, 1893, pp.iv-xxxviii.)

2. *Pam.*, II, 694. Among the other speakers were Ernest Renan as well as Prince Wladyslaw Czartoryski, the son of the famous Prince Adam. See Boleslaw Limanowski to Lucjan Limanowski, letter dated July 10, 1890, BN, MS 2898, fol. 193; *Pam.*, II, 427.

3. *Pam.*, II, 409-10; Boleslaw Limanowski to Lucjan Limanowski, letter dated February 14, 1890, BN, MS 2898, fol. 189; S. Badzynski to Limanowski, letter dated March 11, 1890, BJ, MS 6871, fols. 1-2.

4. Towarzystwo Mlodziezy Polskiej in Zurich to Limanowski, letter dated December 14, 1890, BJ, MS 6872, fol. 75; *Pam.*, II, 428-29; Boleslaw Limanowski to Lucjan Limanowski, letter dated December 22, 1890, BN, MS 2898, fols. 194-95.

5. *Pam.*, II, 429; Badzynski to Limanowski, letter dated April 9, 1891, BJ, MS 6871, fols. 3-4; Towarzystwo Mlodziezy Polskiej in Zurich and Towarzystwo Polskie in Zurich to Limanowski, letter dated April 8, 1891, BJ, MS 6872, fols. 76-77; M. Rudnicki, Association des Anciens Elèves de L'Ecole Polonaise to Limanowski, letter dated April 15, 1891, BJ, MS 6872, fol. 3.

6. Towarzystwo Polskie in Zurich and Towarzystwo Mlodziezy Polskiej in Zurich to Limanowski, letter dated March 30, 1891, BJ, MS 6872, fol. 78; *Pam.*, II, 433-35.

7. Boleslaw Limanowski to Lucjan Limanowski, letters dated July 10, 1890, May 12, 1891 and April 21 as well as June 2, 1892, BN, MS 2898, fols. 193, 196-97, 204-5, 208-9; Maria Wyslouch to Limanowski, letters dated October 5, 1890 and June 19, 1892, BJ, MS 6871, fols.177-78, 189-91; Maria Limanowska to Lucjan Limanowski, letters dated June 20, 1891 and summer 1892, BN, MS 2898, fols. 242-45; *Pam.*, II, 426. Even Limanowski's sociological, i.e., theoretical works, were banned by the censors, who often did not understand them. See Wislicki to Limanowski, letters dated December 13, 1888 and October 11, 1890, BJ, MS 6872, fols. 47-49; *Pam.*, 437-39; Limanowski to Maria Wyslouch, letters dated May 30, August 7 and 19, 1892., Korespondencja Wyslouchow, MS 12,149, fols. 101-3, 117-18, 120.

8. Boleslaw Limanowski to Lucjan Limanowski, letters dated July 15, 1891, September 20, 1892 and January 24, 1893, BN, MS 2898, fols. 198-99, 213-15; *Pam.*, II, 438; J. Krakow to Limanowski, letter dated September 13, 1892, BJ, MS 6871, fols. 115-16; L. Gadon to Limanowski, letter dated October 5, 1892, BJ, MS 6871, fol. 43.

9. Limanowski, *Galicja przedstawiona slowem i olowkiem* (Warsaw-Lvov, 1892); *Pam.*, II, 377-78. Limanowski dealt here with Galicia's geography, ethnography, history, agriculture, trade, industry, transportation as well as political and social institutions. A favorable review of this book by Ivan Franko appeared in Wyslouch's *Lvov Messenger* in 1891. Wislicki published the chapter on industry separately as well, as a supplement to the *Weekly Review* of the first half-year, 1890. See Wislicki to Limanowski, letters dated October 11, 1890, April 1, 1891 and January 31, 1892, BJ, MS 6872, fols. 48-49, 52-53, 54-55; Limanowski to Wislicki, letters dated December 2 and 16, 1891, BP, Korespondencja Adama Wislickiego.

10. S. Szczpanowski, *Nedza Galicji w cyfrach* (Lvov, 1888), p. 13, cited by Limanowski in his "Przemsyl i ludnosc robotnicza przemyslowa," *Dodatek do Przegladu Tygodniowego*, 1st half-year (1890), pp. 296-96, 291-301. Cf. Limanowski, *O kwestii robotniczej* (Lvov, 1871), pp. 17-18, advocating a close alliance of the intelligentsia and manual labor. See also Limanowski, "Przemysl," pp. 291-301. Szczepanowski felt that fostering industrialization should be considered a patriotic duty in Poland. To this end, he wished to combine the "Polish romantic soul with the British brain" by means of an appropriate education of the young people in Poland. See W. Feldman, *Stronnictwa i programy polityczne w Galicji, 1846-1906* (Cracow, 1907), II, 32-34.

11. Limanowski, *Galicja*, pp. 90-95.

12. See *Labor (Praca)*, No. 3, February 14, 1892, cited by W. Najdus, "Ruch robotniczy w Galicji w latach 1890-1900," *PH*, LIII (1962), 93-94. Also see ibid., pp. 87-93.

13. *Pam.*, II, 436, 695-97; "Polskie partie robotnicze i sprawa narodowa," *Przeglad Socjalistyczny*, No. 1 (1892), pp. 1-7. Cf. M. Sc[aevola—J. Lorentowicz], "Nasza krew," *Pobudka*, No. 5 (1892), pp. 1-3, cited in Jan Kancewicz, "Zjazd paryski socjalistow polskich," *Z Pola Walki*, No. 4/20 (1962), pp. 5-6; Badzynski to Limanowski, letter dated February 4, 1892, BJ, MS 6871, fols. 6-7. See Limanowski to Maria Wyslouch, letter dated September 23, 1892, Ossol., Korespondencja Wyslouchow, MS 12,149, fols. 124-25.

14. Limanowski, "Swiadomosc narodowa," *Przeglad Socjalistyczny*, No. 1 (1892), pp. 39-42; *Pam.*, II, 695-97. Also see his "Stanislaw Worcell i jego teoria rozwoju wlasnosci," *Przeglad Socjalistyczny*, No. 2/3 (1893), pp. 63-74.

15. Balicki to Limanowski, letters dated April 19 and May 17, 1892, BJ, MS 6871, fols. 8-10; S. Kieniewicz, *Historia Polski, 1795-1918* (Warsaw, 1968), pp. 401-4.

16. *Pam.*, II, 439-41; Limanowski, "Z moich wspomnien o narodzinach PPS," *W trzydziesta rocznice: Ksiega Pamiatkowa PPS* (Warsaw, 1923). pp. 23-24; L. Wasilewski, ed., "Dokumenty do historii zjazdu paryskiego," *Niepodleglosc*, VIII (1933), 112-120; Limanowski to Maria Wyslouch, letter dated December 2, 1892, Ossol., Korespondencja Wyslouchow, MS 12,149, fols. 128-29.

17. Wasilewski, ed., *op. cit.*, pp. 121-41; Res [Feliks Perl], "Boleslaw Limanowski," *Przedswit*, XXX, No. 11 (1910), pp. 703-6. Cf. M. Zawadka, "Pierwszy zjazd PPS," *Przeglad Socjalistyczny*, No. 12 (1947), p. 40.

18. A. Prochnik, *Ku Polsce socjalistycznej* (Warsaw, 1936), p. 22; Wasilewski, ed., *op. cit.*, pp. 109-13. Cf. Zawadka, *op. cit.*, pp. 38-39; J. Durko, "Zjazd paryski i powstanie PPS," *Przeglad Socjalistyczny*, No. 10/11 (1946), pp. 29-30; Kancewicz, "Powstanie Socjaldemokracji Krolestwa Polskiego i Litwy," pp. 15-29 (Unpublished monograph, University of Warsaw, 1959); Kancewicz, "Zjazd," pp. 4-17; H. Jablonski, *Polityka PPS w czasie wojny, 1914-8* (Warsaw, 1958), p. 14. It is not strictly true that the delegates to the congress were "ten of Limanowski's group and eight members of the 'Proletariat'." (See J. P. Nettl, *Rosa Luxemburg*, London, 1966, I, 61.) There were three members of the *La Diane* group; four delegates of the Workers' Unity; a self-styled representative of the Union of Polish Workers, Grabski, who later on joined the National Democrats; and an "independent" Polish student from Berlin. This takes care of the ten of "Limanowski's group." However, the "eight members of the Proletariat" were now also ideologically akin to Limanowski. Moreover, it was a nationalistic socialist, the *émigré* Lorentowicz, who objected to defining the future Poland as merely "democratic," as proposed by Mendelson and his commission preparing the new program, which consisted of two members of the Proletariat and three delegates of the Workers' Unity group. See Kancewicz, "Zjazd," pp. 17-18. Cf. Prochnik, *Ku Polsce*, p. 23.

19. A) *Political Program*

    1) Direct, universal and secret ballot, both sanctioning and initiatory.

2) Complete equality before the law of all nationalities on the basis of a voluntary federation;

3) Local self-government at the levels of commune and province. Election of administrative personnel;

4) Equality before the law for all citizens irrespective of sex, race, nationality or religious denomination;

5) Complete freedom of speech, press, assembly and association;

6) Free judiciary; elected judges. Civil servants responsible to the courts;

7) Free, compulsory and universal education at all levels; students to be maintained by the state;

8) Abolition of a standing army. National peoples' militia;

9) Progressive income and inheritance taxes. Abolition of tax on food and on articles of first necessity.

See "Szkic programu PPS," *Przedswit*, No. 5 (1893), pp. 5-6. Cf. *Pam.*, II, 684-87; Wasilewski, ed., *op. cit.*, pp. 142-43; J. Mulak, "Z dziejow mysli programowej PPS," *Przeglad Socjalistyczny*, No. 5 (1948), pp. 18-21; Zawadka, *op. cit.*, pp. 41-42; F. Gross, *The Polish Worker* (New York, 1945), p. 116.

20. B. Suchodolski, *Polskie tradycje demokratyczne* (Wroclaw, 1946), p. 126; J. Borejsza, *Emigracja polska po powstaniu styczniowym*, (Warsaw, 1966), pp. 421-22; Jablonski, *Polityka PPS*, p. 12.

21. J. Trabzynski [L. Kulczycki], "Boleslaw Limanowski," *Krytyka*, No. 3 (1900), p. 185; Jablonski, *Polityka PPS*, pp. 12-14.

22. *Dzieje Polski lat ostatnich* (Warsaw, n.d.), p. 57, cited by Jablonski, *Polityka PPS*, p. 510.

NOTES TO CHAPTER VII

1. Limanowski, *Pamietniki*, ed., J. Durko (Warsaw, 1958), II, 441-42 (referred to below as *Pam.*). We should note that Limanowski, in addition to declining the post offered to him on the executive of the ZZSP, had refused at the congress itself in November 1892, to become a member of the drafting commission headed by Mendelson, which prepared the agenda and drafted the new program on the basis of resolutions passed at the congress. See J. Kancewicz, "Zjazd paryski socjalistow polskich (17-23 XI 1892 r.)," *Z Pola Walki*, V, No. 4/20 (1962), p. 17.

2. H. Gierszynski to Limanowski, letters dated March 28 and July 24, 1893, BJ, MS 6871, fols. 60-63.

3. P. Brock, "Boleslaw Wyslouch, tworca ideowy ruchu ludowego w Galicji," Diss., the Jagiellonian University of Cracow, 1950, pp. 282, 285; *Pam.*, II, 451-55.

4. The report of Badeni prepared at the time for the Cracow police, in order to facilitate their identification of Limanowski, described him as "fifty-nine years old, tall, with drawn face, dark eyes and long dark beard streaked with gray." See Badeni to Directorate of Police in Cracow, Memorandum dated December 1, 1893, APKr., St. GKr. 13,No. 1216; *Pam.*, II, 453.

5. Boleslaw Limanowski to Lucjan Limanowski, letter dated December 21, 1893, BN, MS 2898, fol. 226; Limanowski to Centralization, letter dated February 27, 1894, ZHP, APPS, Listy B. Limanowskiego, 1890-1911, AM 1190/7, No. 2; *Pam.*, II, 456-462. In Lvov a certain bookseller and publisher, Gubrynowicz, agitated against Limanowski as the translator of the famous *Elements* by Drysdale (1880). Gubrynowicz charged that Limanowski had demoralized Polish society, and he might well have harmed Limanowski's case with the Galician authorities. See *Pam.*, II, 459-460.

6. *Pam.*, II, 457-58, 465; Maria Wyslouch to Limanowski, letter dated February 23, 1893, BJ, MS 6871, fols. 192-93; Boleslaw Limanowski to Lucjan Limanowski, letters dated November 28 and December 21, 1893, BN, MS 2898, fols. 224-25, 227.

7. X. Y. [Limanowski], *Stuletnia walka narodu polskiego o niepodleglosc* (Lvov, 1894). Revised editions appeared in Lvov—1906 and in Cracow—1916. This book was translated by Limanowski and L. Kulczycki into Russian as *Sto let bor'by Pol'skago naroda za svobodu* (Moscow, 1907). See Boleslaw Limanowski to Lucjan Limanowski, letters dated August 1 and December 21, 1893, BN, MS 2898, fols. 222, 227; K. Czachowski, "Szermierz wolnosci," *Niepodleglosc*, II (1930), 211; Directorate of Police in Lvov to Directorate of Police in Cracow, Memorandum dated October 21, 1895, APKr., St. GKr. 13, No. 772.

8. Limanowski, "Listy szwajcarskie I," *Tydzien, dodatek do Kurjera Lwowskiego*, II, No. 3 (1894), pp. 22-23.

9. Balicki to Limanowski, letter dated January 2, 1894, BJ, MS 6871, fols. 12-13; *Pam.*, II, 466-67.

10. Limanowski, "Wplyw wiedzy stosowanej na rozwoj zycia spolecznego," *Przedswit* (1895): No. 8, pp. 2-8; No. 9, pp. 3-9.

11. *Loc. cit.*

12. Ibid., No. 9, pp. 8-9.

13. Ibid., p. 9.

14. F. Perl to Limanowski, letter dated February 26, 1894, BJ, MS 6872, fol. 96; *Pam.*, II, 466; J. Galezowski to Limanowski, letter dated February 19, 1894, BJ, MS 6872, fol. 33; Res [F. Perl], "Boleslaw Limanowski," *Przedswit,* No. 11 (1910), p. 705; L. Gadon to Limanowski, letter dated October 5, 1892, BJ, MS 6871, fol. 43; J. Krakow to Limanowski, letter dated September 13, 1892, BJ, MS 6871, fols. 115-16; Boleslaw Limanowski to Lucjan Limanowski, letters dated September 20, 1892 and January 24, 1893, BN, MS 2898, fols. 212-15.

15. *Pam.*, II, 469, 473-84; Limanowski, "Pamietniki," BN, MS 6417/14, fol. 641; Limanowski to Maria Wyslouch, letter dated May 31, 1894, Ossol., Korespondencja Wyslouchow, 1890-1914, MS 12,149, fols. 131-33.

16. Boleslaw Limanowski to Lucjan Limanowski, letter dated August 2, 1896, BN, MS 2898, fols. 243-44; also see his letters dated September 2, 1894, August 28 and December 31, 1895, December 25, 1896 and April 23, 1899, BN, MS 2898, fols. 234-35, 239, 241-42, 245-46, 253-54; Jerzejowski to Limanowski, letter dated February 20, 1895, ZHP, APPS, Kopialy Korespondencji ZZSP, 1893-99, AM 698/2, Vol. IV, No. 40; Limanowski to Jedrzejowski, letter dated March 5, 1895, ZHP, APPS, Listy . . . AM 1190/7, No. 14; *Pam.*, II, 492. He seldom wrote to Lucjan from Paris, because he feared to be short of money for carfare; if he were to walk to the office in the morning, he would lose a half-hour of his valuable writing time.

17. *Pam.*, II, 486-87. Limanowski preferred non-partisan duties. Thus, it is not surprising that between 1893 and 1900 he served on the "honorary jury" of the ZZSP, which tried members of this organization for minor misdemeanors. (See *Pam.*, II, 442-45, 698-703.) Moreover, he became the chief cashier of an originally non-partisan fund-raising organization, the "Red Cross," which was founded early in 1894 for the purpose of aiding Polish political prisoners and exiles. (See *Pam.*, II, 555-57.) To mark its tenth anniversary, the "Red Cross" published a collection of autobiographical material on the uprising of 1863, in which Limanowski's reminiscences also appeared. See Limanowski, "Pierwsza manifestacja w Wilnie w 1861 r.," in *Ksiega Zbiorowa w 40^{ta}tocznice powstania styczniowego, 1863-1903* (Lvov, 1903).

18. Limanowski to Jedrzejowski, letters dated May 29, August 27 and September 23, 1894, ZHP, APPS, Listy . . . AM 1190/7, Nos. 6,

8 and 9; Jedrzejowski to Limanowski, letter dated August 22, 1894, ZHP, APPS, Kopialy . . . AM 698/1, Vol. III, No. 66; Boleslaw Limanowski to Lucjan Limanowski, letter dated December 26, 1894, BN, MS 2898, fol. 237.

19. Dmowski to Limanowski, letters dated July 28, December 2, 1895 and February 2, 1897, BJ, MS 6872, fols. 58-64; Myslinski, *op. cit.*, pp. 132-35; *Pam.*, II, 472, 499, 526.

20. A. Prochnik, *Ku Polsce socjalistycznej* (Warsaw, 1936), p. 23. Cf. J. P. Nettl, *Rosa Luxemburg* (London, 1966), I, 41-42.

21. Ibid., II, 845; ibid., I, 90, 94.

22. Ibid., I, 90-92. Rosa Luxemburg went beyond the tactical objections of Warynski (and his political friends) to the struggle for independence; she justified the Russo-Polish connection in terms of an inexorable historical necessity.

23. Cited in Prochnik, *Ku Polsce*, pp. 12-14. For Krauz's views on the Polish question see his *Wybor pism politycznych*, published posthumously in 1907. See also A. Zarnowska, "Kazimierz Kelles-Krauz, 1872-1905," *Z Pola Walki*, II (1958), No. 1.

24. See Limanowski, "Grupa socjalistyczna w parlamencie francuzkim," *Kalendarz Robotniczy na rok 1895*, Cracow, pp. 61-66.

25. Limanowski to Centralization, letters dated September 28 and October 27, 1894, ZHP, APPS, Listy . . . AM 1190/7, Nos. 10 and 11; W. Jodko to Limanowski, letter dated October 12, 1894, ZHP, APPS, Kopialy . . . AM 698/1, Vol. III, No. 185.

26. Limanowski to Centralization, letter dated April 29, 1894, ZHP, APPS, Listy . . . AM 1190/7, No. 5; Wojciechowski to Limanowski, letter dated October 14, 1894, ZHP, APPS, Kopialy . . . AM 698/1, Vol. III, No. 194. At this time the Paris Branch (one of the most active groups of the ZZSP) numbered eleven members; the ZZSP as a whole had seventy-four. In the circumstances of emigration, it was a sizeable figure. See Limanowski, "Z moich wspomnien o narodzinach PPS," *W trzydziesta rocznice: Ksiega Pamiatkowa PPS* (Warsaw, 1923), p. 25.

27. Internal Constitution of the Paris Branch, Preliminary Meeting of November 11, 1894, ZHP, APPS, Materialy Sekcji Paryskiej, AM 1440/8, No. 2; Limanowski to Centralization, letter dated November 15, 1894, ZHP, APPS, Listy . . . AM 1190/7, No. 12; Wojciechowski to Limanowski, letter dated November 17, 1894, ZHP, APPS, Kopialy . . . AM 698/1, Vol. III, No. 325. The Archives of the Paris Branch seem incomplete; also Limanowski apparently attended the monthly meetings irregularly due, perhaps, to frequent illnesses in his family.

28. Limanowski, "Listy szwajcarskie III," *Tydzien, dodatek do Kurjera Lwowskiego*, II, No. 12 (1894), p. 93.

29. Ibid., Plekhanov's pamphlet was first published in Russian as a foreword to the Russian translation of Alphons Thun's book, which was sponsored by the "League of the Russian Revolutionary Social Democracy" (Geneva, 1903). See G. V. Plekhanov, *Sochineniia* (Moscow, n.d.), IX-X, 2, 5-29. In 1898 Limanowski was to speak on behalf of the ZZSP on the occasion of Lavrov's birthday. Two years later he was to request the Centralization to authorize him to speak on behalf of the ZZSP at Lavrov's funeral and to pay him thus his last respects. See Wojciechowski to Limanowski, letter dated May 7, 1898, ZHP, APPS, Kopialy ... AM 698/4, Vol. XIII, No. 7; Limanowski to Centralization, telegram, n.d., AM 1190/7.

30. Limanowski to Centralization, letter dated May 28, 1895, ZHP, APPS, Listy ... AM 1190/7, No. 16.

31. Jedrzejowski to Limanowski, letter dated July 15, 1895, ZHP, APPS, Kopialy ... AM 698/2, Vol. V, No. 444.

32. Limanowski to Jedrzejowski, letter dated July 28, 1894, ZHP, APPS, Listy ... AM 1190/7, No. 17.

33. *Pam.*, II, 487, 489-90.

34. Ibid., II, 494-95. Yet the majority in the SDKPiL appeared to be much less hostile toward Limanowski than was Luxemburg or Warski. See M. Zychowski, *Boleslaw Limanowski, 1835-1935* (Warsaw, 1971), pp. 256-63.

35. Limanowski to Jedrzejowski, letter dated June 22, 1896, ZHP, APPS, Listy ... AM 1190/7, No. 25.

36. *Pam.*, II, 497; Lettre ouverte aux membres du Congrès Socialiste Internationale à Londres, 1896," ZHP, APPS, Listy ... AM 1190/7, No. 26, pp. 1-7.

37. See Limanowski, Rev. of *Czy Europa ma skozaczec? Przyczynek do kwestii wshodniej*, by Wilhelm Liebknecht, *Tydzien, dodatek do Kurjera Lwowskiego*, V, No. 32 (1897), pp. 255-56; Cf. F. Engels, "Preface to the Polish Edition of 1892," *The Manifesto of the Communist Party*, trans. S. Moore (1888; rpt. Moscow, n.d.), pp. 38-41.

38. Lettre ouverte ..., pp. 8-16.

39. Ibid., p. 17.

40. Nettl, *op. cit.*, I, 98-99. Cf. Limanowski, *Odrodzenie i rozwoj narodowosci polskiej na Slasku* (1911; rpt. Warsaw, 1921), pp. 41-42.

41. Z. Klemensiewicz to Limanowski, letter dated May 9, 1896, BJ, MS 6871, fols. 87-90.

42. In the second half of the decade the Galician Social Democracy became the Polish Social Democratic Party of Galicia and Teschen Silesia (PPSD), due to the transformation of the parent Austrian party—hitherto organized on the basis of the territorial principle—into a federation of autonomous ethnic parties. (See W. Najdus, "Ruch robotniczy w Galicji w latach 1890-1900," *PH*, LIII, 1962, 109-11.) The *Right of the People* appeared semi-monthly from 1896 to 1902. Revived in 1904 as a weekly organ of the PPSD for peasants, by 1912-13 it reached a circulation of 3800-4100 copies. See J. Buszko, "Polskie czasopisma socjalistyczne w Galicji," *Zeszyty Prasoznawcze*, II, Nos. 1-2 (1961), p. 33; Klemensiewicz to Limanowski, letters dated December 24, 1896 and July 28, 1897, BJ, MS 6871, fols. 91-93.

43. Limanowski, "Chlopi i socjalizm," *Naprzod*, V. No. 31 (1896), p. 3.

44. Limanowski, "Chlopi i socjalizm," *Kalendarz Robotniczy na rok 1897*, pp. 34-36.

45. Limanowski, Rev. of *Czy Europa ma skozaczec?* by Wilhelm Liebknecht, p. 256. Cf. Limanowski, Rev. of *Pisma wojskowo-polityczne*, by General W. Chrzanowski, *Gazeta Literacka*, No. 9 (1871), pp. 10-11.

46. Jerzejowski to Limanowski, letter dated January 22, 1897, ZHP, APPS, Kopialy . . . AM 698/3, Vol. VIII, No. 96; Wasilewski to Limanowski, letter dated January 27, 1898, ZHP, APPS, Kopialy . . . AM 698/4, Vol. XII, No. 193.

47. Limanowski, "1848-1898," *Przedswit*, XVIII, No. 3 (1898), pp. 2-5. See also his "Co bylo glownym powodem upadku rewolucji 1848 r.," *Trybuna*, II, No. 12 (April 15, 1907), pp. 6-16.

48. Maria Wyslouch to Limanowski, letter dated April 16, 1898, BJ, MS 6872, fols. 194-95; Limanowski to Maria Wyslouch, letter dated April 28, 1898, Ossol., Korespondencja Wyslouchow, MS 12, 149, fols. 135-37.

49. Limanowski, "Wielka wojna w przyszlosci," *Tydzien, dodatek do Kurjera Lwowskiego*, VI, No. 20/21 (1898), pp. 160-62; Limanowski, "Co bylo glownym . . . ," pp. 6-16. In 1907 Limanowski was to publish his last article on Mickiewicz; in it he stressed that the poet professed socialism of an "internationalist variety." According to Limanowski, "what social science tells us today, Mickiewicz had already divined." See "Mickiewicz jako publicysta," *Trybuna*, II, No. 6 (January 15, 1907), pp. 6-11.

50. The Annual Report by Paris Branch for 1895, December 10, 1895, ZHP, APPS, Okolniki Centralizacji, AM 1050/1, No. 10.

51. The Centralization usually agreed with Limanowski. In this instance, too, they opposed Krauz and refused to heed his warnings. See ibid., above; J. Mulak, "Z dziejow mysli programowej PPS," *Przeglad Socjalistyczny*, IV, No. 5 (1948), p. 22; Limanowski to Centralization, letters dated January 21, March 5 and December 4, 1895, ZHP, APPS, Listy ... AM 1190/7, Nos. 13, 14 and 22; Minutes of Meetings of the Paris Branch, January 12 and January 26, 1895, ZHP, APPS, Materialy Sekcji Paryskiej, AM 1440/8, Nos. 15 and 17.

52. Motion by Paris Branch, March 26, 1896, ZHP, APPS, Okolniki Centralizacji, AM 1050/1, No. 1.

53. Motion by Paris Branch, May 7, 1899, ZHP, APPS, Okolniki Centralizacji, AM 1050/1, No. 6; Paris Branch to Centralization, letter dated July 21, 1899, ZHP, APPS, Materialy Sekcji Paryskiej, AM 1440/8, No. 67.

54. Limanowski to Centralization, letter dated December 4, 1895, ZHP, APPS, Listy ... AM 1190/7, No. 22; Circular dated December 13, 1895, ZHP, APPS, Okolniki Centralizacji, AM 1050/1, No. 9.

55. Motion by Centralization dated December 18, 1898, ZHP, APPS, Okolniki Centralizacji, AM 1050/1, No. 20; Motion by Paris Branch dated June 24, 1899, ZHP, APPS, Okolniki Centralizacji, AM 1050/1, No. 10; Limanowski to Centralization, letter dated December 4, 1895, ZHP, APPS, Listy ... AM 1190/7, No. 22; Circular dated December 13, 1895, ZHP, APPS, Okolniki Centralizacji, AM 1050/1, No. 9.

56. The Annual Report by Paris Branch for 1895, December 10, 1895, ZHP, APPS, Okolniki Centralizacji, AM 1050/1, No. 10; "Motion by B. Limanowski," 1899, ZHP, APPS, Okolniki Centralizacji, AM 1050/1, No. 8; Limanowski to Foreign Committee, letter dated May 21, 1900, ZHP, APPS, Listy ... AM 1190/7, No. 45; *Pam.*, II, 520-21.

57. Circular dated February 7, 1899, Congress of ZZSP, January 10-11, 1899, ZHP, APPS, Okolniki Centralizacji, AM 1050/1, No. 2; Jedrzejowski to Limanowski, letter dated January 29, 1900, BJ, MS 6872, fol. 114.

NOTES TO CHAPTER VIII

1. Protocol of the Polish Delegation, the Preliminary Meeting, September 23, 1900; Declaration of the Polish Delegation, September 27, 1900, the International Socialist Congress in Paris, 1900, ZHP, APPS, Miedzynarodowe Kongresy Socjalistyczne, AM 1440/ 16, Nos. 2 and 14; Limanowski, *Pamietniki*, ed., J. Durko (Warsaw, 1958), II, 531-34, 725-26 (referred to below as *Pam.*).

2. Declaration by B. Limanowski, September 1900, the International Socialist Congress in Paris, 1900, ZHP, APPS, Miedzynarodowe Kongresy Socjalistyczne, AM 1440/16, Nos. 27 and 11.

3. Limanowski, "Lettre ouverte au Citoyen Guesde," dated February 28, 1901, ZHP, APPS, Materialy Seckji Paryskiej, AM 1440/9, No. 4.

4. Appeal to Compatriots, Paris, February 1900, ZHP, APPS, Listy B. Limanowskiego, Jubileusz 1900 r., AM 1190/7.

5. "Obchody Jubileuszowe, 1900," in Limanowski, *Socjalizm-Demokracja-Patriotyzm* (Cracow, 1902), pp. 210-11, also 205-10 (referred to below as *SDP.*); Jubilee of B. Limanowski, the Jubilee Committee, 1900, ZHP, APPS, Druki Ulotne Sekcji Paryskiej, AM 1440/9, Nos. 11, 24 & 28; *Pam.*, II, 535-37.

6. Zjednoczenie Towarzystw Mlodziezy Polskiej Zagranica to Limanowski, letter dated February 11, 1900, BJ, MS 6872, fols. 85-90; *Pam.*, II, 534-35, 538-39, 726; "Obchody Jubileuszowe," *SDP*, pp. 214-15; Zwiazek Mlodziezy Postepowej in Zurich to Limanowski, letter dated January 30, 1900, BJ, MS 6872, fol. 92. The socialist student union sponsored the second edition of Limanowski's *Powstanie narodowe 1863 i 1864 r.* (Lvov, 1900), which was printed in London in 2925 copies. See Redakcja i Administracja *Swiatla*, ZHP, APPS, Drukarnia i Ksiegarnia PPS, AM 1050/13, Nos. 111-12.

7. "Obchody Jubileuszowe," in *SDP*, pp. 215-16; *Pam.*, II, 537-40.

8. *Socjalizm-Demokracja-Patriotyzm* (Cracow, 1902). This book consisted chiefly of: a biographical sketch by Limanowski's friend, the historian and philosopher W. M. Kozlowski; Limanowski's autobiographical essay; and some of his pamphlets which had already appeared separately. As of January 1905, 2000 copies had been sold, of which 500 were in Russian Poland. (See ZHP, APPS, Listy B. Limanowskiego, 1890-1911, AM 1190/7; Jedrzejowski to Limanowski, letter dated September 11, 1901, Kopialy Korespondencji Komitetu Zagranicznego PPS, AM 699/2, Vol. IV, No. 347; AM 699/4, Vol. XII, No. 235.) There was still considerable demand for his earlier

writings, like the biographical sketch of General Wroblewski in *Zywot Generala Jaroslawa Dabrowskiego* (1878), by W. Rozalowski, and Limanowski's *Socjalizm jako konieczny objaw dziejowego rozwoju* (1879). See Ksiegarnia Polska to J. Kaniowski [B. A. Jedrzejowski], September 26, 1899, ZHP, PPS, Korespondencja i rachunki wplywajace do Redakcji *Swiatla*, AM 1050/10, No. 33.

9. K. Dunin-Wasowicz, "Promien i Promienisci," *Plomienie*, III (October, 1947), 20–23. The organ of the *"Promienisci"* was the *Ray (Promien)*, founded back in February, 1899, by the Union of Polish Student Societies Abroad.

10. "Boleslaw Limanowski," *Swiatlo*, III, No. 9 (1900), pp. 2-6. In the *Workers' Calendar* appeared Limanowski's photograph, a biographical article, as well as a poem about Limanowski by the poet Wladyslaw Orkan, of which one stanza may be translated as follows:

> Our hearts beat faster,
> When we meet those
> Who left the battlefields in time,
> Before they became cemeteries,
> To "march ahead with the living,"
> Toward the next field of battle. . . .

See "Boleslaw Limanowskiemu," *Kalendarz robotniczy na rok 1901*, p. 5.

11. According to the liberal Positivist Warsaw school of historiography, the partitioning powers bore the prime responsibility for the fall of Poland, in effect interrupting her political, social and moral regeneration which took place in the period of the so-called Polish Enlightenment. This "optimistic" view was partly grounded in the republican conceptions of Lelewel; it saw in the "pessimism" of the Cracow school kinship with the hostile historiography of the partitioning powers and a reflection of loyalist politics which were both monarchical and anti-democratic. (See M. H. Serejski, ed., *Historycy o historii*, Warsaw, 1963, pp. 135-36.) Limanowski agreed, in principle, with the Positivist Warsaw school and its "objectivist" approach to Polish history. While favoring idealization of Polish history as a means of renewing sagging spirits and enhancing national pride, he feared at the same time that it could act as a sanction of social injustice in the past (and present). Actually, his approach to history was a dual one—Positivist and Romantic. He considered history a science (aimed at enunciation of general laws), as well as a literary discipline (aimed at recreation of the past both vividly and faithfully); and he felt that historians should teach the young public morality by drawing biographical sketches of distinguished men and

women on the model of the *Lives* of Plutarch. See his Rev. of *Literatura Polska. Kurs historii wiekow srednich* (1872), by T. Korzon, *Gazeta Narodowa*, April 9, 1873, p. 1. Limanowski, "Z dziedziny nauki o spoleczenstwie," *Niwa*, IX (1876), 674; Limanowski, Rev. of *Historia starozytna* (1876), by T. Korzon, *Szkice spoleczne i literackie*, II (April 8, 1876), 183-84; K., "Boleslaw Limanowski," *Tydzien, dodatek do Kurjera Lwowskiego*, VIII, No. 35 (1900), pp. 273-74.

12. Minutes by police in Cracow, dated August 22, 1901, APKr., Czesciowe sprawozdanie z dzialalnosci PPS w liscie T. T. Jeza, St. GKr. 256, Komitet Narodowy, No. 664; Maria Wyslouch to Limanowski, letter dated November 25, 1900, BJ, MS 6872, fol. 197; Limanowski to Wyslouch, letter dated October 23, 1900, Ossol., Papiery Wyslouchow, Vol. V, MS 7179, fol. 483.

13. Limanowski to Wyslouch, letter dated July 22, 1901; also see letters dated February 18 and June 27, 1900, Ossol., Papiery Wyslouchow, Vol. V, MS 7179, fols. 487-88, 477, 479-81; *Pam.*, II, 495-96; 499, 544-46, 564; Jedrzejowski to Limanowski, letters dated January 20, August 29 and September 9, 1901, ZHP, APPS, Kopialy . . . AM 699/2, Vol. V, No. 105 and AM 699/2, Vol. IV, Nos. 314 and 347.

14. Jedrzejowski to Limanowski, letters dated April 30 and November 16, 1902, ZHP, APPS, Kopialy . . . AM 699/2, Vol. VI, No. 31; AM 699/3, Vol. VII, No. 467. Limanowski to Wilhelm Feldman, letter dated September 1, 1902, Korespondencja W. Feldmana, Ossol., MS 12, 280, fol. 535; Limanowski to Jedrzejowski, letter dated February 1, 1903, ZHP, APPS, Listy . . . AM 1190/7, No. 78.

15. In 1922 Limanowski extended his *History of Polish Democracy* (which ended in 1848) to cover the insurrection of 1863. This edition was reissued twice after the Second World War, in 1946 and in 1957. Contemporary Polish historians—both at home and abroad—still value this pioneering work, due to the wealth of factual material which it contains. See Lidia and Adam Ciolkosz, *Zarys dziejow socjalizmu polskiego* (London, 1966), I, 501. Cf. H. Jablonski, "Wstep," in *Pam.*, II, xi; *Pam.*, II, 469. A review, favorable to Limanowski, did eventually appear in *Forward*.

16. Jedrzejowski to Limanowski, letter dated October 21, 1901, ZHP, APPS, Kopialy . . . AM 699/2, Vol. IV, No. 441; Limanowski to Jedrzejowski, letter dated November 2, 1901, ZHP, APPS, Listy. . . AM 1190/7, No. 55.

17. Limanowski to Jedrzejowski, letters dated April 26, 1902 and March 15, 1903, ZHP, APPS, Listy ... AM 1190/7, Nos. 65 and 82.

18. Limanowski, "Glos francuzki o przyszlosci narodu polskiego," [Rev. of "L'avenir de la nationalité polonaise," by Anatol Leroy-Beaulieu, in *Revue des revues*, No. 2 (1900)] , *Tydzien, dodatek do Kurjera Lwowskiego*, VIII, No. 8 (1900), p. 57.

19. Limanowski to Jedrzejowski, letters dated December 28, 1902 and March 10, 1903, ZHP, APPS, Listy ... AM 1190/7, Nos. 76 and 81; *Pam.*, II, 558.

20. Limanowski, "Nacjonalizm a socjalizm," *Kalendarz Robotniczy na rok 1903*, pp. 17-20; Limanowski to Boleslaw Wyslouch, letter dated May 23, 1904, Ossol., Papiery Wyslouchow, Vol. V, MS 7179, fols. 499-501.

21. Limanowski to Jedrzejowski, letters dated September 30 and October 28, 1903, and February 8, 1904, ZHP, APPS, Listy ... AM 1190/7, Nos. 91, 93 and 98.

22. A. Malinowski to Limanowski, letter dated December 18, 1903, ZHP, APPS, Kopialy ... AM 699/3, Vol. IX, Nos. 188-189; Limanowski to Malinowski, letter dated January 2, 1904, ZHP, APPS, Listy ... AM 1190/7, No. 96; Motion by Paris Branch dated January 21, 1904, ZHP, APPS, Okolniki OZ PPS, AM 1050/1, No. 4.

23. *Pam.*, II, 556-57; Z. Zygmuntowicz, "Boleslaw Limanowski w swietle akt autryjackich," *Niepodleglosc*, XI (1935), 135.

24. Boleslaw Limanowski to Lucjan Limanowski, letters dated September 12, 1900 and January 15, 1902, BN, MS 2898, fols. 257-58, 267-68.

25. Boleslaw Limanowski to Lucjan Limanowski, letter dated July 31, 1902, BN, MS 2898, fols. 273-74.

26. Limanowski to Foreign Committee, letter dated March 13, 1904, ZHP, APPS, Listy ... AM 1190/7, No. 100.

27. Limanowski, "Przyszlosc Polski w ludzie polskim," *Kalendarz Robotniczy na rok 1904*, pp. 68-70.

28. *Pam.*, II, 570-72.

29. Boleslaw Limanowski to Lucjan Limanowski, letter dated May 19, 1904, BN, MS 2898, fols. 291-92.

30. *Pam.*, II, 572-73, 582-83, 643. Since Debski rejected Limanowski's proposal, it is quite unlikely that he would have spoken about it to Pilsudski (whom Limanowski was to meet personally for the first time only in 1907). By the time Limanowski repeated it to

Jedrzejowski and Wasilewski in August, 1904, Pilsudski was already
in Japan, negotiating with the Japanese authorities. Thus, it would
appear that Pilsudski conceived the idea of an alliance with Japan
independently of Limanowski. Pilsudski seemed primarily concerned
with raising funds in Japan for the purpose of an anti-Russian insurrection at home, though the idea of a Polish legion formed on Japanese soil was also entertained by the leadership of the PPS (according
to testimony of Stanislaw Wojciechowski). Dmowski, however, had
arrived in Japan prior to Pilsudski, in mid-May 1904; he succeeded
in persuading the Japanese that an uprising in Poland at this time
was bound to be cut down by Russia aided by Germany, without
benefitting Japan itself. It is interesting to note that Pilsudski's trip
was financed by the Japanese General Staff, whereas Dmowski's
expenses were paid by a Polish industrialist.(See M. K. Dziewanowski, "The Polish Revolutionary Movement and Russia, 1904-7,"
Harvard Slavic Studies, IV, 1957, 383n. Cf. M. Kulakowski, ed.,
Roman Dmowski w swietle listow i wspomnien, London, 1968, I,
39.)All Pilsudski gained were some funds to buy Japanese ammunition, in exchange for intelligence regarding Russian troop movements. See S. Kieniewicz, *Historia Polski, 1795-1918* (Warsaw, 1968),
pp. 434-35.

31. *Pam.*, II, 573-76; J. Grabiec, in *Pamieci Wilhelma Feldmana*
(Cracow, 1922), p. 102; J. Myslinski, *Grupy polityczne Krolestwa
Polskiego w Zachodniej Galicji, 1895-1904* (Warsaw, 1967), pp. 242-
45, 251. Due to the outbreak of revolution in 1905 and the untimely death of Krauz himself in that same year, the "university" soon
ceased to function.

32. Myslinski, *op. cit.*, p. 248; *Pam.*, II, 587-88, 590-91; Boleslaw
Limanowski to Lucjan Limanowski, letter dated October 9, 1904,
BN, MS 2898, fol. 293. However, Limanowski's stay in Galicia was
marred by the memory of the recent suicide of his third son, Witold,
who shot himself on November 9, 1903. Witold, having lost his
mother when he was only four years old, became perhaps the least
emotionally stable of Limanowski's four boys. At the time of his
suicide he was a bank clerk in Lvov, about to be promoted. This
would have enabled him to complete his high school education. The
Wyslouchs arranged his burial. It was a truly traumatic experience
for Limanowski to learn of Witold's death, and now he was much
saddened by his first visit to his son's grave. See *Pam.*, II, 363, 569,
588-91; Maria Limanowska to Boleslaw Wyslouch, letter dated
December 2, 1903, Ossol., Papiery Wyslouchow, Vol. V, MS 7179,

fols. 463-65; Limanowski to Wyslouch, letter dated November 22, 1903, Ossol., ibid., fol. 489.

33. Limanowski, *Rozwoj przekonan demokratycznych w narodzie polskim* (Cracow, 1906); *Pam.*, II, 600-1; Ciolkosz, *op. cit.*, I, 501.

34. Limanowski, *Rozwoj*, pp. 38-39.

35. Ibid., pp. 30-32.

36. Ibid., pp. 38-39.

37. *Narod i panstwo; Studium socjologiczne* (Cracow, 1906). Like Limanowski's lectures in Zakopane, the book was published by the PPS's cooperative publishing firm "Book *(Ksiazka),*" founded by Jedrzejowski in Galicia in 1904. Malinowski to Limanowski, letter dated January 20, 1904, ZHP, APPS, Kopialy ... AM 699/3, Vol. IX, No. 227; W. Pobog-Malinowski, *Najnowsza historia polityczna Polski*, 2nd ed. (London, 1963), I, 613.

38. Limanowski, *Narod i panstwo*, pp. 94-95.

39. Ibid., pp. 96-99; Limanowski, Rev. of *Bau und Leben des Sozialen Körpers* (1875-78), by Albert Schäffle, *Ateneum*, 1882, III, 583-84.

40. Boleslaw Limanowski to Lucjan Limanowski, letter dated October 9, 1904, BN, MS 2898, fols. 293-94.

41. Boleslaw Limanowski to Lucjan Limanowski, letter dated January 2, 1905, BN, MS 2899, fols. 1-2.

42. W. Gumplowicz, *Kwestia polska a socjalizm* (Cracow, 1908), p. 27.

43. *Pam.*, II, 594-96; Boleslaw Limanowski to Lucjan Limanowski, letter dated April 24, 1905, BN, MS 2899, fols. 3-4.

44. In April 1904, several parties met in Geneva under the auspices of that ambiguous figure priest Gapon (i.e., the PPS, the Russian Socialist Revolutionary Party, and the socialist revolutionary Georgian, Armenian, Latvian, Finnish and Belorussian parties). Pilsudski represented the PPS at this conference, which resulted in two resolutions. The first called for a *sovereign* Constituent Assembly for both Finland and Poland, whereas the second envisioned cultural autonomy for the component nationalities in the remainder of the Russian Empire. Thus the Russian party pledged itself to a separate Constituent Assembly in Warsaw as part of its revolutionary strategy. This represented the first and the only instance when a Russian party expressed full solidarity with the aspirations of the PPS. See L. Plochocki [Leon Wasilewski] , *Rosyjskie partie polityczne*

i ich stosunek do sprawy polskiej (Cracow, 1905), p. 133; *Pam.*, II, 599; "Przemowienie Boleslawa Limanowskiego," *Kalendarz Robotniczy za rok 1905*, pp. 33-35.

45. Ibid., pp. 36-37; BJ, Fragment Papierow Boleslawa Limanowskiego, Add. 245/62.

46. Limanowski to Foreign Committee, letters dated February 12, April 16 and May 16, 1905, ZHP, APPS, Listy. . . AM 1190/7, Nos. 120, 125 and 129. On November 4, 1905, all charitable institutions aiding the revolutionaries in Russian Poland united to form the Union to Aid Victims of the Revolution which had similar functions to those of the "Red Cross." It provided all types of aid to revolutionaries and their families; for instance, food, clothing, money and legal aid. See Manifesto of the Foreign Union to Aid Political Victims, dated November 4, 1905, BJ, MS 6872, fol. 116; Circular of March 4, 1906, ZHP, APPS, Czerwony Krzyz—Druki ulotne, AM 1141/7, No. 12.

47. [H. Kamienski], *Wojna ludowa* przez X.Y.Z., Bendlikon, 1866; Jedrzejowski to Limanowski, letter dated October 23, 1905, ZHP. APPS, Kopialy. . . AM 699/4, Vol. XII, No. 119; *Pam.*, II, 607-8.

48. *Pam.*, II, 608.

49. Boleslaw Limanowski to Lucjan Limanowski, letters dated October 16, 1905, and February 18 and May 23, 1906, BN, MS 2899, fols. 10-11, 18-19, 21.

50. Limanowski, "Przebudowanie Rosji" [Rev. of *Russlands Wiederaufbau* (1906), by Alexander Ular], *Trybuna*, I (December 1, 1906), 26-27. The *Tribune* was an organ of the right-wing PPS, edited by T. Bobrowski. It appeared in Cracow during the revolutionary years of 1906-7.

51. *Pam.*, II, 611-12.

52. Boleslaw Limanowski to Lucjan Limanowski, letter dated October 16, 1905, BN, MS 2899, fols. 10-11.

53. Limanowski, "Mickiewicz jako publicysta," *Trybuna*, II (January 1, 1907), 7-9. Late in 1906, Wladyslaw Mickiewicz, the son of the poet, informed Limanowski that he had been awarded a pension by the Society to Honor and Feed Emigré Leaders (Stowarzyszenie "Czci i Chleba"), a fraternal organization founded in 1862 to aid distinguished Polish exiles. It has never received any official subsidies and depended entirely on voluntary contributions. Although the grant was only 250 francs a year, Limanowski was overjoyed

to receive it; he felt that it symbolized a certain measure of recognition for Polish socialism. See *Pam.*, II, 635-36; Boleslaw Limanowski to Lucjan Limanowski, letter dated January 16, 1907, BN, MS 2899, fol. 33; W. Mickiewicz, "Emigracja Polska, 1860-1890," BN, MS 5302, fols. 150-54. (A published version appeared in Cracow, in 1908.)

54. I. Daszynski, *Pamietniki* (Cracow, 1925-26), I, 227-28; *Pam.*, II, 603.

55. A. Prochnik, *Ku Polsce socjalistycznej* (Warsaw, 1936), pp. 23-25. Cf. J. Mulak, "Z dziejow mysli programowej PPS," *Przeglad Socjalistyczny*, No. 6 (1948), pp. 23-25; A. Zarnowska, *Geneza rozlamu w Polskiej Partii Socjalistycznej, 1904-6* (Warsaw, 1965), pp. 350-56, 469-72; *Pam.*, II, 632. Prior to the revolution, the PPS appealed mainly to intelligentsia (of gentry and middle class social origin and of Polish, Lithuanian, German and Jewish ethnic origin). In addition to its regular members, the PPS had sympathizers outside the party, organized in the student "Union *(Spojnia)*" in Paris, the *émigré* Union of Progressive Youth, and the "Radiant Ones" in Galicia. The revolution produced mass parties. Appealing to workers in heavy industry, the PPS by 1906 influenced—directly or indirectly—about 50% of workers in Russian Poland. The numerical strength of the PPS around the time of the split was 55,000; the SDKPiL then had around 30,000 members. Thus, in 1906 Polish socialists, including the Jewish Bund, totalled about 100,000. This compared favorably with the contemporary strength of the united French socialist party (52,000 in 1907). See Zarnowska, *Geneza*, pp. 457, 465n; Zarnowska, "Zasieg, wplyw i baza spoleczna PPS w przeddzien Rewolucji 1905 r.," *KH*, LXVII (1960), 367-85.

56. Boleslaw Limanowski to Lucjan Limanowski, letters dated October 8 and November 4, 1906, BN, MS 2899, fols. 28-30; Limanowski to Boleslaw Wyslouch, letter dated April 21, 1907, Ossol., Papiery Wyslouchow, Vol. V, MS 7179, fols. 513-16; Zygmunt Limanowski to Daszynski, letter dated October 17, 1906, ZHP, Listy do I. Daszynskiego, 1899-1930, MS 70/II/1, No. 71; *Pam.*, II, 636-37; Zygmuntowicz, *op. cit.*, pp. 136-37.

57. Following his return to Galicia, Limanowski had to renew his permit to stay each year, but he encountered no difficulty in doing so. (See Zygmuntowicz, *op. cit.*, p. 137; Limanowski to Daszynski, letter dated November 9, 1906, ZHP, Listy . . . MS 70/II/1, No. 72; Boleslaw Limanowski to Lucjan Limanowski, letter dated October 8, 1906, BN, MS 2899, fol. 29; Viceroy to Chief of Police in Cracow,

letter dated August 13, 1907, APKr., St. GKr. 4, No. 1303.) This document and others pertaining to Limanowski's activities in Galicia in the period from 1907 to 1910 are appended to *Pam.*, III, 603-18.

58. Limanowski to Wyslouch, letter dated April 21, 1907, Ossol., Papiery Wyslouchow, Vol. V, MS 7179, fols. 513-15; Helena Limanowska to Boleslaw Limanowski, letter dated April 3, 1907, MS 2896, fols. 16-17; *Pam.*, II, 652, also 638-39.

59. Limanowski handed Huysmans ten copies of an article, which was an abridgement of his treatise *Nationality and State*, for distribution to delegates to the forthcoming international socialist congress at Stuttgart. (See Limanowski, "Nation et Etat," *Courrier Européen*, November 1906.) The artifle received a favorable review from editors of this socialist periodical appearing in Paris. See Boleslaw Limanowski to Lucjan Limanowski, letter dated November 4, 1906, BN, MS 2899, fol. 31; *Pam.*, II, 653, 655-56.

60. *Pam.*, II, 655, 658-66; Boleslaw Limanowski to Lucjan Limanowski, letter dated June 9, 1907, BN, MS 2899, fols. 42-43.

61. *Pam.*, II, 661; Limanowski, "Spostrzezenia i uwagi z pobytu w Londynie," *Naprzod*, September 24, 1907, p. 1.

62. *Loc. cit.*

63. Limanowski, "Spostrzezenia," *Naprzod*, September 27, 1907, p. 1.

64. Limanowski, "Spostrzezenia," *Naprzod*, September 28, 1907, p. 1.

65. *Loc. cit.*

66. *Loc. cit.*

67. Limanowski, "Spostrzezenia," *Naprzod*, September 27, 1907, p. 1; September 28, 1907, p. 1.

68. Limanowski, "Spostrzezenia," *Naprzod*, October 1, 1907, p. 2.

69. *Pam.*, II, 667-72.

NOTES TO CHAPTER IX

1. Boleslaw Limanowski to Lucjan Limanowski, letter dated September 7, 1907, BN, MS 2899, fols. 46-47.

2. In the latter instance, under the auspices of the open "People's University" in memory of Adam Mickiewicz, founded in 1898. During the peak years of its activity (1901–2), this institution arranged 631 lectures for about 160,649 people. See *Dziesieciolecie Uniwersytetu Ludowego* im. *Adama Mickiewicza w Krakowie, 1899-1909* (Cracow, 1909), p. 7, cited in J. Miaso, *Uniwersytet dla Wszystkich: 1906-1913* (Warsaw, 1960), p. 71; Limanowski, *Pamietniki*, ed., J. Durko (Warsaw, 1958), II, 727-28.

3. Boleslaw Limanowski to Lucjan Limanowski, letter dated September 15, 1907, BN, MS 2899, fols. 48-49. *Forward* was the most important socialist paper in Poland before 1918, founded in 1892 as a semi-monthly. Surviving frequent confiscations and even steadily increasing the number of its subscribers, due to its appeal to workers as well as the tenacity of its editors, in 1900 *Forward* became a daily. By 1912, its subscribers numbered between 6000 and 7000. See J. Buszko, "Polskie czasopisma socjalistyczne w Galicji," *Zeszyty Prasoznawcze*, II (1961), 31-35; F. Perl and Z. Zaremba, eds., *Z dziejow prasy socjalistycznej w Polsce* (Warsaw, 1919), pp. 36-37.

4. "Kobiety i polityka," *Naprzod,* May 31, 1908, p. 5.

5. *Loc. cit.* Also see Limanowski's "Hasla dnia Kobiet," *Naprzod*, March 19, 1911, p. 1.

6. For instance, in *Forward* he translated speeches given in the third Duma (1907-12), as well as proclamations issued by the Russian Orthodox priest Petrov, who was a champion of the working classes in the Duma. (Petrov wrote a letter to the Russian Metropolitan Antonius, stating his political views, and was promptly expelled from the Church.) Also in *Forward* Limanowski summarized short stories by Maxim Gorky and reviewed a book on Polish socialism by a contemporary Russian author, A. L. Pogodin, *Glavnye techeniia pol'skoi politicheskoi mysli, 1863-1907* (St. Petersburg, 1907); Limanowski, "List Popa Pietrowa," *Naprzod*, February 1, 1908, No. 51; Limanowski, *Pamietniki*, ed., J. Durko (Warsaw, 1961), III, 37, 620 (referred to below as *Pam.*).

7. Limanowski, Rev. of *Historia ruchu rewolucyjnego w Rosji* (1908), by Feliks Kon, *Naprzod*, August 11, 1908, p. 2; August 12, 1908, p. 2.

8. *Loc. cit.* He was now convinced that when a revolution threatened Tsardom, the Russian government instituted reforms in Poland, to be rescinded as soon as the danger had passed. Thus, the Tsar took back concessions granted to Poland as a result of the Russo-Japanese War and the revolution of 1905, that is, schools in Polish and the right to form trade unions. Limanowski felt that Russia could not afford to liberalize its institutions in Poland, for every step toward democratization only enhanced Polish patriotic activity. Accordingly, every reform in Russian Poland from 1864 was but a temporary measure destined to become eventually a dead letter, as was the case with the rural self-government in this region. He was unconditionally opposed to Polish participation in the Russian Duma (1907-17), for he believed that it fostered servility toward the Tsarist

regime and, at the same time, Polish deputies in it were helpless to protect the Polish national interest. See Limanowski, *Studwudziestoletnia walka narodu polskiego o niepodleglosc* (Cracow, 1916), pp. 383-85, 439-40.

9. Limanowski, "Konwencja angielsko-rosyjska," *Naprzod*, October 20, 1907, p. 2.

10. No. 6 (August, 1908) in *Wybor Pism*, II, 114-66. See J. P. Nettl, *Rosa Luxemburg* (London, 1966), II, 848; Limanowski, Rev. of *Kwestia polska a socjalizm* (1908), by W. Gumplowicz, *Naprzod*, September 27, 1908, pp. 2-3.

11. Limanowski, "Spoleczne rozczlonkowanie narodow austryjackich," *Naprzod*, October 13, 1907, p. 2.

12. Limanowski, "Listy o wspolczesnej literaturze francuzkiej," *Przegkad Tygodniowy*, XX (1885), 72; W. Feldman, *Stronnictwa i programy polityczne w Galicji, 1846-1906* (Cracow, 1907), pp. 151-54; Limanowski, Rev. of *Przyczynek do kwestii rolnej w Krolestwie Polskim* (1908), by Piotr Gurkowski [W. Gumplowicz], *Naprzod*, October 16, 1908, p. 2 and October 17, 1908, p. 2. Limanowski may have been confirmed in his antipathy toward the peasant proprietor while revising that year a translation of Wilhelm Blos's Marxist history of the French Revolution *(Die Französische Revolution,* Stuttgart, 1889). He seems to have been suspicious of the vested peasant interest—both as an obstacle to political democratization and to the struggle for independence. Yet he recognized that it was the emancipated French peasantry who saved the French Revolution.

13. The authors of the preamble to the PPS peasant program of 1905 detected no capitalist concentration in Polish agriculture. They recognized that peasant prejudice impeded agitation in the village but, at the same time, they considered private property in land incompatible with socialism. Having classified the Polish peasant into four categories—the landless, dwarf holders, self-sufficient farmers, and rich farmers using hired labor—they hoped fully to appeal to the first two categories only. Part I of this program dealt with I) protection of labor: (a) labor legislation, (b) freedom of association, (c) social security for accident and old age; II) agrarian reforms: (a) nationalization of state and church lands as well as of private land where there was no direct heir, (b) nationalization of forests, lakes and rivers, (c) elimination of the "checkerboards *(szachownice),*" (d) state insurance of grain, (e) state aid to cooperatives, (f) state schools of agronomy; III) protection of rural population as a whole: (a) self-government, (b) state levy of taxes for rural schools, roads

and poor relief, (c) free medical and legal aid for peasants. See "Projekt Programu Rolnego," *Przedswit*, Nos. 6-8 (1905), pp. 248-252; A. Laski, "Poglady przedrozlamowej PPS w kwestii rolnej," *Z Pola Walki*, II, No. 4/8 (1959), pp. 915-16.

14. P. Gurkowski [W. Gumplowicz], *Przyczynek do kwestii rolnej w Krolestwie Polskim* (Cracow, 1908), pp. 118-19, cited by Laski, *op. cit.*, pp. 918-19; A. Wronski [W. Jodko], *Program rolny PPS* (Cracow, 1910), pp. 4-5.

15. Wronski, *op. cit.*, pp. 28-35; Limanowski, Rev. of *Przyczynek...*, by P. Gurkowski, *Naprzod*, October 16, 1908, p. 2.

16. Limanowski, Rev. of *Przyczynek...*, by P. Gurkowski, *Naprzod*, October 17, 1908, p. 2; Limanowski, "Dzieje socjalizmu w Stanach Zjednoczonych Polnocnej Ameryki" [Rev. of *History of Socialism in United States*, by M. Hillquit], *Naprzod*, September 13, 1908, No. 253; Limanowski, *Historia demokracji polskiej w epoce porozbiorowej*, 4th ed. (Warsaw, 1957), pp. 44-45. Actually, the "Great Depression" had resulted in a revival of small peasant proprietorship in Western Europe. Due to the competition from the American and Russian wheat growers, large-scale cultivation—which was best suited for wheat—became less profitable there than small-scale, diversified farming. See W. E. Lunt, *History of England*, 4th ed. (New York, 1956), p. 767.

17. Pp. 9-10, 15-18, 36, 93-124, cited by Z. Zechowski, *Socjologia Boleslawa Limanowskiego* (Poznan, 1964), p. 119.

18. Limanowski, "Monizm w historii," *Krytyka*, No. 11 (1908), pp. 334-43. Feldman's *Criticism* was a very stimulating monthly, which appeared in Cracow from 1899 to 1914 and was sympathetic to the right-wing PPS. See *Zarys historii prasy polskiej*, ed. by B. Krzywoblocka and A. Slisz, Part II, No. 2 (Warsaw, 1959), p. 7; J. Wilhelmi, "Wstep," in W. Suchodolski, *Krytyka, 1899-1914: Bibliografia zawartosci* (Wroclaw, 1953), pp. 5-8.

19. Limanowski, "Monizm w historii," p. 336.

20. Ibid., pp. 336-39; J. B. Bury, *The Idea of Progress* (1932, rpt. New York, 1955), p. 310. Limanowski was unaware of an inconsistency in his thinking. While attacking his Marxist opponents, he professed the ideal of a Positivistic history. However, when stating his own views, he argued from an idealistic premise; this was quite compatible with his assuming the primacy of the economic "base" in primitive society. As mentioned above, he espoused two ideals of history— the Positivistic and the Romantic. He considered the first a scientific discipline aimed at enumerating general laws, and to the second he

attributed the role of an idealistic moralizer. See Limanowski, "Z dziedziny socjologicznej," *Niwa*, IX (1876), 674.

21. Limanowski, "Narod, panstwo i miedzynarodowosc," *Krytyka*, No. 4 (1908), p. 339. See also W. Najdus, "Ruch robotniczy w Galicja w latach 1890-1900," *PH*, LIII (1962), 112-13; S. Kieniewicz, *Historia Polski, 1795-1918* (Warsaw, 1968), pp. 431-32.

22. Limanowski, "Narod, panstwo i miedzynarodowosc," pp. 335-36; Limanowski, "Ze wspomnien o Wilhelmie Feldmanie," *Pamieci Wilhelma Feldmana* (Cracow, 1922), pp. 138-40.

23. Limanowski considered the Brno resolutions the most important pronouncement on "the liberation of nationality from its bondage to the [alien] state" since the Prague Slav Congress's Manifesto of 1848. See Limanowski, *Stanislaw Worcell* (1910; rpt. Warsaw, 1948), p. 353; Limanowski, "Narod, panstwo i miedzynarodowosc," pp. 341-43; On the Brno Congress and further developments on the nationality issue in the Austrian party see *Pam.*, II, 521; Najdus, "Ruch robotniczy," pp. 112-13; R. Pipes, *The Formation of the Soviet Union* (Harvard University Press, 1954), pp. 24-27; Robert A. Kann, *The Multinational Empire* (New York, 1950), II, 155-57.

24. Limanowski, "Narod, panstwo i miedzynarodowosc," pp. 342-43.

25. Limanowski, *Stanislaw Worcell* (Cracow, [1910]). In contrast to Limanowski, Milkowski considered Worcell a true democrat of "the Mazzini and Lelewel type," rather than a socialist. See Milkowski to Limanowski, letter dated October 16, 1908, BN, MS 6870, fols. 7-8; *Pam.*, III, 373-74; Lidia i Adam Ciolkosz, *Zarys dziejow socjalizmu polskiego* (London, 1966), I, Preface and pp. 447n-448n; P. Brock, "The socialists of the Polish 'Great Emigration'," in Asa Briggs and John Saville, eds., *Essays in Labour History* (London, 1960), p. 146.

26. Limanowski, *Stanislaw Worcell* (1910; rpt. Warsaw, 1948), p. 423.

27. Ibid., p. 425.

28. Ibid., p. 414. For instance, in the early 1890s a group of Polish student socialists in Paris honored Worcell by assuming his name for their organization. And in 1905-6 a People's University was founded in Galicia in memory of Worcell.

29. The award given in memory of P. Barczewski—who was a conservative landowner and philanthropist from the Polish Ukraine—went to Jan Antoniewicz for his biography of the Polish patriot and painter, Grottger. See *Korespondencja Sekretariatu Generalnego*

Akademii Umiejetnosci, 1911/107, Archives of BPAN, MS PAU I 73. Cf. E. Haecker, "Boleslaw Limanowski," *Naprzod*, November 20, 1910, p. 3; H. Nowina, "Ksiazka Limanowskiego o Stanislawie Worcellu," *Przedswit*, No. 11 (1910), pp. 710-16; *Pam.*, III, 145-46.

30. Limanowski to J. N. Janowski, letter dated June 4, 1883, BN, Korespondencja J. N. Janowskiego, MS 4281, fol. 102; Janowski to Limanowski, letter dated June 8/9, 1883, BJ, MS 6870, fol. 3; Limanowski to Maria Wyslouch, MS 12,149, fols. 106-7; Limanowski, *Stanislaw Worcell*, pp. 7-9; Boleslaw Limanowski to Lucjan Limanowski, letters dated February 8 and May 9, 1907, MS 2899, fols. 34, 40-41.

31. Limanowski, *Stanislaw Worcell*, pp. 200-1. On the one hand, Limanowski considered Worcell's theory too idealistic and, on the other (as we have seen) he had criticized Krauz's materialism. Limanowski did not object to Krauz's "Law of Social Introspection" as such, but to Krauz's deriving it from the "economic relations in primitive societies." Krauz maintained that class society was not suddenly created by an ethnic conquest, but that it had evolved gradually. See Limanowski, "Monizm w historii," pp. 340-42. Cf. Zechowski, *op. cit.*, pp. 138-39.

32. Limanowski, *Stanislaw Worcell*, p. 194. However, Limanowski agreed with Lelewel that the privileged landed nobility was an alien element in Polish history. This conviction probably lay at the basis of his faith in democracy. (See Boleslaw Limanowski to Lucjan Limanowski, letter dated December 25, 1896, BN, MS 2898, fols. 245-46.) Usually Limanowski treated the idea of an ethnic conquest in Poland as axiomatic; yet, in some instances, he considered it an unproven hypothesis. See Limanowski, "Narod, panstwo i miedzynarodowosc," pp. 337-38. Cf. his *Historia demokracji polskiej*, I, 11-12.

33. Limanowski, "Z dziedziny nauki o spoleczenstwie," *Niwa*, IX (1876), 448-55, 655. This treatise was based *inter alia* on works of the following authors: *First Principles* (1862), by H. Spencer; *Physics and Politics* (1872), by W. Bagehot; *The Origin of Civilization and the Primitive Condition of Man* (1870), by Sir John Lubbock; *The Descent of Man* (1871), by Charles Darwin; *Vorlesungen über die Menschen-und Tierseele* (1863), by W. Wundt; *Du Système social et des Lois qui le régissent* (1848), by L. A. J. Quetelet. See also Limanowski, "Z dziedziny socjologicznej," *Przeglad Tygodniowy*, XXII (1887), 172.

34. Limanowski, "Demokracja w Polsce," Pamphlet in a Series *Latarnia*, II, No. 3 (Cracow, 1903), pp. 1-2; *Stanislaw Worcell*, pp. 356-57. Limanowski believed that progress hitherto had benefitted mainly the privileged classes. He argued that the status of both the urban and rural worker had actually declined since the sixteenth century. (Like the authors of the *Communist Manifesto*, he assumed that England was the classic example of the economic rise of the bourgeoisie in Europe.) Limanowski's analysis of working class conditions in England was based on Engels's account in *Die Lage der arbeitenden Klassen in England* (Leipzig, 1846). See Limanowski, Rev. of *Socjalizm jako objaw choroby spolecznej* (1878), by Dr. Juliusz Au, *Tydzien*, No. 56 (1878), pp. 316-17.

35. George Woodcock, *Anarchism* (1962; rpt. Cleveland and New York, 1967), p. 25.

36. Limanowski, *Stanislaw Worcell*, pp. 358-60.

37. As an instance of the kind of works Limanowski studied in the history of Western Europe, we might point to Augustin Thierry's *Histoire de la conquête de l'Angleterre par les Normands* (1825). (See his *Narod i panstwo*, Cracow, 1906, p. 89.) According to the Polish sociologist J. S. Bystron, Limanowski may be considered as "one of the most interesting social theorists in Poland," who was the first European author to tackle sociologically the idea of antagonism between nationality and state. This idea was likely to occur only to a member of a subjugated nationality, like Limanowski. See J. S. Bystron, "Pojecie narodu w socjologii polskiej," *Rok Polski*, I (1916), 35-37; "Rozwoj problemu socjologicznego w nauce polskiej," *Archiwum Komisji do badania Historii Filozofii w Polsce*, I, Part II (1917), 199-200.

38. Limanowski, *Stanislaw Worcell*, pp. 343-44; *Narod i panstwo*, p. 82. In postulating an antagonism between nationality and state, Limanowski opposed the theory of social organicism, which seemed to preclude a revolutionary, voluntaristic action like struggle for independence and asserted an evolutionary development of state defined as "social organism." By stressing, instead, the "organicism" of the subjugated nationality in contrast to the "mechanism" of the alien state, Limanowski produced what seemed to be a unique concept of nationality. Though his theory was too schematic and had little scientific value—his distinction between nationality and state was due to an arbitrary juxtaposition of spontaneity and organization— Bystron considers that it represented perhaps the most empirical approach possible within the context of sociological organicism. As a

social theorist, writes Bystron, Limanowski was not only a pioneer among Poles, but he was also an independent thinker; for he was willing to sacrifice his ideological constructs to truth, as demonstrated by his gradual abandonment of sociological organicism, as well as by his implicit rejection of Comte's concept of sociology as a natural science of society. Because Limanowski had always, albeit incorrectly, identified nationality with society, from the beginning his moralistic patriotic idealism was incompatible with his adoption of Comte's concept of sociology governed by morally neutral laws, i.e., empirical generalizations. See Bystron, "Pojecie," pp. 34-39; "Rozwoj," pp. 200-2.

39. Limanowski, *Narod i panstwo*, pp. 83-87. According to the Polish-American sociologist Florian Znaniecki (*Modern Nationalities*, Urbana, Ill., 1952, p. xiii), "nationalism," derived from the word "nation," suggests solidarity of a people composing a nation. In English "nation" is usually defined as "a totality of citizens in a sovereign state." Due to the political connotation of "nation" and the ambiguity of this term (e.g., Welsh "nation"), some investigators of modern nationalism use the word "nationality" to define "a collectivity of people with certain common and distinctive cultural characteristics, sometimes racial traits, and a definite geographical location." Terms "nation" or "nationality" have been employed in this book to denote either the political or the cultural aspect of nationalism.

40. Limanowski, *Stanislaw Worcell*, pp. 345-56.

41. Ibid., p. 346; Limanowski, *Socjalizm jako konieczny objaw dziejowego rozwoju* (Lvov, 1879), p. 65.

42. Limanowski, *Socjologia* (1919; rpt. Cracow, 1921), II, 110-11; *Historia demokracji polskiej*, I, 11-12. In his *Narod i panstwo* Limanowski implied that Slavdom may have "patterned" itself on the Western model of state! (See pp. 84-89.) Apparently, he espoused Bagehot's idea of "unconscious imitation" as being an important factor in the development of political societies, at times confusing Lelewel's concept of a "cultural conquest" in Poland with Szajnocha's idea of an "ethnic conquest" of Slavs by the Normans.

43. Limanowski, *Socjologia*, II, 124-25; Res [Feliks Perl], *Dzieje ruchu socjalistycznego w zaborze rosyjskim do powstania PPS* (1910; rpt. Warsaw, 1958), p. 126; E. Haecker, "Boleslaw Limanowski," *Naprzod*, November 20, 1919, p. 2.

44. M. Charnay, "Benoît Malon, 23 VI 1841 - 13 IX 1893," *La Grande Encyclopedie*, n.d., XXII, 1068-69. Malon appears more

modern than Limanowski; in his definition of "integral socialism,"
Malon stressed the contemporary economic premises of socialism,
that is: "1) the new modes of production and exchange; 2) the role
of an industrial proletariat in transforming both the State and the
Commune, socialization of capital, and organization of labor." See
Benoît Malon, *Le socialisme integral* (Paris, [1890]), I, 17-18; Li-
manowski, "Socjalizm utopijny i naukowy," *Przeglad Wilenski*, I,
No. 27 (1912), p. 6.

45. Limanowski, *Stanislaw Worcell*, p. 144.

46. *Loc. cit.*

47. Limanowski, *Stanislaw Worcell*, pp. 145-49, 419; Limanowski, "Charakterystyka J. Supinskiego," in Jozef Supinski, *Szkola
polska gospodarstwa spolecznego* (Warsaw, 1911), p. 23. Limanowski had come to feel that, in addition to a *Weltanschauung*, each educated individual needed a religious belief. And religion ought not to
clash with one's "world-outlook," because philosophy and religion
shared the same "main truths." In Limanowski's opinion, a man who
was ready to sacrifice his self-interest for the great cause of "ennoble-
ment and liberation of man" had perforce strong religious feelings,
whether he belonged to a religious denomination or not. Accordingly, religion should disdain vulgar materialism and yet seek fulfillment
of man on earth. And he agreed with the disciple of Saint-Simon,
Enfantin, that the "dogma of the Fall" had been discredited; "faith
in Progress" should become the religion of the future. Notwithstanding his assignment of religion to the realm of emotions (faith and
will), and of philosophy and science to the realm of intellect, by the
beginning of the twentieth century Limanowski realistically departed
from his earlier and more rigid position when, writing in 1878, he
had placed science and religion in two completely disparate compart-
ments of human experience. He now assumed that for an educated
man religion formed but part of his *Weltanschauung*. See Limanowski, Rev. of *Religiöse Weltanschauung* (1903), by Albert Kalthoff,
Tydzien, dodatek do Kurjera Lwowskiego, XIII, Nos. 46 and 47
(1905), pp. 363-77. Cf. Limanowski, "Wiedza i wiara," *Tydzien*,
No. 44 (1878), pp. 113-14. Albert Kalthoff (1850-1906) was a distinguished Protestant theologian.

48. Limanowski to Feldman, letter dated February 1, 1903,
Korespondencja Wilhelma Feldmana, Ossol., MS 12,280, fol. 538.

49. Limanowski, *Narod i panstwo*, pp. 90-93; *Stanislaw Worcell*,
pp. 152-57; "Narod, panstwo i miedzynarodowosc," pp. 336-39.
Like the authors of the *Communist Manifesto* Limanowski considered France the classic example of the rise of bourgeoisie as a political
force in Europe.

50. Marx did not claim to have originated the idea of class struggle; he adapted to his purpose a theory created by French "middle class historians" to explain the French Revolution. In a letter to Engels he referred to Augustin Thierry as "the father of class struggle in French historical writing." (See his letter dated July 27, 1854, *Marx-Engels Correspondence, 1846-1895*, p. 71, cited in G. Sabine, *A History of Political Theory*, 3rd ed., New York, 1962, pp. 767-68.) Limanowski in producing his *Nationality and State (Narod i panstwo)* of 1906, which was a brief popular history of the French Revolution as well as a "sociological" treatise, had based himself on contemporary works of Jaurès and Aulard as well as on Augustin Thierry's famous book on the development of the "Third Estate" in France: Augustin Thierry, *Essai sur l'histoire de la formation et progrès du Tiers Etat*, 1853. (See Limanowski to Feldman, letter dated February 1, 1903, Ossol., Korespondencja Wilhelma Feldmana, MS 12,280, fols. 537-38.) As a follower of Thierry, Limanowski in a sense was a "middle class historian"; his theory did not explain the rift within the "Third Estate" and the rise of the "Fourth Estate." See also Haecker, "Boleslaw Limanowski," *Naprzod*, November 20, 1910, p. 2; Res, *Dzieje ruchu socjalistycznego*, pp. 125-26.

51. Ibid., pp. 126-27.

52. He was also popular with a dissenting group of newcomers to the National Democratic Party, led by the Germanophile publicist Studnicki-Gizbert. (By 1909 the latter had evolved ideologically from membership in the leftist PPS toward the right, *via* Wyslouch's Peasant Party.) See *Pam.*, III, 99, 626, also 49; Boleslaw Limanowski to Lucjan Limanowski, letters dated December 4, 1907 and January, 1908, BN, MS 2899, fols. 67-68, 74.

53. Boleslaw Limanowski to Lucjan Limanowski, letter dated July 1, 1908, BN, MS 2899, fols. 88-89; *Pam.*, III, 61-62.

54. Boleslaw Limanowski to Lucjan Limanowski, letter dated September 1, 1909, BN, MS 2899, fols. 121-22; *Pam.*, III, 119-21.

55. *Pam.*, III, 156-58, 162-64; Compte rendu des organisations socialistes polonaises, Copenhagen, 1910, ZHP, APPS, AM 1440/16, No. 2; Limanowski to Wyslouch, letter dated September 13, 1910, Ossol., Korespondencja Wyslouchow, Vol. V, MS 7179, fols. 543-46.

56. Boleslaw Limanowski to Lucjan Limanowski, letter dated November 4, 1910, BN, MS 2899, fols. 145-46; *Pam.*, III, 173; Archiwum Uniwersytetu Jagiellonskiego, MS 116/I-111, cited in J. Buszko, *Ruch socjalistyczny w Krakowie, 1890-1914, na tle ruchu robotniczego w Zachodniej Galicji* (Cracow, 1961), p. 271. For an

account by police of the student unrest in Cracow see Z. Zygmuntowicz, "Boleslaw Limanowski w swietle akt austriackich," *Niepodleglosc*, XI (1935), 137-38; Police Report, dated October 30, 1910,
APKr., St.GKr. 27, L.2726/pr; APKr., St.GKr. 4, passim. Documents
on Limanowski's Jubilee of 1910 had been appended to *Pam.*, III,
606-18.

57. Zygmuntowicz, *op. cit.*, pp. 138-39; Report by Police Official
Dr. Jurkowski, dated November 21, 1910, APKr., St.GKr. 4, re-
printed in *Pam.*, III, 617-18; "Jubileusz Boleslawa Limanowskiego,"
Naprzod, November 22, 1910, p. 1.

58. I. Daszynski, *Pamietniki* (Cracow, 1925-26), II, 66. Writing
these memoirs in independent Poland, Daszynski was to remark at
this point: "I am so happy that this wonderful old man is still alive
and active!" See ibid., pp. 66-67; *Pam.*, III, 176-191.

59. Boleslaw Limanowski to Lucjan Limanowski, letter dated
January 2, 1911, BN, MS 2899, fols. 149-50; "Jubileusz Boleslawa
Limanowskiego we Lwowie," *Zycie*, I (1910): No. 5, p. 73; No. 6,
pp. 84-85; No. 9, pp. 132, 141; No. 11, p. 164; G. Danilowski,
"Boleslaw Limanowski," *Zycie*, I, No. 7 (1910), 103-4.

60. Boleslaw Limanowski to Lucjan Limanowski, letter dated
December 1, 1910, BN, MS 2899, fols. 147-48; Haecker, "Boleslaw
Limanowski," *Naprzod*, November 21, 1910, p. 3; Res, "Boleslaw
Limanowski," *Przedswit*, XXX, No. 11 (1910), pp. 697-704; M.
Kukiel, "Limanowski jako historyk," *Zycie*, I, No. 7 (1910), 105-6;
W. F. [Wilhelm Feldman], "Boleslaw Limanowski," *Krytyka*, No.
11 (1910), p. 191.

61. Limanowski, *Powstanie narodowe 1863 i 1864 r.*, 2nd ed.
(Lvov, 1900), p. 19; A. Warski, "Jubileusz szlachetnego socjalisty
utopijnego," in his *Wybor pism i przemowien* (Warsaw, 1958), I,
469-75.

62. Ibid., pp. 475-77. Unlike the Bolsheviks, the PPS Militant Or-
ganization never robbed private institutions, but solely "confiscated"
the enemy government funds. See W. Rupniewski, ed., "Z archiwow
Ochrany—Sprawa Walerego Slawka," *Niepodleglosc*, II (1930), 357n.
Also see BN, Papiery Wyslouchow, Vol. XXIX, Mfm 30099, No. 353
(Ossol. MS 7203).

63. *Pam.*, III, 192-93, 218-19, 224, 228-29; Limanowski, "Krakowianom," *Pamiatka Majowa*, 1911, p. 2; Boleslaw Limanowski to
Lucjan Limanowski, letters dated February 1, March 1, April 3,
August 1, November 4 and December 2, 1911; February 1 and June
1, 1912, BN, MS 2899, fols. 151-56, 164-64, 170-73, 176-77, 184-
85; Limanowski, "Katorga: Przemowienie wygloszone w Salinach
Wielkich," *Naprzod*, July 26, 1911, pp. 1-2.

64. As an instance of this, we might mention a letter dated March 4, 1912 from a certain Georgian Toumanoff, advising Limanowski that he had been granted an honorary membership in an international association devoted to the organizing of an annual ceremony in memory of Herzen, held at the latter's monument in Nice. See *Pam.*, III, 238-39; Boleslaw Limanowski to Lucjan Limanowski, letters dated February 1 and August 11, 1912, BN, MS 2899, fols. 176-77, 188-89.

65. Limanowski, *Szermierze Wolnosci* (Cracow, 1911). This work is still considered valuable by historians. (See Ciolkosz, *op. cit.*, I, 501.) In 1891 Limanowski published three of his sketches in Wyslouch's *Week* (i.e., of Szaniecki, Zaliwski and Heltman); in 1893 he sent one (i.e., of Stolzman) to the Cracow *New Reform (Nowa Reforma)*; finally, the remaining previously published biographical articles on Emilia Plater, Worcell and Mieroslawski have already been mentioned above.

66. Limanowski, *Szermierze*, pp. 128-30.

67. Limanowski, *Historia ruchu rewolucyjnego w 1846 r.* (Cracow, 1913), Preface. This last major work by Limanowski was the first history dealing with events of 1846 in the whole of Poland. It was the most thoroughly researched of Limanowski's books on Polish history (according to a private interview this author had with Professor S. Kieniewicz). Limanowski utilized here much material hitherto untouched by historical research. See also Ciolkosz, *op. cit.*, I, 503.

68. H. Jablonski, *Polityka PPS w czasie wojny 1914-8* (Warsaw, 1958), pp. 51-53; *Pam.*, III, 252, 256-58; 637, No. 32; Boleslaw Limanowski to Lucjan Limanowski, letter dated September 9, 1912, BN, MS 2899, fols. 190-91; Kieniewicz, *Historia*, pp. 478-80; Daszynski, *Pamietniki*, II, 123.

69. Congress of Polish Irredentists, ZHP, Akta TKSSN, MS I/AAJ/I, cited in Jablonski, *Polityka*, pp. 52, 518.

70. *Pam.*, III, 265.

71. Boleslaw Limanowski to Lucjan Limanowski, letter dated December 4, 1912, BN, MS 2899, fols. 196-97; *Pam.*, II, 637-38, No. 42; Kieniewicz, *Historia*, p. 480; Daszynski, *Pamietniki*, II, 119-23.

72. *Pam.*, III, 287-89.

73. See E. Bobrowski, "Pamietniki, 1912-8," I, 1-6, 31-35, BN, Mfm 32371. Original typsecript at Ossol., MS 12004.)

74. *Pam.*, III, 267-68, 638, No. 46.

75. J. Buszko, *Spoleczno-polityczne oblicze Uniwersytetu Jagiellonskiego w dobie autonomicznej Galicji, 1869-1914* (Cracow, 1963), pp. 92-93; *Pam.*, III, 271-72; 275; Boleslaw Limanowski to Lucjan Limanowski, letters dated January 8 and February 7, 1913, BN, MS 2899, fols. 198-201.

76. Boleslaw Limanowski to Lucjan Limanowski, letters dated April 2, June 2 and July 6, 1914, BN, MS 2899, fols. 231-32, 235, 237-38.

77. *Pam.*, III, 323-24; Slawek do przedstawiciela Instytutu Jozefa Pilsudskiego, sten. H. Zaleskiej, AZHP, cited in Jablonski, *Polityka*, pp. 77-78, 523; Z. Zychowski, *Boleslaw Limanowski, 1835-1935* (Warsaw, 1971), pp. 327-28; Limanowski, "Pamietniki," BN, MS 6418/III, fols. 63-64; Hans Roos, *A History of Modern Poland*, trans. J. R. Foster (New York, 1966), pp. 17-18.

78. Roos, *op. cit.*, pp. 16-18; *Pam.*, III, 642-43, Nos. 6 and 17.

79. J. Maternicki, "Historia i historycy polscy w latach pierwszej wojny swiatowej," Diss., University of Warsaw, 1966, pp. 31, 102. *Inter alia*, he contributed to the conservative *Polish News (Wiadomosci Polskie)* and the right-wing PPS *Peasant Cause (Chlopska Sprawa)*. See also Limanowski, "Pamietniki," BN, MS 6418/III, fol. 93.

80. *Pam.*, III, 412-13, No. 11; Maternicki, *op. cit.*, pp. 175-76. Eventually, Limanowski published an article on the original *Legiony* in *Polish Culture (Kultura Polski)* which was the organ of the League for Polish Independence *(Liga Niezawislosci Polski)* formed in September 1917, by Polish populists, right-wing socialists and progressive democrats (PSL, PPSD, PPS and PSP). See *Pam.*, III, 658, No. 48. Like Mickiewicz, Limanowski held a mystic belief that "blood spilled in the cause of freedom is never wasted." He argued that, without Dabrowski's *Legiony*, there would have been no Duchy of Warsaw which provided a basis for an independent Polish nationhood. See Limanowski, "Z powodu rocznicy powstania Kosciuszkowego," *Chlopska Sprawa*, No. 7 (April, 1915), p. 3; "Jan Henryk Dabrowski, tworca pierszych legionow polskich," *Kultura Polski*, II (1918), 340-43.

81. Limanowski, *Studwudziestoletnia walka narodu polskiego o niepodleglosc* (Cracow, 1916), pp. 383-84, 431, 441; Maternicki, *op. cit.*, pp. 173-74; *Pam.*, III, 484-85. The occupation of Russian Poland by the Central Powers in the summer of 1915 facilitated circulation of Limanowski's books in this region; his works were now read by the lower classes, especially by peasants. See Limanowski, "Pamietniki," BN, MS 6814/V, fol. 29, *Pam.*, III, 559.

82. Maternicki, *op. cit.*, pp. 175-76; Limanowski, *Studwudzie-stoletnia walka*, pp. 349, 382; "Z powodu powstania Kosciuszko-wego," pp. 1-3; "Sprawa polska i wloscianie," *Wiadomosci Polskie*, No. 48 (1915), pp. 5-6.

83. *Dzieje Litwy* (Warsaw-Cracow, 1917), p. 68; H. Gierszynski to Limanowski, letter dated January 1, 1894, BJ, MS 6871, fols. 64-65.

84. Limanowski, *Dzieje Litwy*; "Lacznosc Polski i Litwy w wal-kach o wolnosc," *Wiadomosci Polskie*, No. 56 (1915), p. 3; "Odrod-zenie sie narodu litewskiego," [Rev. of *Litwa, Studium o odrodzeniu sie narodu litewskiego* (1908), by Michal Römer], *Naprzod*, Febru-ary 2, 1908, p. 3.

85. Limanowski, *Dzieje Litwy*, p. 66; *Studwudziestoletnia walka*, pp. 2-3; Maternicki, *op. cit.*, pp. 134-35.

86. Limanowski, *Stanislaw Worcell*, p. 64. Yet, back in 1893, in his brief account of the second partition, Limanowski had "justified" Russia's position to some extent. Russia, allied with one antagonist in the Polish civil war, demanded in payment for its services pro-vinces which were only in part ethnically Polish. But Prussia, the ally of the other antagonist in this civil war, and allegedly Russia's enemy, had turned traitor to its Polish ally and taken those provinces which constituted the "cradle" of the Polish nation; the loss of this terri-tory by Poland was far more detrimental than that of the eastern borderlands. See his *1793-1893: W stuletnia rocznice drugiego roz-bioru Polski* (Lvov, 1893), p. 16.

87. Limanowski, "Pamietniki," BN, MS 6418/III, fols. 46, 86.

88. *Pam.*, III, 529-30; Limanowski, "Passiv warten oder tätig ein-greifen?" *Polnische Blätter*, III (April, 1916), 88-94; "Ze wspomnien o Wilhelmie Feldmanie," *Pamieci Wilhelma Feldmana* (Cracow, 1922), pp. 139-40. In April 1916, the German Chancellor Bethmann-Hollweg—due to the pressure exerted on him by the Chief of Staff Falkenhayn, who desired Polish divisions in the struggle against Russia, as well as due to representation made by German trade associations—abandoned the old idea of annexing Russian Poland to Austria. In-stead, both governors of the Central Powers in Russian Poland, i.e., the German governor in Warsaw and the Austrian one in Lublin, on November 5, 1916, proclaimed on behalf of their respective monarchs the creation of an "independent" Polish state "linked to the two allied Powers." See Roos, *op. cit.*, pp. 20-21.

89. *Pam.*, I, 185; *Pam.*, III, 470; Limanowski, "Udzial Polakow w walkach wolnosciowych w dobie porozbiorowej," in F. Konieczny, ed., *Polska w kulturze powszechnej*, Part I (Cracow, 1918), pp. 88-89.

NOTES TO CHAPTER X

1. Hans Roos, *A History of Modern Poland*, trans. J. R. Foster (New York, 1966), pp. 64-65, 92, 96-97, 132-39; A. J. Groth, "Dmowski, Pilsudski and Ethnic Conflict in the Pre-1939 Poland," *Canadian Slavic Studies*, III (Spring, 1969), 69-91.

2. J. Grzybowski, "Parlament Polski, 1919-1939," *Swiat*, XVIII (October 13, 1968), 6-7.

3. Ibid., p. 6; Roos, *op. cit.*, pp. 100-1.

4. Roos, *op. cit.*, p. 110.

5. Ibid., pp. 115-16. For a recent study of Pilsudski's *coup* see Joseph Rothschild, *Pilsudski's Coup d'état* (New York, 1966).

6. Roos, *op. cit.*, pp. 138-41.

7. In 1918 Limanowski published his final treatise on nationalization of industry and socialization of agriculture, titled *The Nationalized State and Nationalization of Land (Panstwo unarodowione i unarodowienie ziemi*, Warsaw, 1918). In it he defined democratization of a state as reconciliation of an individual interest with the general interest, economically as well as politically. Valuing highly the formal aspects of democracy, e.g., universal suffrage and freedom of association, he argued that it was the latter which had enabled European socialists (for instance, in Belgium and Austria) to organize and to create workers' educational centers as well as cooperatives. Cooperatives, he maintained, had proved economically self-sufficient in these countries and had also financed other socialist activity. He presumed that both producers' cooperatives and the recent trend to concentration of capital were bound to lead to nationalization of industry. In addition, he felt (writing during the final months of the war) that "war socialism" was likely to become a means of nationalizing distribution in Europe. When Germany occupied Russian Poland in August 1915, she created both wholesale and retail government outlets for distributing food; this development, Limanowski assumed, had been a result of socialist propaganda and he hoped it might become a basis for a very gradual nationalization of both retail and wholesale trade in Poland, beginning with food and vital commodities like coal. See ibid., pp. 7-15.

8. Ibid., pp. 13-14, 19.

9. Ibid., pp. 20-24.

10. Ibid., pp. 21-25. The very moderate agrarian Reform Bill which passed in the Sejm on July 15, 1920—while the Bolsheviks threatened Warsaw—envisaged a gradual expropriation of large estates

for the benefit of the peasants. Half of the peasant families in the former Congress Poland and four-fifths in former Galicia then cultivated less than twelve acres. At the same time, the big estates constituted one-fifth of arable land in central and southern Poland, and in the west and east as much as one-third of such land. The main aims of the Agrarian Bill of July 15, 1920 were: 1) expropriation of estates of over 450 acres (with compensation), to be divided into allotments; 2) consolidation of scattered strips; and 3) improvement of the land. On December 28, 1925, the Sejm ratified a second Bill which remained in force, with minor modifications, up to World War II. Its provisions were less advantagous to the poorer peasants than those of the previous Bill had been. (Thirty-five senators, including Limanowski, opposed the amended Bill, and 60 senators favored it.) By 1938 one-tenth of Poland's arable land was distributed by the government among peasants. However, the amount of land available was too small to satisfy peasants' land hunger without extensive industrialization, which interwar Polish politicians considered impossible, due to the poverty of the country. See Roos, *op. cit.*, pp. 105-7; M. Zychowski, *Boleslaw Limanowski, 1835-1935* (Warsaw, 1971), pp. 383-84.

11. Limanowski, *Panstwo unarodowione*, pp. 25-26.

12. Ibid., pp. 26-27.

13. Ibid., pp. 27-28, 30-32.

14. Z. Zechowski, *Socjologia Boleslawa Limanowskiego* (Poznan, 1964), pp. 153-54.

15. Limanowski, "Jedno z najwazniejszych zadan pracy socjalistycznej u nas," *Trybuna*, III, No. 2 (January 15, 1921), pp. 38-39.

16. Limanowski, "Pamietniki, 1919-1928," p. 670 (unpublished diary in typescript, BN, MS 6418; referred to below as "Pam".), p. 17 in the published version *Pamietniki (1919-1928)*, ed. J. Durko (Warsaw, 1973); referred to below as *Pam.*

17. Ibid., p. 698.

18. Ibid., p. 701, *Pam.*, p. 45.

19. Ibid., pp. 703-5, *Pam.*, p. 48. On August 5, 1921, Limanowski was decorated with the Order Polonia Restituta, Class III "for serving the Polish Commonwealth as a civilian in the period of struggles for [Polish] independence." (See ibid., p. 709, *Pam.*, p. 52; Private Papers of Limanowski, 1885-1923, BN, MS 2905, p. 4.) Later, in 1923, Limanowski was assigned the military rank of "private" for the purpose of receiving a government pension. See Polish Ministry of Defense to Limanowski, letter dated April 12, 1923, BN, Private Papers of Limanowski, 1885-1923, MS 2905.

20. *Bolszewickie panstwo w swietle nauki* (Warsaw, 1921).

21. Ibid., pp. 20-21; Roos, *op. cit.*, pp. 72-73. Between 1918 and 1920 Limanowski published five biographical sketches of prominent Polish patriots. He entitled the series "Polish Plutarch." Three of them dealt with Kosciuszko, Staszic and Kollataj, the liberal reformers in the era of Polish Enlightenment. Limanowski argued in the first of these sketches that the government of Moraczewski had in fact implemented Kosciuszko's political program by formally granting equal civil rights to all adult citizens in Poland. In the second pamphlet he presented Staszic (a forerunner of Warsaw Positivism) as a scientist, pioneer of industrialization, philanthropist, social philosopher and social reformer. Staszic, the son of a mayor of the town of Pila in western Poland (which the Prussians renamed Scheidemühl), had founded a model cooperative rural community on his Hrubieszow estate, and an Employment Exchange to aid the city poor in Warsaw. And now Limanowski advocated that these institutions of Staszic be revived to serve as an inspiration to the Ministry of Labor and Social Security in independent Poland. In the last of the three pamphlets, Limanowski—mindful of the Russian revolutions in 1917 and using the words of Kollataj—argued that the oppressed must be liberated before they became aware of being exploited: "No human being is as terrible as an enlightened slave. . . a man, who is the offspring of oppression and captivity and who becomes overwhelmed by the passion of revenge, is surely the most predatory and deadliest creature imaginable." See *Tadeusz Kosciuszko* (Warsaw, [1920]), pp. 98-99; *Stanislaw Staszic* (Warsaw [1920]), pp. 39, 96-97; Hugo Kollataj, Listy do Stanislawa Malachowskiego, III, 184, cited in Limanowski, *Hugo Kollataj* (Warsaw, [1920]), pp. 34-35.

22. Limanowski, *Bolszewickie panstwo*, p. 21.

23. Ibid., pp. 21-22.

24. The gathering was held in Warsaw on May 7-8, 1922. (It included a representative of British Labour Party youth.) For Limanowski's address see "Przemowienie powitalne honorowego przewodniczacego Zjazdu," *I Zjazd Niezaleznych Socjalistycznych Zwiazkow Akademickich: Deklaracja Ideowa* (Warsaw, 1922), pp. 1-7.

25. The outcome of the November elections of 1922, showing the more conservative composition of the Senate as compared to the Sejm, in approximate percentages was as follows:

	Right	Center	Left	Minorities
Sejm	28%	29.9%	22.1%	20%
Senate	36%	24.4%	13.4%	24.3%

(See H. Swoboda [Adam Prochnik], *Pierwsze pietnastolecie Polski niepodleglej, 1918-1933: Zarys dziejow politycznych*, Warsaw, 1933, p. 114, 123. Cf. A. J. Groth, "Polish Elections, 1919-1928," *Slavic Review*, XXIV, 1965, 653-665.) In 1922, the principal Polish political groupings in parliament were as follows: 1) the Right included the National Democratic Party; 2) the Center was composed of the "Piast" peasant party, the National Workers' Party, the Polish Center and the Citizens' Center; and 3) the Left consisted of the PPS, the peasant "Liberation" party and several other lesser leftist populist parties, as well as the Communist Party. See ibid., p. 654.

26. "Boleslaw Limanowski," *Robotnik*, November 8, 1922, pp. 1-2.

27. Speech of November 28, 1922, *Sprawozdania stenograficzne z posiedzen Senatu Rzeczpospolitej Polskiej*, 1922-27, 1/3-1/5; *Pam.*, pp. 221-23.

28. H. Jablonski, "Wstep," in Limanowski, *Pamietniki*, ed. J. Durko (Warsaw, 1958), II, xii.

29. Swoboda, *op. cit.*, p. 130. Limanowski did not report in his diary what must have been a very painful incident in his life.

30. Limanowski, "W sprawie polityki narodowosciowej," *Robotnik*, May 16, 1924, in Appendix No. 2 to "Pam.," pp. 1-5, *Pam.*, pp. 228-31. Moreover, Limanowski became much concerned with the welfare of political prisoners in interwar Poland, irrespective of their political orientation or national origins. See for instance Zychowski, *op. cit.*, pp. 384-85.

31. Speech of July 25, 1924, *Sprawozdania stenograficzne* ... 1922-27, LXVIII/85-91; S. Kieniewicz, *Historia Polski, 1795-1918* (Warsaw, 1968), pp. 473-74.

32. *Centralizm i federalizm, panstwo narodowe i panstwo narodo-wosciowe* (Lvov, 1926).

33. Ibid., p. 6.

34. "Akademia ku czci tow. senatora Boleslawa Limanowskiego," *Robotnik*, December 9, 1924, pp. 1-2; "Pam.", December 8, 1924, p. 785, *Pam.*, p. 121.

35. "Pam.", pp. 808-11, *Pam.*, pp. 143-45. *Inter alia*, in February 1926, he wrote an article about a distinguished PPS female activist, Maria Paszkowska, for the *Voice of Women (Glos Kobiet)*. In the hospital, Limanowski read a great deal, including Shaw's *Saint Joan*. Around this time, he was approached by a Czech admirer, Jaroslav Cechacek, for permission to translate into Czech his *History of Social Movements*, as well as *Sociology* (composed between 1888

and 1900, but published for the first time only in independent Poland, in two editions of 1919 and 1921). Cechacek had asked Limanowski to provide a foreword and a photograph toward the translation of the former study. At the XX congress of the PPS, held on December 31, 1925, the Czech delegate and Limanowski's admirer of long standing, Frantisek Soukup, referred to him, perhaps somewhat exaggeratedly, as "the teacher of socialism among the Czechs." See ibid., p. 811, *Pam.*, p. 145.

36. Jablonski, "Wstep," in Limanowski, *Pamietniki*, II, x-xi.

37. L. Krzywicki, *Wspomnienia* (Warsaw, 1959), III, 278; J. Wyka, "Koloryt socjalny jednego wieku" [Rev. of Limanowski, *Pamietniki*, III], *Tworczosc*, XVIII (1962), 154.

38. Limanowski, *Pamietniki*, ed. J. Durko (Warsaw, 1961), III, 389.

39. Ibid., p. 401.

40. Ibid., p. 411.

41. Ibid., p. 507.

42. Ibid., p. 434.

43. Limanowski, "O charakterze kultury polskiej," *Kultura Polski*, I (1917), pp. 5-7; Limanowski, *Pamietniki*, III, 465.

44. "Pam.", p. 818, *Pam.*, p. 152.

45. "Pam.", p. 819, *Pam.*, p. 152. Limanowski—like the parliamentary club of the PPS—had voted for Pilsudski as a candidate for the Presidency. However, when the latter resigned in favor of Moscicki, Limanowski—again like the club—decided that the Sejm had been insulted.

46. "Pam.", pp. 821, 832, *Pam.*, pp. 154, 164.

47. Roos, *op. cit.*, p. 117.

48. "Pam.", pp. 850, 853, *Pam.*, 180, 183. See article by Nowaczynski in the Warsaw *Gazeta Warszawska Poranna*, July 23, 1927.

49. "Pam.", pp. 865-66, *Pam.*, p. 195.

50. "Pam.", pp. 867-69, *Pam.*, pp. 196-97.

51. Limanowski, "W sprawie krzywdzenia bratnich slowianskich narodow w naszej Rzplitej," *Robotnik*, March 4, 1928, p. 3.

52. Limanowski, "Boleslaw Limanowski wzywa cala Polske pracujaca by 4 i 11 marca glosowali na listy PPS (2)," *Robotnik*, March 2, 1928, p. 1; "Pam.", pp. 869-70, *Pam.*, pp. 197-98.

53. "Pam.", p. 871, *Pam.*, pp. 198-99. Pilsudski's *coup* possibly interrupted an evolution toward a well-functioning two-party parliamentary democracy in Poland. (See Grzybowski, *op. cit.*, p. 7.) The

PPS might well have emerged as the principal opposition in a potential two-party system. Between 1919 and 1928 the number of supporters of the PPS in elections to the Sejm rose from 515,062 to 1,481,279. (See Groth, "Polish Elections," p. 656.) According to Prochnik, the outcome of the elections in approximate percentages was as follows:

	BBWR (Government Bloc)	Right	Center	Left	Minorities
Sejm	34%	9%	12%	25%	20%
Senate	36.9%	9%	10.8%	18%	25.3%

See Swoboda, *op. cit.*, p. 252. Cf. Groth, "Polish Elections," pp. 655-58.

54. "Pam.", p. 872, *Pam.*, pp. 199-200; Swoboda, *op. cit.*, pp. 256-57.

55. Jablonski, "Wstep," in Limanowski, *Pamietniki*, II, xiii.

56. "Pam.", 872, *Pam.*, pp. 199-200. Swoboda, *op. cit.*, p. 255. The victory of Daszynski represented the last free election in interwar Poland. (See Grzybowski, *op. cit.*, p. 7.) Due to the inadequacy of the land reform, the radical peasant "Liberation" party gained ground. The whole left, led by the PPS, rallied together to defeat Pilsudski's candidate for the Speaker, Bartel. See Roos, *op. cit.*, pp. 117-18.

57. "Pam.", pp. 873, 877, 880, 882; *Pam.*, pp. 203-4, 205-6; H. Moscicki and W. Dzwonkowski, ed., *Parliament Rzeczpospolitej Polskiej, 1919-1927* (Warsaw, 1928), pp. 284-85.

58. "Pam.", pp. 885-86, *Pam.*, p. 211. Before recording the daily happenings in his life, Limanowski assembled both written and printed documents for the given year, i.e., letters, notes and the like, in order to enhance the accuracy of his account.

59. Limanowski, *Rozwoj polskiej mysli socjalistycznej* (Warsaw, 1929), pp. 19-21.

60. Ibid., p. 21; Grzybowski, *op. cit.*, p. 7.

61. "Pam.", p. 890, *Pam.*, p. 215; Jablonski, "Wstep," in Limanowski, *Pamietniki*, II, xii; Swoboda, *op. cit.*, pp. 263-64.

62. "Pam.", p. 890, *Pam.*, p. 215.

63. Ibid, pp. 891-92, *Pam.*, pp. 216-17.

64. Limanowski to Emil Haecker, letter dated December 16, 1928, AZHP, Papiery Emila i Franciszki Haecker, AM 692; Zychowski, *op. cit.*, pp. 402-6, 431-32.

65. Boleslaw Limanowski to Ignacy Moscicki, letter dated August 17, 1929, CKW PPS, Okolnik No. 12: "W sprawie konfiskaty listu

tow. Senatora Boleslawa Limanowskiego," No. 1845/1388; *Pam.*,
pp. 242-43; Roos, *op. cit.*, p. 118; Zychowski, *op. cit.*, p. 412.

66. Roos, *op. cit.*, p. 119. Actually, the conception of this Bloc
originated with the prominent ideologist of the PPS Mieczyslaw
Niedzialkowski, who was the principal author of the new party's
program of 1937. Niedzialkowski, the editor of the *Worker* since the
death of Perl in 1927, was executed by the Nazis during the Second
World War. See Adam Ciolkosz, *Trzy Wspomnienia* (London, 1946),
p. 34.

67. Roos, *op. cit.*, pp. 119-20.

68. Swoboda, *op. cit.*, pp. 314-16.

69. Ibid., pp. 317-18.

70. Pilsudski paid a very high price for his victory. The Brest-
Litovsk affair aroused indignation throughout the country and the
brutal suppression of the left provoked opposition even in the government camp. Pilsudski demoralized the administration of justice,
alienated the bulk of the nation, and hence had to rely on fellow
officers for support. See Roos, *op. cit.*, pp. 120-22; Zychowski, *op.
cit.*, p. 417; Letter to the author from Mrs. Zofia Eiblowa (Emil
Haecker's daughter), dated August 29, 1968.

71. Jablonski, "Wstep," in Limanowski, *Pamietniki*, II, xiii.

72. Pilsudski justified his *coup* of 1926 by alleging a need to
cleanse Polish political life; his rule was called "Sanacja," that is, a
regime whose purpose was to cleanse the nation morally.

73. Limanowski to Emil and Franciszka Haecker, letter dated
January 11, 1931, AZHP, Papiery Emila i Franciszki Haecker, AM
692. Almost two years later, in December, 1932, he again wrote—
this time to Emil alone—regretting his inability to attend the PPS
commemoration of the fortieth anniversary of the party, which was
to take place in Cracow. He assured Haecker that he was "united
wholeheartedly in spirit with those at the meeting." And, he went
on, "I hope that our socialist party will continue to develop for the
benefit of the entire working class. My best wishes go out to the
editorial board of *Forward*, and especially to comrade Emil Haecker,
in gratitude for his forty years of hard and sacrificial toil as its editor-
in-chief. I wish you all may live to see a more propitious era for the
independent Polish press." See Limanowski to Emil Haecker, letter
dated December 2, 1932, in ibid.

74. *Ksiega Jubileuszowa PPS, 1892-1932* (Warsaw, n.d.), p. 275.

75. Roos, *op. cit.*, pp. 122-24.

76. M. Zielenczyk, Rev. of Limanowski's *Pamietniki*, I, *Polityka*, I (August 14-20, 1957), 5; Jablonski, "Wstep," in Limanowski, *Pamietniki*, II, xiv; Zychowski, *op. cit.*, pp. 420-22.

77. *Sprawozdanie z dzialalnosci Wydzialu Humanistycznego Uniwersytetu Warszawskiego za rok 1934/5*, pp. 130-31.

78. "Boleslaw Limanowski," *Czlowiek w Polsce*, II (1934), 5.

79. M. Handelsman, "Boleslaw Limanowski jako badacz historii powszechnej," *PH*, XXXIII (1936), 331-32; St. Zakrzewski, "Boleslaw Limanowski—30 X 1835—1 II 1935," *KH*, XLIX (1935), 262; Zychowski, *op. cit.*, pp. 427-28.

80. Speech of February 27, 1935, *Sprawozdania stenograficzne z posiedzen Senatu Rzeczpospolitej Polskiej, 1931-1935*, col. 3; Jablonski, "Wstep," in Limanowski, *Pamietniki*, II, xiv. Limanowski survived his two younger sons. As mentioned above, Witold committed suicide in 1903. The youngest son Stanislaw, a medical doctor, died prematurely in 1927. Zygmunt, a professor of Administrative Statistics and Actuarial Science, died in 1943; finally, the oldest and most brilliant of Limanowski's four sons, Mieczyslaw, who was a distinguished geologist, died in 1948.

81. "Boleslaw Limanowski," *Mysl Narodowa*, XV (February 10, 1935), 95.

82. Zychowski, *op. cit.*, pp. 429-30. Limanowski had willed his savings and books to socialist youth organizations. (See ibid., p. 432.) In Cracow, the street renamed in honor of Limanowski is located appropriately in Podgorz, a working class district in this city. Podgorz coincidentally was also the name of his native estate in Livonia.

83. Roos, *op. cit.*, pp. 138-41, 144-45.

NOTES TO CHAPTER XI

1. "Czem moze sie stac Liga Narodow," *Trybuna*, II, No. 10 (March 13, 1920), p. 293.

2. *Odrodzenie i rozwoj narodowosci polskiej na Slasku* (1911; rpt. Warsaw, 1921), pp. 57-58.

3. *Mazowsze Pruskie* (Warsaw, 1925), p. 19.

4. *Panstwo unarodowione i unarodowienie ziemi* (Warsaw, 1918), pp. 17-18.

5. "Czem moze sie stac Liga Narodow," p. 294.

6. *Rzut oka na rozwoj polskiej mysli socjalistycznej* (Chicago, 1922), p. 14; "Konferencja pokojowa w Paryzu" [Rev. of *The Peace Conference* (1919), by E. J. Dillon], *Naprzod*, January 26, 1921, p. 3.

7. "Jedno z najwazniejszych zadan pracy socjalistycznej u nas," *Trybuna*, III, No. 2 (January 15, 1921), pp. 37-38.

8. *Mazowsze Pruskie*, pp. 5-6.

9. Ibid., p. 7.

10. "Czem moze sie stac Liga Narodow," pp. 291-92.

11. Ibid., pp. 292-93; "Konferencja pokojowa w Paryzu," p. 3.

12. "Pamietniki, 1919-1928," pp. 708-9 (Typescript of diary, BN, MS 6418, referred to below as "Pam."), pp. 51-52 in the published version: *Pamietniki (1919-1928)*, ed. J. Durko (Warsaw, 1973), referred to below as *Pam*.

13. Limanowski's comments on Latvia, however, apparently remained in manuscript. See his "W Podgorzu po wojnie po 43 latach," BN, Papiery Wyslouchow, Vol. XXIX, Mfm 30099, fols. 367-375. (Original at Ossol., MS 7203.) Cf. Limanowski, *Pamietniki*, ed. A. Prochnik (Warsaw, 1957), I, 225-26. Limanowski defined federalism as a natural gravitation of kindred peoples toward a common political organization, based on the principle of mutualism of rights and obligations. Federalism, he believed, had been the earliest principle expressed by man as a social animal. In modern times, the idealism of the French Revolution was becoming gradually embodied in political institutions. Hence liberalization of both individuals and peoples. And, Limanowski went on, with the continuous spiritual progress of each people, egoistic nationalism was being superseded by the federal principle (i.e., a voluntary association of nationalities for the sake of mutually guaranteed autonomous development). He assumed a gradual demise of religious as well as ethnic differences between the peoples of Europe. See Limanowski, *Centralizm i federalizm, panstwo narodowe i panstwo natodowosciowe* (Lvov, 1926), pp. 1-6; *Rozwoj polskiej mysli socjalistycznej* (Warsaw, 1929), pp. 27-28; *Panstwo unarodowione*, p. 17.

14. *Studwudziestoletnia walka narodu polskiego o niepodleglosc* (Cracow, 1916), pp. 400-1.

15. *Odrodzenie*, p. 18.

16. *Mazowsze Pruskie*, pp. 16-17.

17. *Polityka wynaradawiajaca rzadu pruskiego w stosunku do ludnosci polskiej* (Czestochowa, 1919).

18. Ibid., p. 16.

19. See Note 2. In this book Limanowski surveyed the history of Silesia as 1) a Polish land—up to the fourteenth century; 2) a fief of

Bohemia; 3) a Habsburg land; and 4) a Prussian land. He also historically traced the development in recent times of socialism and nationalism in both Austrian and German Silesias. In the 1921 edition,
he included an extra chapter titled "Silesia becomes part of Poland."
Limanowski explained why the Polish element had survived in Upper
Silesia, whereas this had not been the case in Lower Silesia; in the
latter, during the disastrous Thirty Years' War, Polish peasants had
either perished or been compelled to emigrate elsewhere. See *Odrodzenie*, pp. 16-17.

20. Ibid., pp. 17-18, 25-27.

21. Ibid, pp. 22-27, 38.

22. Ibid, pp. 38-49.

23. See for instance, Dr. Paul von Nieborowski, *Oberschlesien Po-
len und der Katholicismus* (Berlin, 1919), referred to by Limanowski
in his *Odrodzenie*, p. 64. See also ibid., pp. 61-64.

24. *Odrodzenie*, pp. 50-57.

25. Ibid., p. 61.

26. *Mazowsze Pruskie* (Warsaw, 1925), circulated in 3000 copies.
Limanowski had dedicated the pamphlet to "the courageous Polish
young people of Allenstein." See also "Pam.", pp. 747-79, 804,
Pam., pp. 86-88, 111-15, 139-40; *Mazowsze*, pp. 2-4.

27. Under the oppressive rule of the Teutonic Knights, many
Poles and Lithuanians perished. In West Prussia, too, intensive German colonization resulted in creating a predominantly German gentry and bourgeoisie. However, the Germans in the Prussias rebelled
against the hated regime of the Knights in 1454, forming the so-
called "Lizard Conspiracy," and swore allegiance to the Polish King
Kazimierz Jagiellonczyk. Thereupon, Kazimierz fought a lengthy
war with the Knights, which was terminated by the Peace of Torun
(Thorn) in 1466. Thereby, the King acquired West (i.e., Royal)
Prussia and Ermeland outright. East Prussia, however, became merely a fief of Poland. Thus the Mazurians, who were settled in the
southern parts of East Prussia, remained under Prussian rule. When
the Order became secularized in 1525, its former Grand Master,
Albert of Hohenzollern, swore fealty to the Polish King, Zygmunt
the Old. In 1618, one of the descendants of the former died child-
less; thereupon, the Polish King Zygmunt Waza III gave East Prussia
in vassalage to the Elector of Brandenburg, in exchange for the
latter's promise to aid him in his bid for the Swedish crown. Finally,
in 1687 Poland ceded all its rights in East Prussia in favor of the incumbent Hohenzollern of Brandenburg. In 1701, his descendant was

to crown himself King in Prussia. See *Mazowsze*, pp. 9-15; *Polityka*, p. 5.

28. *Mazowsze*, pp. 1-2; *Polityka*, p. 5.

29. *Mazowsze*, pp. 19-21, 24-25.

30. *Mazowsze*, pp. 9, 25-26; "Mazury. Z wrazen wakacyjnych," *Kalendarz Robotniczy PPS na rok 1924*, "Pam.", Appendix No. 1, p. 1 (*Pam.*, pp. 224-27); "W sprawie Mazurskiej," *Robotnik*, December 23, 1924, "Pam.," Appendix No. 3, p. 1 (*Pam.*, pp. 235-36). Like Dmowski, Limanowski favored maximum expansion of Poland in the west. East Prussia had been vital to Dmowski's plans as head of the Polish Delegation to the Paris Peace Conference. In order to give Poland access to the sea and to avoid the creation of a dangerous "corridor," at Versailles Dmowski demanded that those parts of East Prussia, which were not to be incorporated in the new Polish-Lithuanian state should be separated from Germany to form an independent republic. See Hans Roos, *A History of Modern Poland*, trans. J. R. Foster (New York, 1966), pp. 25, 48-49, 52-53.

31. "Mazury. Z wrazen," "Pam.", Appendix No. 1, pp. 1-5 (*Pam.*, pp. 224-27); *Mazowsze*, pp. 28-28.

32. *Mazowsze*, p. 23; "W sprawie Mazurskiej," "Pam.", Appendix No. 3, pp. 1-3 (*Pam.*, pp. 235-36); "Z Prus Wschodnich," *Robotnik*, July 10, 1925, "Pam.", Appendix No. 6, pp. 1-3 (*Pam.*, pp. 240-41); "Czy Prusy Wschodnie moga sie uwazac za kraj niemiecki?" *Robotnik*, January 1, 1925, "Pam.", Appendix No. 4, pp. 1-3 (*Pam.*, pp. 237-39).

33. "Pamietajcie o Mazurach," *Jedniodniowka PPS*, May 1, 1925, in "Pam.", Appendix No. 5 (*Pam.*, p. 239).

34. *Mazowsze*, p. 23.

35. Ibid., pp. 60-62.

36. Ibid., pp. 64-69. In 1922 an umbrella Polish Union came into being in Berlin; it united Polish organizations in four centers of Polish population in Germany: 1) east of the Vistula; 2) west of the Vistula; 3) Silesia; and 4) Westphalia. Limanowski argued that by 1923, in Marienwerder and Ermeland alone, the Union numbered around 4000 members.

37. Ibid., p. 67.

38. Limanowski believed that a minority might well secure its civil and political rights within a decentralized and democratic multi-national state. In the case of political trials, he insisted that the defendant be guaranteed due process of law, that is, proceedings in the

language of the accused, and that half of the jury consist of the latter's compatriots. He argued that the most sensitive area of minority rights was educational policy, advocating separate boards to administer religion and education for each religious or ethnic group in a multi-national state, to be financed from the overall state budget proportionally to the population numbers in a given ethnic or religious group. See *Panstwo unarodowione*, pp. 16-17.

39. *Mazowsze*, pp. 62-64. Limanowski had been impressed by S. Srokowski's book on East Prussia, *Z Krainy Czarnego Krzyza: Uwagi o Prusiech Wschodnich* (Poznan, 1925). In it the author alleged instances of economic separatism in East Prussia. Limanowski cited two such cases: 1) a certain von Hippel, who was vice-president of the Chamber of Commerce in Tilsit, allegedly had said in Koenigsberg in 1920 that good relations with Poland would be in the economic interest of the country; and 2) a certain Dr. Hans Zint, who was the chairman of a district court in Danzig, writing in a socialist paper *Die Glocke* around that time had advocated a "By-Vistula Federation," in order to reconcile the respective economic interests of Poland, Germany and the Free City of Danzig. See Limanowski, *Mazowsze*, pp. 69-70.

40. *Mazowsze*, p. 70.

41. *Socjologia* (1919; rpt. Cracow, 1921), II, 74-77.

42. Ibid., pp. 83-84.

43. Ludwig Gumplowicz (1838-1909) was a pioneering Polish sociologist of Jewish descent; he became one of the first European scholars to establish sociology as a social science in its own right. The founder of the "sociological" school of jurisprudence, from 1875 until his death he was professor of public law at the University of Graz, Austria. Gumplowicz defined sociology as a study of group interaction. He believed that group conflict—whether between nationalities or classes—provided the motive force of social evolution. He owed this thesis to Joseph A. de Gobineau (whose writings-as we shall see below—were at least in part familiar to Limanowski as well), and also to his observing the ethnic strife in Austria-Hungary. Like Limanowski, Gumplowicz was influenced to some extent by Comte and Spencer; he accepted only the general import of Darwin's theory, that is, the concept of struggle for existence. Limanowski had obviously been impressed by Gumplowicz's well-documented theory of the origins of the state as a by-product of an ethnic conquest—especially, his *Rasse und Staat* (1875)—which has gained wide assent in recent times. Nevertheless, some modern sociologists feel

that Gumplowicz underrated the element of consensus in the formation of some states. Limanowski, in postulating—like Lelewel—an alternate theory of the rise of the state as a federation of kindred tribes, formed in defense against an alien conquest, has remedied the onesidedness of Gumplowicz's theory. The latter proved unpopular in Poland, due to its implications and the fact that Gumplowicz wrote in German. See "Ludwig Gumplowicz," *International Encyclopedia of the Social Sciences*, ed., David L. Sills (1968), VI, 293-95. Cf. Z. Zechowski, *Socjologia Boleslawa Limanowskiego* (Poznan, 1964), pp. 49-50.

44. Limanowski, "Z dziedziny nauki o spoleczenstwie," *Niwa*, IX (1876), 451-54; *Historia demokracji polskiej w epoce prozbiorowej*, 4th ed. (Warsaw, 1957), I, 9-12 and II, 380-81; *Rozwoj przekonan demokratycznych w narodzie polskim* (Cracow, 1906), pp. 30-32; *Narod i panstwo* (Cracow, 1906), pp. 74, 89-93; "Narod, panstwo i miedzynarodowosc," *Krytyka*, No. 4 (1908), p. 337; *Panstwo unarodowione i unarodowienie ziemi* (Warsaw, 1918), pp. 1-3.

45. Limanowski, *Historia demokracji*, II, 380-81; *Socjologia*, II, 80-83. Unlike Gumplowicz, who believed that the European middle classes were formed from foreign stocks, Limanowski argued that usually they represented a mixture of the gentry and the common people (i.e., peasants), where the foreign element was numerically insignificant. See his *Socjologia*, II, 70-74. Cf. *International Encyclopedia*, VI, 293-95.

46. *Socjologia*, I, 251.

47. Ibid., II, 21-24, 49, 125-27. Limanowski viewed the study of sociology under three aspects: 1) he distinguished "social statics" from "social dynamics" which to him represented anatomy and physiology respectively as applied to social relations. 2) Moreover, he recognized a "social dynamics of the philosophy of history," which dealt with conscious process of human development, i.e., progress. Gumplowicz dismissed the latter as mere metaphysics. (See ibid., II, 48-49.) 3) Like Comte, Limanowski viewed history as a source of concrete data serving as a basis for making sociological generalizations. Yet he also defined history comprehensively as a "record of human life in all of its manifestations," and he endorsed Saint-Simon's concept of history as a "series of commentaries on the march of civilization." He appeared to believe that history aimed at synthesizing all of the social sciences. Thus, he unwittingly identified history with sociology as defined by Comte. See ibid., I, 255.

48. *Socjologia*, II, 78-79.

49. Ibid., II, 123-24.

50. "Spoleczna mysl religijna," *Naprzod*, April 20, 1924, p. 3.

51. *Loc. cit.* Cf. "Socjalizm utopijny i naukowy," *Przeglad Wilenski,* I, No. 27 (1912), p. 7; *Rozwoj polskiej mysli socjalistycznej*, pp. 9-10, 14; *Socjologia*, II, 124; *Rzut oka*, pp. 15-16.

52. *Bolszewickie panstwo w swietle nauki* (Warsaw, 1921), pp. 22-23, 9-13; Limanowski, *Rozwoj polskiej mysli*, pp. 8-9.

53. Limanowski, "Pan Majewski jako krytyk *Kapitalu* Marxa," *Przedswit*, Nos. 9-10 (Sept. and Oct. 1919), pp. 10-15. This is a comment on a review by a Polish writer of the third volume of Marx's *Capital*.

54. *Socjologia*, I, 205-6.

55. Ibid., I, 207.

56. Ibid., I, 208-14.

57. Ibid., II, 38-39, 42-43, 52.

58. Ibid., II, 5.

59. *Essai sur L'inégalité des races humaines* (Paris, 1853-55).

60. *Socjologia*, I, 202-4. In his political theory Limanowski stressed the concept of "race" more than Gumplowicz had done. The latter distinguished between class struggle and ethnic strife (and his "tribes" were historico-ethnic groups rather than Limanowski's sometimes "natural" racial entities). In contrast, Limanowski could not conceive of a social conflict which was not originally a struggle between divergent ethnic stocks. Thus, he was less modern and realistic as a sociologist than Gumplowicz had been. It is interesting to note how Limanowski and Gumplowicz, having started from the same premise of ethnic conquest, arrived at completely dissimilar philosophies of history. Gumplowicz had been a pessimist and a rigid determinist in his outlook. He defined class struggle as a blind, merciless force, and he rejected the idea of progress in favor of a "cyclical theory" of man's development. He believed that the state was a natural organism; and laws were but a means of subjecting a weaker group to rule by a stronger one. The idea of a "just" law seemed to him preposterous! History, according to Gumplowicz, was but the history of strife between interest and ethnic groups. He rejected social organicism, and he did not believe in the advent of a "welfare state," considering socialist ideology but a harmful illusion. Ideology represented to Gumplowicz nothing but cant, a mere rationalization of self-interest. And he dismissed patriotic history as a "poetic clap-trap," though as

a young man he allegedly fought in the 1863 uprising. Gumplowicz was too cynical, whereas Limanowski was too Utopian. See L. Gumplowicz, *Filozofia spoleczna*, trans. S. Posner, (Warsaw-Lvov, 1909), passim; "Socjologiczne pojmowanie historii," in M. H. Serejski, ed., *Historycy o historii*,(Warsaw, 1963),pp. 432-44; Limanowski, *Socjologia*, II, 146-47.

61. *Socjologia*, II, 40-41. I assume this was a product of Limanowski's revision in the years 1912 to 1914. The manuscript of *Socjologia* was completed in 1900, prior to *Stanislaw Worcell*. In the latter work published in 1910, he considered nationality to be more "spiritual" than the tribe had been. See *Stanislaw Worcell* (1910; rpt. Warsaw, 1948), pp. 343-48.

62. *Socjologia*, II, 70-74. Cf. Alexander J. Groth, "Dmowski, Pilsudski and Ethnic Conflict in Pre-1939 Poland," *Canadian Slavic Studies*, II (Spring, 1969), p. 74. But in his *History of Polish Democracy*, I, 11-17, first published in 1901, Limanowski offered a more sophisticated explanation of Poland's fall. In it he argues that in the long run it was Poland's eastern expansion which resulted in the partitions. It gave rise to a powerful nobility in the eastern lands and brought strife and anarchy to Poland. Gentry equality in Poland became more and more fictitious, while peasants were gradually transformed into serfs. Why did the king in Poland fail to check the nobility by aligning himself with the barons, as had been the case in Western Europe? Because, Limanowski answers, continuous wars with Muscovy, as well as with Sweden and Turkey, resulted in the preponderance in Poland of the warrior class. The acquisition of Lithuania and part of Livonia by Poland had averted from the fifteenth century on an imminent conflict between the landless gentry (i.e., the potential middle class) and the landed nobility, since the former were now settled on newly acquired and underpopulated lands. Meanwhile, the recovery of Danzig (Gdansk) by the Peace of Thorn (Torun) in 1466 had facilitated export of grain to Europe by the nobility. Faced with shortage of labor, because the peasants escaped to the new lands, the nobility tied them to the soil. Moreover, serfdom became aggravated by depopulation and destruction resulting from constant warfare. Without the influx of peasants or gentry, the towns became almost exclusively Jewish; this facilitated the crushing of their autonomy. The Commonwealth, due to both strife created by its powerful Lithuanian families, and to religious fanaticism—for which Limanowski blamed the Jesuits (who were supported by Polish kings since the end of the sixteenth century)—

became an easy prey to foreign intervention. Moreover, its predominantly rural economy, as well as extreme decentralization, rendered it defenseless in a struggle against all three, or any one, of its powerful neighbors. (See also *Historia Litwy*, Chicago and Paris, 1895, pp. 44-51.) The import of Jews by Polish kings appears to have been an effect rather than a cause of the weakness of the Polish middle class.

63. *Rozwoj polskiej mysli*, pp. 17-18.

64. Ibid., p. 19.

65. Ibid., pp. 7-8; *Rzut oka*, p. 15.

66. *Rozwoj polskiej mysli*, pp. 21-23.

67. *Rzut oka*, p. 16.

68. See his letters to Emil and Franciszka Haecker, written in the early 1930s and discussed in the previous chapter.

NOTES TO CHAPTER XII

1. See Introduction by E. H. Carr to N. G. Chernyshevsky, *What is to be done?* trans. B. R. Tucker (revised and abridged by Ludmilla B. Turkevich; New York, 1961), pp. xi-xii.

2. S. Hook, *Marx and the Marxists: The Ambiguous Legacy* (New York, 1955), pp. 68-71.

3. See for instance Z. Zechowski, *Socjologia Boleslawa Limanowskiego* (Poznan, 1964), pp. 29-30.

4. When Limanowski died, at least one of his obituaries failed to give him the recognition which he deserved as a pioneering Polish historian. See W. Borowy, "Obituary: Boleslaw Limanowski," *SEER*, XIV (1935/6), 429-30. Cf. W. J. Rose in ibid., pp. 425-27. See also W. M. Kozlowski, "Zycie i praca Boleslawa Limanowskiego," in *Socjalizm-Demokracja-Patriotyzm* (Cracow, 1902), p. 14. Cf. Zechowski, *op. cit.*, pp. 28-29.

5. Zechowski, *op. cit.*, pp. 169, 172-74. Cf. Z. F., "Socjologia Boleslawa Limanowskiego. Czesci dwie," *Przedswit*, Nos. 9-10 (1919), pp. 40-47. This anonymous reviewer argues that Limanowski failed to distinguish the "historical" (or "cultural") disciplines from the natural sciences, as the German scholar Windelband had done. Limanowski did not keep up with modern trends; and in his bibliographical sketch he entirely omitted mentioning the "economic school" of sociology (i.e., Marx, Engels, Kautsky, Petr Struve). Though Limanowski made some changes in the manuscript of his *Sociology* between

1912 and 1914, he admitted that this revision was a superficial one "because I was aware that creativity diminishes in old age." (See his *Socjologia*, 1919; rpt. Cracow, 1921, I, 19.) Limanowski realized that his work was dated. He felt that some of his ideas might have been considered original, if his book had been published thirty years earlier; for instance, he pointed to organicism, now discredited. He was convinced that "development of knowledge itself leads to certain conclusions and, consequently, in a given period of history similar ideas germinate in many minds," expecting that this phenomenon would become a common occurrence upon "democratization of knowledge," that is increased production of scholars. See his *Socjologia*, I, 18.

6. Z. Zaremba, "Social Transformation in Poland, Part II," *JCEA*, XII (1952/3), 289.

BIBLIOGRAPHY

I. MANUSCRIPT PRIMARY SOURCES

A. *Collections of Private Origin*

(1) Biblioteka Jagiellonska (Cracow)

Korespondencja J. N. Janowskiego. MS 4281.

Fragment korespondencji Boleslawa Limanowskiego. MS 6870.

Korespondencja Boleslawa Limanowskiego z lat 1862-1914, Vol. I. MS 6871.

Korespondencja Boleslawa Limanowskiego z lat 1862-1914, Vol. II, MS 6872.

Fragment papierow Boleslawa Limanowskiego. Add. MS 245/62.

Korespondencja Mariana Dubickiego, Vol. III. Add. MS 10/67.

(2) Biblioteka Narodowa (Warsaw)

Korespondencja Limanowskich, 1843-1934. MS 2896-2903.

Notatki i papiery Boleslawa Limanowskiego, 1845-1929. MS 2904.

Papiery osobiste Boleslawa Limanowskiego, 1885-1923. MS 2905.

Legitimacje i fotografie Boleslawa Limanowskiego. MS 2906.

B. Limanowski, Pamietniki, 1835-1907. MS 6417.

B. Limanowski, Pamietniki, 1835-1907; Dziennik, 1914-1928. MS 6418.

Fragment korespondencji Marii Konopnickiej, 1891-1909. MS 2667.

Korespondencja Danilowskich, 1893-1945. MS 2984.

Korespondencja T. Korzona, Vol. I. MS 5937.

Listy do Juliana Lukaszewskiego, 1886-7. Microfilm copy No. 28062. (Original at Ossol. MS 4397.)

Listy do Stanislawa Kozminskiego, 1908. Microfilm copy No. 28685. (Original at Ossol. MS 6714.)

Papiery Wyslouchow, Vol. XXIX. Microfilm copy No. 30099. (Original at Ossol. MS 7203.)

(3) Biblioteka Polskiej Akademii Nauk (Cracow)

Korespondencja T. Lenartowicza, 1886, Vol. IV. MS 2028.

Archiwum Biblioteki Polskiej Akademii Nauk, Dziennik Podawczy Akademii Umiejetnosci w Krakowie, 1873-1891.
MS PAU I 206.

Korespondencja Sekretariatu Generalnego Akademii Umiejetnosci w Krakowie. 1873-1891. MS PAU I 73, 1911/107.

(4) Biblioteka Publiczna (Warsaw)
    Korespondencja Adama Wislickiego, 1879-1894.

(5) Ossolineum (Wroclaw)
    Papiery Wyslouchow, 1895-1922, Vol. V. MS 7179.
    Korespondencja Wyslouchow, 1890-1914. MS 12,149.
    Korespondencja Wilhelma Feldmana, 1898-1919. MS 12,280.

B. Public Documents

(1) Archiwum Glowne Akt Dawnych (Warsaw)
    Kancelaria Gubernatora Warszawskiego. Referat I Tajny,
       3/1874. Polski Komitet Narodowy we Lwowie.

(2) Archiwum Panstwowe Krakowa. Akta Dyrekcji Policji, 1852-
    1918.
    Obchody 50^{lecia} pracy naukowej i spolecznej Boleslawa
       Limanowskiego, 1900-1913. St. GKr.4.
    Obchody rocznic powstania styczniowego w Krakowie,
       1885-1914. St. GKr.5.
    Zakaz rozprzestrzeniania w Galicji broszur Boleslawa Li-
       manowskiego, 1878-1895. St. GKr. 13.
    Czesciowe sprawozdanie z dzialalnosci PPS w liscie T. T.
       Jeza, 1901. St. GKr. 256.
    Starania o zezwolenie na pobyt w Galicji, 1904-1915.
       St. GKr. 311.
    Socjalno-rewolucyjne organizacje Polakow, 1889. St.
       GKr. 508/9/10.
    Agitacja socjalistyczna w kraju i na emigracji, 1886. St.
       GKr. 534.
    Rozprzestrzenianie broszur o tresci socjalistycznej w
       Galicji, 1881-2. St. GKr. 622.
    Konfiskata "Politycznej a spolecznej rewolucji" Boleslawa Li
       manowskiego, 1883. St. GKr. 740.
    Konfiskata w Krakowie *Rownosci*, 1879-1881. St. GKr. 809.

(3) Archiwum Zakladu Historii Partii (Warsaw)
    Archiwum PPS
       Kopialy korespondencji ZZSP, 1893-1899. AM 698/1-4.
       Kopialy korespondencji OZ PPS, 1900-7. AM 699/1-4.

Materialy pierwszych polskich organizacji socjalistycznych i niepodleglosciowych. B. Limanowski: Druki ulotne, 1881-1892. AM 877/1.

Odpisy dokumentow CK MSZ w Wiedniu i Dyrekcji Policji w Krakowie, 1877-1890. AM 877/2.

Zjazd Paryski, 1892. 305/11/1.

ZZSP i OZ PPS. Protokoly posiedzen, 1893-1898. Okolniki Centralizacji, 1893-1899; Okolniki KZ PPS, 1900-9. AM 1050/1.

Korespondencja od socjalistow polskich, 1891-1914. AM 1050/2-3.

Sprawy honorowe, 1891-1908. AM 1050/7.

Korespondencja i rachunki wplywajace do redakcji i administracji *Swiatla*, 1898-1900. AM 1050/10-11-12-13.

Czerwony Krzyz, 1894-1908. AM 1141/7.

Korespondenja czlonkow i dzialaczy PPS. Listy Boleslawa Limanowskiego, 1890-1911. AM 1190/7.

Wydzial Zagraniczny PPS. Sekcja Paryska, 1904-1914. AM 1314/6; Komitet Jubileuszowy ku czci Boleslawa Limanowskiego, 1910. AM 1314/7.

Materialy Sekcji ZZSP i OZ PPS. Sekcja Paryska, 1894-1909. AM 1440/8-9.

Druki ulotne, 1893-1909. Sekcja Paryska, 1900-1909. AM 1440/9.

Wykazy imienne czlonkow i kierownictwa. AM 1440/12-13.

Miedzynarodowe Kongresy Socjalistyczne, 1889-1910. AM 1440/16.

Listy do I. Daszynskiego, 1899-1930. 70/11/1.

Papiery Emila i Franciszki Haecker, 1894-1938. AM 692.

C. Personal Correspondence addressed to the author

Zofia Eiblowa (daughter of Emil Haecker). Letter dated August 29, 1968.

II. PUBLISHED PRIMARY SOURCES*

A. Works of Boleslaw Limanowski.

(1) Monographs, Pamphlets and Essays

Bolszewickie panstwo w swietle nauki. Warsaw, 1921.

Centralizm i federalizm, panstwo narodowe i panstwo narodowosciowe, Lvov, 1926.

"Charakterystyka Henryka Kamienskiego," in Henryk Kamienski, *Filozofia ekonomii materialnej ludzkiego spoleczenstwa.* Abridged by Dr. Z. Daszynska-Golinska. Warsaw, 1911.

"Charakterystyka Jozefa Supinskiego," in Jozef Supinski, *Szkola polska gospodarstwa spolecznego.* Abridged by by Z. Daszynska-Golinska. Warsaw, 1911.

Dwaj znakomici Komunisci: Tomasz Morus i Tomasz Campanella i ich systematy, Utopia i Panstwo Sloneczne. Studium Socjologiczne. Lvov, 1873.

Ferdynand Lassalle i jego polemiczno-agitacyjne pisma. Geneva, 1882.

Foreword, in *Artykuly polityczne Adama Mickiewicza.* Cracow, 1893.

Foreword, in Adam Prochnik, *Demokracja Kosciuszkowska.* 1920; rpt. Warsaw, 1946.

Galicja przedstawiona slowem i olowkiem. Illustrations by W. Tetmajer. Warsaw, 1892.

Historia demokracji polskiej w epoce porozbiorowej. 2 vols., 4th ed. Warsaw, 1957. 1st edition: Zurich, 1901; 2nd revised edition: Cracow, 1922; 3rd edition: Warsaw, 1946. Abridged popular PPSD edition: *Demokracja w Polsce,* Cracow, 1903.

Historia Litwy. Chicago and Paris, 1895. 2nd revised edition: *Dzieje Litwy.* Warsaw, 1917.

Historia ruchu narodowego od 1861 do 1864 r. 2 vols. Vol. I (1861-63); *Vol. II (1863-64). Published anonymously in Lvov, 1882, ·and reprinted under the title *Historia powstania narodu polskiego 1863 i 1864 r.* Lvov, 1894. A revised edition bearing this title appeared in Lvov in 1909.

Historia ruchu rewolucyjnego w Polsce w 1846 r. Cracow, 1913.

Historia ruchu spolecznego w drugiej polowie XVIII stulecia. Lvov, 1888.

*Books by Limanowski and secondary source materials given below, which also include original documents or excerpts therefrom, are marked with *.

Historia ruchu spolecznego w XIX stuleciu. Lvov, 1890. Czech edition: *Dejiny socialniho hnuti XIX stol.* Translated by Ant. Hojn and Ab. Hojn. Praha, 1891.

"Kilka uzupelniajacych uwag i krotki zyciorys J. F. Beckera," in *Manifest do ludnosci rolniczej (Manifesto to Farmers).* Proclaimed by the Geneva Branch of the International and written by J. P. Becker. Translated from the German by K. Sosnowski. Geneva, 1883.

"Krakowianom," in *Pamiatka Majowa.* Cracow, 1911.

Boleslaw Limanowski to Ignacy Moscicki, letter dated August 17, 1929. Published by the Central Executive Committee of the PPS as Circular No. 12.

Losy narodowosci polskiej na Slasku. Lvov, 1874. Revised editions: *Odrodzenie i rozwoj narodowosci polskiej na Slasku.* Warsaw, 1911 and 1921.

Mazowsze Pruskie, Warsaw, 1925.

Narod i panstwo: Studium Socjologiczne. Cracow, 1906.

O kwestii robotniczej. Lvov, 1871.

O powstaniu polskiem 1863-4 roku: Dwa odczyty popularne. Cracow, 1913.

Ostatnie lata dziejow powszechnych od 1846 do dni dzisiejszych. Lvov, 1881. A revised Polish edition of the German universal history by Wilhelm Müller, published anonymously, with new chapters on the history of Poland written by B. Limanowski, and extended to cover events up to the year of publication. 1st Polish edition covers the period from 1848 to 1875: Lvov, 1878. Translated by B. Limanowski.

Panstwo unarodowione i unarodowienie ziemi. Warsaw, 1918.

Patriotyzm i Socjalizm. 2nd revised edition. Paris, 1888. 1st edition: Geneva, 1881.

Plakan, Janko [B. Limanowski]. *Dziewczyna nowego swiata: Obrazek z sennej rzeczywistosci.* Lvov, 1872.

Polityka wynaradawiajaca rzadu pruskiego w stosunku do ludnosci polskiej. Czestochowa, 1919.

Powstanie narodowe 1863 i 1864 r. Lvov, 1888 and 1900.

"Prawa Platona," in *Sobotka: Ksiega zbiorowa na uczczenie 50go jubileuszu Seweryna Goszczynskiego.* Lvov, 1875.

Plutarch Polski:
Hugo Kollataj. Warsaw, [1920].
Romuald Traugutt. Warsaw, [1920].
Stanislaw Staszic. Warsaw, [1920].

Tadeusz Kosciuszko. Warsaw, [1920].

Walerjan Lukasinski. Warsaw [1918].

"Przemowienie powitalne hon. przewodniczacego Zjazdu," *I Zjazd Niezaleznych Zwiazkow Akademickiej Mlodziezy Socjalistycznej: Deklaracja Ideowa.* Warsaw, 1922.

Rozwoj polskiej mysli socjalistycznej. Warsaw, 1929.

Rozwoj przekonan demokratycznych w narodzie polskim. Cracow, 1906.

Rzut oka na rozwoj polskiej mysli socjalistycznej. Chicago, 1922.

Socjalizm jako konieczny objaw dziejowego rozwoju. Lvov, 1879.

Socjologia. 2 vols. Cracow, 1919 and 1921.

Socjologia Augusta Comte'a. Lvov, 1875.

Stanislaw Worcell: Zyciorys. Cracow, 1910 and Warsaw, 1948.

Studwudziestoletnia walka narodu polskiego o niepodleglosc. Cracow, 1916. Revised edition of *Stuletnia walka narodu polskiego o niepodleglosc:* Lvov, 1894 and 1906. Russian edition: *Sto let bor'by pol'skago naroda za svobodu.* Translated by B. Limanowski and L. Kulczycki, Moscow, 1907.

Szermierze wolnosci. Cracow, 1911.

"Udzial narodu polskiego w rewolucjach 1848 i 1849 r.," in Karol Marks, *Rewolucja i kontrewolucja w Niemczech.* Translated from the German by B. Limanowski. London, 1897.

"Udzial Polakow w walkach wolnosciowych w dobie porozbiorowej," in *Polska w kulturze powszechnej.* Edited by Feliks Konieczny. Cracow, 1918.

1793-1893. W stuletnia rocznice drugiego rozbioru Polski. Lvov, 1893.

"Wilhelm Liebknecht," in W. Liebknecht, *W obronie prawdy (Zu Trutz und Schutz).* Translated from the German by B. Limanowski. Geneva, 1882.

(2) Periodicals Containing Limanowski's
Articles and Reviews

Ateneum, Warsaw, 1882/III, 1886/I, 1886/III, 1887/II, 1887/IV, 1888/III, 1889/III.

Bluszcz, Warsaw, 1872, 1873.

Chlopska Sprawa, Warsaw, No. 7 (April, 1915).

Gazeta Literacka, Lvov, Nos. 6-9 (1871).

Gazeta Narodowa, Lvov, 1872, 1873, 1876.

Glos, Warsaw, 1887, 1888.

Kalendarz Robotniczy, Cracow, 1895, 1896, 1897, 1901, 1903, 1906, 1909; Chicago, 1904; Warsaw, 1924.

Krytyka, Cracow, Nos. 4 and 11 (1908).
Kwartalnik Historyczny, Lvov, 1902.
Na Dzis, Cracow, 1872.
Naprzod, Cracow, 1896, 1907, 1908, 1910, 1911, 1912, 1917, 1921, 1924.
Niwa, Warsaw, 1876.
Opiekun Domowy, Warsaw, July 24, 1867.
Pobudka (La Diane), Paris, 1889, 1890, 1891.
Prawo Ludu, Cracow, 1908.
Przedswit (L'Aurore), London, 1895, 1898; Warsaw, 1919.
Przeglad Tygodniowy, Warsaw, 1869, 1872, 1873, 1878, 1879, 1880, 1881, 1885, 1886, 1887, 1893.
Przeglad Socjalistyczny, Paris, 1892, 1893.
Przeglad Spoleczny, Lvov, 1886, 1887.
Przeglad Wilenski, Vilna, 1912.
Przyjaciel Domowy, Lvov, July 1, 1875.
Rekodzielnik, Lvov, November 19, 1871.
Robotnik, Warsaw, 1924, 1925, 1928.
Rownosc, Geneva, 1879.
Szkice Spoleczne i Literackie, Cracow, April 8, 1876.
Towarzysz Pilnych Dzieci, Lvov, Nos. 3-10 (1877).
Trybuna, Cracow, 1906, 1907.
Trybuna, Warsaw, 1920, 1921.
Tydzien Literacki, Artystyczny, Naukowy i Spoleczny, Lvov, 1877, 1878, January 5, 1879.
Tydzien Polski, Lvov, 1879.
Tydzien, dodatek Literacki do Kurjera Lwowskiego, Lvov, 1894, 1895, 1897, 1898, 1900, 1905.
Wiadomosci Polskie, Teschen-Piotrkow, Nos. 48 and 56 (1915).
Zycie, Lvov, 1911.

(3) Memoirs, Autobiographical Essays and Autobiographical Articles

"Garsc wspomnien z pobytu w Genewie," *Kalendarz Robotniczy PPS na rok 1922,* pp. 82-87.
"Jak stalem sie socjalista," in *Socjalizm-Demokracja-Patriotyzm.* Cracow, 1902, pp. 83-137.
"Kartka z mego zycia," *Naprzod*, November 20, 1910, pp. 1-2.
"Kartka z mego zycia," *Przedswit*, No. 11 (1910), pp. 706-9.
"Ludwik Mieroslawski," *Swiatlo*, III (1900), 49-62.

Pamietniki, 4 vols: Vol. I (1835-1870), edited with a Foreword by A. Prochnik, Warsaw, 1937; reprinted 1957; Vol. II (1870-1907), edited by J. Durko with a Foreword by H. Jablonski, Warsaw, 1958; Vol. III (1907-1919) and IV (1919-1928), edited by J. Durko, Warsaw, 1961 and 1973.

"Pierwsza manifestacja w Wilnie 1861 r.," in *W czterdziesta rocznice powstania styczniowego*. Lvov, 1903. (Ossol., MS 8033.)

Plakan, Janko [B. Limanowski]. "Wybrzeza Morza Bialego," *Klosy*, X (March-April, 1870), 207-10, 223-26.

"Rok uchwalenia swieta robotniczego," *Jednodniowka Majowa*. Warsaw, May 1, 1920, pp. 3-5.

"Wspomnienia z pobytu w Galicji," *Przedswit*, XXI, No. 11 (1901), pp. 401-7; *Z Pola Walki*, London, 1904.

"Z moich wspomnien o narodzinach PPS," in *W trzydziesta rocznice: Ksiega Pamiatkowa PPS*. Warsaw, 1923, pp. 23-25.

"Ze wspomnien o Wilhelmie Feldmanie," in *Pamieci Wilhelma Feldmana*. Cracow, 1922, pp. 138-41.

B. *Biographical Sketches, Obituaries and Reminiscences about Limanowski.*

"Akademia ku czci tow. senatora Boleslawa Limanowskiego," *Robotnik*, December 9, 1924, pp. 1-2.

Bezmaski, H. [Stanislaw Posner]. "Portret Boleslawa Limanowskiego," *Robotnik*, October 13, 1923, p. 2.

Bobrowski, E. "Pamietniki, Vol. I: 1912-1918." Microfilm copy: BN, Mfm 32371. (Original at Ossol. MS 12,004.)

"Boleslaw Limanowski," *Czlowiek w Polsce*, II (1934), 5.

"Boleslaw Limanowski" *Kalendarz Robotniczy na rok 1901*, pp. 6-8.

"Boleslaw Limanowski," *Kalendarz Robotniczy na rok 1918*, pp. 92-94.

"Boleslaw Limanowski," *Mysl Narodowa*, XV (February 10, 1935), 95.

"Boleslaw Limanowski," *Robotnik*, November 8, 1922, pp. 1-2.

Borowy, W. "Obituary of B. Limanowski," *SEER*, XIV (1935/6), 429-33.

Chalasinski, J. "Swietej Pamieci Boleslaw Limanowski," *Ruch Prawniczy Ekonomiczny i Socjogiczny*, XV (1935), 281-83.

Czachowski, K. "Szermierz wolnosci," *Niepodleglosc*, II (1930), 193-213.

Danilowski, G. "Boleslaw Limanowski," *Zycie*, I (November 19, 1910), 103-4.

Daszynski, I. *Pamietniki*, 2 vols. Cracow, 1925-26.

Dabrowski, J. [Jozef Grabiec]. "Wielki Starzec demokracji polskiej, Boleslaw Limanowski," *Kultura Slowianska*, I (1925), 4-5.

"Dwa Jubileusze. 86-lecie Boleslawa Limanowskiego; 70-lecie Karola Kautsk'ego," *Kalendarz Robotniczy PPS na rok 1925*, pp. 72-76.

Haecker, E. "Boleslaw Limanowski," *Naprzod*, November 20, 1910, pp. 1-3.

Handelsman, M. "Boleslaw Limanowski," *Swiat*, XXX (January 5, 1935), 5.

————. "Boleslaw Limanowski jako badacz historii powszechnej," *PH*, XXXIII (1935), 328-32.

"Jubileusz Boleslawa Limanowskiego,"*Zycie* (October-November 1910), 55-56, 73, 84-85, 132.

"Jubileusz Boleslawa Limanowskiego we Lwowie," *Zycie*, I (December 10, 1910), 164.

K. "Boleslaw Limanowski," *Tydzien, dodatek do Kurjera Lwowskiego*, VIII No. 34 (August 26, 1900), pp. 265-67; Nos. 35 (September 2, 1900), pp. 273-74.

Kozlowski, W. M. "Zycie i praca Boleslawa Limanowskiego," in *Socjalizm-Demokracja-Patriotyzm*. Cracow, 1902.

Krzywicki, L. *Wspomnienia*. 3 vols. Warsaw, 1957-1959.

Kukiel, M. "Limanowski jako historyk," *Zycie*, I (1910), 105-6.

Lopacinski, W. "Boleslaw Limanowski," *PH*, XXXIII (1936), 324-25.

Minkiewicz, R. "Obnazcie glowy," *Zycie*, I (1910), 141.

Moscicki, H. and W. Dzwonkowski, eds. *Parlament Rzeczypospolitej Polskiej, 1919-1927*. Warsaw, 1928, pp. 284-85.

Orkan, W. "Boleslawi Limanowskiemu," *Kalendarz Robotniczy na rok 1901*, p. 5.

Os...rz, S. [Leon Wasilewski]. *Boleslaw Limanowski*. Cracow, 1910.

Poniecki, W. "Boleslaw Limanowski," in *Mysliciele i bojownicy*. Warsaw, 1935, pp. 153-58.

Prochnik, A. *Boleslaw Limanowski*. Warsaw, 1934.

————. "Boleslaw Limanowski," *KH*, XLIX (1935), 263-70.

————. "Boleslaw Limanowski," *Wiadomosci Literackie*, No. 8 (February 24, 1935), p. 3.

Res [Feliks Perl]. "Boleslaw Limanowski," *Przedswit*, XXX, No. 11 (1910), pp. 697-706.

Rzepeccy, T. and K. *Sejm i Senat Rzeczypospolitej Polskiej, 1928-1933*. Poznan, 1933, pp. 199-200.

Rzepeccy, T. and W. *Sejm i Senat Rzeczypospolitej Polskiej, 1922-1927*. Poznan, 1933, pp. 440-41.

Starza, P. "Boleslaw Limanowski," *Swiatlo*, III (1900), 1-6.

Studnicki-Gizbert, W. "Boleslaw Limanowski," in *Z okolic Dzwiny: Ksiega zbiorowa na dochod czytelni polskiej w Witebsku*. Wilno, 1912, pp. 93-96.

Tokarz, W. "Boleslaw Limanowski, jako badacz historii Polski," *PH*, XXXIII (1936), 326-27.

Trabczynski, J. [Ludwik Kulczycki], "Boleslaw Limanowski," *Krytyka*, No. 3 (1900), pp. 179-86.

W. F. [Wilhelm Feldman] "Boleslaw Limanowski," *Krytyka*, No. 11 (1910), pp. 185-91.

Wasilewski, L. *Boleslaw Limanowski w setnym roku zycia*. Warsaw, 1934.

Wojtkowski, A. "Boleslaw Limanowski," *Awangarda Panstwa Narodowego*, XIII (March 1935), 38.

Zakrzewski, K. "Apostol rewolucji polskiej," *Pion*, III (1935), 2.

Zakrzewski, St. "Boleslaw Limanowski 30 X 1835—1 II 1935," *KH*, XLIX (1935), 262.

*Zygmuntowicz, Z. "Boleslaw Limanowski w swietle akt austriackich," *Niepodleglosc*, XI (1935), 130-39.

"Zyciorys Boleslawa Limanowskiego," *Swiatlo*, No. 10-11 (May 1, 1920), p. 5.

C. Other Sources.

B. "Un livre socialiste polonais," *La Revue Socialiste*, IX (January-June 1889), pp. 190-208.

Chrzanowski, I. ed. *Joachim Lelewel, czlowiek i pisarz*. Introduction by S. Pigon. Warsaw, 1946.

Debogorii-Mokrievich, V. *Vospominaniia*. Paris, 1894-98.

Drahomanov, M. "Ukrainski hromadivtsi pered pol'skim socializmom i pol'skim patriotismom," *Hromada*, No. 5 (1882), pp. 231-41.

Gieysztor, J. *Pamietniki z lat 1857-1865*. Foreword by T. Korzon. 2 vols. Vilna, 1913.

Glabinski, S. Rev. of *Historia ruchu spolecznego w drugiej polowie XVIII stulecia*, by B. Limanowski, *KH*, III (1889), pp. 351-56.

————. Rev. of *Historia ruchu spolecznego w XIX stuleciu*, by B. Limanowski, *KH*, V (1891), 434-42.

Gumplowicz, L. *Filozofia spoleczna*. Translated by S. Posner. Warsaw-Lvov, 1909.

————. "Socjologiczne pojmowanie historii," in M. H. Serejski, ed., *Historycy o historii, 1775-1918*. Warsaw, 1963, pp. 432-47.

Kieniewicz, S. ed. *Galicja w dobie autonomicznej, 1850-1914*. Wroclaw, 1952.

*Kosciuszko, T. *Napomknienia wzgledem poprawy losu wloscian i Uniwersal Polaniecki*. Edited by K. Baranowski. Warsaw, 1917.

Krzywicki, L. "B. Limanowski: Historia ruchu spolecznego," in his *Dziela*, Vol. III. Warsaw, 1959, pp. 601-4.

Kulakowski, M. ed. *Roman Dmowski w swietle listow i wspomnien.* London, 1968.

Lassalle, F. *The Workingman's Programme*. Translated with an Introduction by Edward Peters. New York, 1899.

Lisicki, H. Rev. of *Powstanie narodowe 1863 i 1864 r.* (1888), by B. Limanowski, *KH*, III (1889), 596-98.

Lukasinski, W. ed. *Postepowa publicystyka emigracyjna, 1831-1846.* Warsaw-Cracow-Wroclaw, 1961.

Malon, B. *Le Socialisme integral*, Vol. I. Paris [1890].

Manifest Rzadu Narodowego Rzeczypospolitej Polskiej do Narodu Polskiego. Cracow, February 22, 1846.

Manifest Towarzystwa Demokratycznego Polskiego. Paris, December 4, 1836.

Marx, K. *Critique of the Gotha Program*. Moscow, 1947. Based on the latest Soviet edition (1945) and re-checked against the German original.

Marx, K. and F. Engels. *Manifesto of the Communist Party*. Reprinted from the English edition by S. Moore, 1888. Moscow, n.d.

Milkowski, Z. *Od kolebki przez zycie. Wspomnienia*, Vol. III. Cracow, 1936-37.

Molska, A. ed. *Pierwsze pokolenie marksistow polskich. Wybor pism i materialow zrodlowych z lat 1878-1886*. 2 vols. Warsaw, 1962.

Nowina, H., "Ksiazka Limanowskiego o Stanislawie Worcellu," *Przedswit*, XXX, No. 11 (1910), pp. 710-17.

Perl, F. and Zaremba, Z. eds. *Z dziejow prasy socjalistycznej w Polsce*. Warsaw, 1919.

Plekhanov, G. U. "O sotsial'noi demokratii v Rossii" (1893), in his *Sochineniia*. Moscow, n.d., IX-X, 5-29.

Plomienczyk, A. [A. Skwarczynski], "Prace Limanowskiego o emigracji po r. 1831," *Krytyka*, No. 2 (1912), pp. 78-86.

Pobog-Malinowski, W. ed. "Umowa Ligi Narodowej [sic] z Paryska Gmina Narodowo-Socjalistyczna," *Niepodleglosc*, VII (1933), 432-34.

"Projekt programu rolnego," *Przedswit*, XXV, Nos. 6-8 (1905), pp. 248-52.

Przeglad Spoleczny, 1886-7. A collection edited by K. Dunin-Wasowicz. Bibliography prepared by J. Czachowska. Wroclaw, 1955.

Przyborowski, W. Rev. of *Historia powstania narodu polskiego 1863 i 1864 r.* (1909), by B. Limanowski, *KH*, XXIV (1910), 320-24.

La Revue Socialiste, XIV (August 1891), 199, 254.

Romaniukowa, F. ed. *Radykalni demokraci polscy, 1863-1875.* Warsaw, 1960.

Rupniewski, W. ed. "Program niepodleglosciowy *Podbuki,*" *Niepodleglosc*, II (1930), 352-54.

Smolenski, W. *Szkoly historyczne w Polsce.* Warsaw, 1898. Abridged in M. H. Serejski, ed. *Historycy o historii (1775-1918).* Warsaw, 1963, pp. 345-56.

Sprawozdania stenograficzne z posiedzen Senatu Rzeczypospolitej Polskiej, 1922-1927; 1931-1935.

Sprawozdanie z dzialalnosci Wydzialu Humanistycznego Uniwersytetu Warszawskiego za rok akademicki 1934/5. Warsaw, 1936.

Straszewicz, L. "Spoleczne kierunki w teorii i w zyciu," *Prawda*, III (1883), 472-73, 482-83, 499-500, 507-8, 521-22, 531-33, 545-46.

Swietochowski, A. *Utopie w rozwoju historycznym.* Warsaw, 1910.

Szajnocha, K. "Lechicki poczatek Polski," in his *Dziela*, Vol. IV, Warsaw, 1876, pp. 85-291.

"Szkic programu PPS," *Przedswit*, XIII, No. 5 (1893), pp. 2-7.

Tygodnik Wielkopolski, II, No. 14 (April 6, 1872), pp. 192-93. Rev. of Limanowski's *Dziewczyna nowego swiata* (1872).

Warski, A. "Jubileusz szlachetnego socjalisty utopijnego," in his *Wybor pism i przemowien*, Vol. I. Warsaw, 1958, pp. 469-76.

Wasilewski, L. ed. "Do historii zjazdu paryskiego," *Niepodleglosc*, VIII (1933), 107-52.

Wronski, A. [W. Jodko-Narkiewicz] . *Program rolny PPS.* Cracow, 1910.

Z. L. S. [W. Przyborowski] . Rev. of *Powstanie narodowe 1863 i 1864 r.* (1900), by B. Limanowski, *KH*, XVI (1902), 134-38.

Zasady nauki spolecznej przes Doktora Medycyny [G. R. Drysdale, *The Elements of Science or Physical, Sexual and Natural Religion* (London, 1854)] . Translated by Limanowski. Geneva, 1880.

Z. F. "Socjologia Boleslawa Limanowskiego. Czesci dwie," *Przedswit*, XXXVIII, Nos. 9-10 (September-October 1919), pp. 40-47.

III. SECONDARY SOURCES

A. Articles

*Bardach, J. "Inflanty, Litwa, Bialorus w tworczosci Boleslawa Limanowskiego," *PH*, No. 3 (1974), LXV, 479-503.

Bardach, J. "Nieznana autobiografia Boleslawa Limanowskiego." *Slowianie w dziejach Europy.* Festschrift for H. Lowmianski. Poznan, 1974, pp. 277-88.

*Borejsza, J. W. and A. Garlicki, "Boleslaw Limanowski" [Rev. of *Pamietniki*, 3 vols.] *KH*, LXIX (1962), 740-44.

Brock, P. "Socialism and Nationalism in Poland, 1840-1846," *Canadian Slavonic Papers*, IV (1959), 121-46.

————. "The Birth of Polish Socialism," *JCEA*, XIII (1953), 213-31.

————. "The Polish Revolutionary Commune in London," *SEER*, XXXV (1956), 116-28.

————. "The Socialists of the Polish 'Great Emigration'," in Asa Briggs and John Saville, eds. *Essays in Labour History.* London, 1960, pp. 140-73.

————. "Zeno Swietoslawski (1811-1875), a Forerunner of the Russian Narodniki," *ASEER* (1954), 566-87.

Buszko, J. "Polskie czasopisma socjalistyczne w Galicji," *Zeszyty Prasoznawcze*, II (1961), 29-35.

Bystron, J. S. "Rozwoj problem socjologicznego w nauce polskiej," *Archiwum Komisji do badania Historii Filozofii w Polsce*, Vol. I, Part II (1917), pp. 189-260.

————. "Pojecie narodu w socjologii polskiej," *Rok Polski*, I, No. 4 (1916), pp. 33-48.

Charnay, M. "Malon, Benoît, 1841-1893," *La Grande Encyclopedie*, XXII, 1068-69.

"Comte, Auguste. 1798-1857." *Wielka Encyklopedia Powszechna PWN* (1963), II, 590-91.

Dunin-Wasowicz, K. "Polska mlodziez akademicka a socjalizm," *Przeglad Socjalistyczny*, No. 5 (1946), pp. 35-37.

————. "Promien i Promienisci," *Plomienie*, III, Nos. 5-6 (October 1947), pp. 20-23.

Durko, J. "Zjazd paryski i powstanie PPS," *Przeglad Socjalistyczny*, Nos. 10-11 (1946), pp. 29-31.

Dziewanowski, M. "The Polish Revolutionary Movement and Russia, 1904-7," *Harvard Slavic Studies*, IV (1957), 375-94.

Gross, Z. "Poczatki ruchu zawodowego w zaborze austryjackim," *Robotniczy Przeglad Gospodarczy*, No. 11 (1947), pp. 19-21.

Groth, Alexander J. "Dmowski, Pilsudski and Ethnic Conflict in Pre-1939 Poland," *Canadian Slavic Studies*, III (Spring, 1969), 69-91.

————. "Polish Elections, 1919-1928," *Slavic Review*, XIV (1965), 653-65.

Grzybowski, K. "Parlament Polski, 1919-1939," *Swiat*, XVIII (October 13, 1968), 6-7.

Bibliography

Jablonski, H. "U kolebki polskiego socjalizmu," *Wiedza i zycie*, XV (1946), 372-76.

Jasnowski, J. "Poland's Past in English Historiography (XVII-XIX centuries)," *Polish Review*, III (1958), 21-35.

Kancewicz, Jan. "Zjazd paryski socjalistow polskich 17-23 XI 1892 r., jego geneza, przebieg i znaczenie," *Z Pola Walki*, V, No. 4/20 (1962), pp. 3-34.

Kieniewicz, S. "Chlopi a walki powstancze o niepodleglosc. Rola klas posiadajacych i ich ocena," *KH*, LXIV (1957), 195-200.

———. "Historiografia polska wobec powstania styczniowego," *PH*, XLIV (1953), 1-35.

Kusinski, S. "Teoria spoleczna E. Abramowskiego," *Mysl Wspolczesna*, No. 1/44 (1950), pp. 26-60.

Lechicki, C. Rev. of *Pamietniki*, Vol. II, by B. Limanowski, *Prasa Wspolczesna i Dawna*, II (1959), 230-32.

Laski, A. "Poglady przedrozlamowej PPS w kwestii rolnej," *Z Pola Walki*, II, No. 4/8 (1959), pp. 32-55.

Mathes, W. L. "The Origins of Confrontation Politics in Russian Universities: Student Activism, 1855-1861," *Canadian Slavic Studies*, II (Spring, 1968), 28-45.

Mikos, S. "W sprawie skladu spolecznego i genezy ideologii Gromad Ludu Polskiego w Anglii, 1835-1846," *PH*, LI (1960), 663-81.

"Mill, John Stuart, 1806-1873," *Encyclopaedia Britannica* (1966) XV, 460-63.

Miller, I. S. "Dokola genezy Gromady Rewolucyjnej Londyn," *PH*, L (1959), 815-30.

*Mulak, J. "Z dziejow mysli programowej PPS," *Przeglad Socjalistyczny*, No. 5 (1948), pp. 17-23; No. 6 (1948), pp. 23-27.

Myslinski, J. "Uniwersytet Wakacyjny w Zakopanem w r. 1904," *Przeglad Historyczno-Oswiatowy*, VI, No. 1 (1963), pp. 7-21.

Najdus, W. "Poczatki socjalistycznego ruchu robotniczego w Galicji (lata siedemdziesiate-osiemdziesiate XIX w.)," *Z Pola Walki*, III, No. 1/9 (1960), pp. 3-34.

———. "Ruch robotniczy w Galicji w latach 1890-1900," *PH*, LIII (1962), 86-114.

Orthwein, K. "Teoria i praktyka pierwszych socjalistow polskich," *Kultura i spoleczenstwo*, II (1958), 119-38.

P. L. "Dwa kongresy socjalistyczne w Paryzu," *Pobudka*, Nos. 7/8/9 (1889), pp. 46-52.

Pietrzak-Pawlowska, I. Rev. of *Pamietniki*, Vol. I, by B. Limanowski, *Studia Zrodloznawcze*, V (1960), 174-77.

345

Romaniukowa, F. "Dalsze dokumenty do historii Gromady Rewolucynej Londyn," *PH*, LI (1960), 548-56.

———. "Zagadnienie wlasnosci gminnej i spoldzielczo-zrzeszeniowej w pogladach radykalnych demokratow polskich," *Zeszyty Naukowe SGPiS*, No. 19 (1960), pp. 119-38.

Sokulski, J. "Stosunki Gillera z B. Limanowskim." Unpublished article, BPAN, MS 6683.

Tracy, M. "Agriculture in Western Europe: The Great Depression, 1880-1900," in Charles K. Warner, ed. *Agrarian Conditions in Modern European History*. New York-Toronto, 1966, pp. 98-112.

Turski, M. "Spor od pokolen," *Polityka*, XI (June 15, 1968), 3-4.

Tyrowicz, M. Rev. of *Pamietniki*, Vol. I, by B. Limanowski, *Roczniki Dziejow Spolecznych i Gospodarczych*, VI (1937), 338-41.

Walczak, S. "Rola genewskiego osrodka emigracyjnego polskich socjalistow w ksztaltowanie sie swiadomosci socjalistycznej w kraju," *Ze Skarbca Kultury*, No. 1/7 (1955), pp. 22-60.

Wandycz, P. S. "The Polish Precursors of Federalism," *JCEA*, XII (1952/3), 346-55.

Wereszycki, H. Rev. of *Pamietniki*, Vol. I, by B. Limanowski, *Niepodleglosc*, XV (1937), 304-7.

Wojenski, T. "Postepowe tradycje publicystyki polskiej na przelomie XIX i XX wieku," *Wiedza i zycie*, XIX (1952), 622-27.

Wyka, J. "Koloryt socjalny jednego wieku" [Rev. of *Pamietniki*, Vol. III, by B. Limanowski], *Tworczosc*, XVIII, No. 4 (1962), pp. 153-55.

Zaremba, Z. "Social Transformation in Poland, Part II," *JCEA*, XII (1952/3), 276-89.

Zawadka, M. "Pierwszy zjazd PPS,"*Przeglad Socjalistyczny*, No. 12 (1947), pp. 38-42.

Zielenczyk, M. "Pamietniki Boleslawa Limanowskiego," *Polityka*, I (August 14-20, 1957), 5.

Zlotorzycka, M. "Dzialalnosc Zwiazku Ludu Polskiego w Anglii, 1872-77," *Niepodleglosc*, XIII (1936), 165-97.

Zarnowska, A. "Zasieg, wplyw i baza spoleczna PPS w przeddzien Rewolucji 1905 r.," *KH*, LXVII (1960), 351-89.

———. "PPS u progu rozlamu (VII Zjazd PPS, Rada Czerwcowa 1905 r.)," *Z Pola Walki*, IV, No. 1/13 (1961), pp. 3-34.

B. Books and Unpublished Dissertations and Monographs

Baczko, B. ed. *Z dziejow polskiej mysli filozoficznej i spolecznej. Wiek XIX*. Vol. III. Warsaw, 1957.

Blit, L. *The Origins of Polish Socialism: The History and Ideas of the First Polish Socialist Party, 1878-1886*. Cambridge, 1971.

Bobrowska, B. "Socjalista romantyk, Emil Bobrowski," (Unpublished monograph, Ossol., MS 12,005.)

Bochenski, A. *Dzieje glupoty w Polsce. Pamflety dziejopisarskie*. Warsaw, 1947.

Borejsza, J. "Oblicze polityczne polskiej prasy emigracyjnej na Zachodzie Europy, 1864-1870." (Unpublished doctoral dissertation, University of Warsaw, 1962.)

———. *Emigracja polska po powstaniu styczniowym*. Warsaw, 1966.

———. *W kregu wielkich wygnancow, 1848-1895*. Warsaw, 1963.

Brock, P. "Boleslaw Wyslouch, tworca ideowy ruchu ludowego w Galicji." (Unpublished doctoral dissertation at the Jagiellonian University of Cracow, 1950. AZHP MS 76/I/2.)

Bury, J. B. *The Idea of Progress: An Inquiry into its Origins and Growth*. 1932; rpt. New York, 1955.

Buszko, J. *Ruch socjalistyczny w Krakowie 1890-1914 na tle ruchu robotniczego w Zachodniej Galicji*. Cracow, 1961.

———. *Spoleczno-polityczne oblicze Uniwersytetu Jagiellonskiego w dobie autonomii galicyjskiej, 1869-1914*. Cracow, 1963.

Carr, E. H. *What is History?* 1961; rpt. London, 1967.

Chernyshevsky, N. G. *What is to be done?* Translated by B. R. Tucker. Revised and abridged by Ludmilla B. Turkevich. New York, 1961.

Ciolkosz, L. and A. *Zarys dziejow socjalizmu polskiego*. Vol. I. London, 1966.

Cole, G. D. H. *The Second International, 1889-1914*. Part I. London, 1963.

Dzieciol, W. *The Origins of Poland*. London, [1966].

Feldman, W. *Dzieje polskiej mysli politycznej, 1864-1914*. 2nd ed. Edited by Jozef Feldman, with an Introduction by Leon Wasilewski. Warsaw, 1933.

———. *Stronnictwa i programy polityczne w Galicji, 1846-1906*. Vol. II. Cracow, 1907.

Gross, F. *The Polish Worker*. New York, 1945.

Haecker, E. *Historia socjalizmu w Galicji i na Slasku*. Vol. I (1846-1882). Cracow, 1933.

Hook, Sidney. *Marx and the Marxists: The Ambiguous Legacy*. New York, 1955.

Jablonski, H. *Miedzynarodowe znaczenie polskich walk narodowwyzwolenczych XVIII i XIX w*. 2nd ed. Warsaw, 1966.

———. *Polityka PPS w czasie wojny 1914-8*. Warsaw, 1958.

Kancewicz, J. "Powstanie SDKPiL jako rewolucyjnej partii proletariatu polskiego Kongresowki na tle rozlamu w polskim ruchu robotniczym." (Unpublished monograph, Warsaw, 1959.)

Kieniewicz, S. *Historia Polski, 1795-1918*. Warsaw, 1968.

————. *Ruch chlopski w Galicji w 1846 r.* Wroclaw, 1951.

————. *Sprawa wloscianska w postaniu styczniowym.* Wroclaw, 1953.

Ksiega Jubileuszowa PPS, 1892-1932. Warsaw, 1932.

Kukiel, M. *Czartoryski and European Unity*. Princeton, 1955.

Leslie, R. F. *Polish Politics and the Revolution of November 1830.* London, 1956.

————. *Reform and Insurrection in Russian Poland, 1856-1865.* London, 1963.

Maternicki, J. "Historia i historycy polscy w latach pierwszej wojny swiatowej." (Unpublished doctoral dissertation at the University of Warsaw, 1966.)

Mazowiecki, M. [L. Kulczycki]. *Historia ruchu socjalistycznego w zaborze rosyjskim.* Cracow, 1903.

Miaso, J. *Uniwersytet dla wszystkich*. Warsaw, 1960.

Mickiewicz, W. "Emigracja Polska, 1860-1890." (Unpublished monograph at Biblioteka Narodowa, MS 5302.) Published in Cracow, 1908.

Mlynarski, Z. *Szkice z dziejow rewolucyjnej prasy w Polsce.* Warsaw, 1963.

Mogilska, H. *Wspolna wlasnosc ziemi w polskiej publicystyce lat 1835-1860.* Warsaw, 1949.

Myslinski, J. *Grupy polityczne Krolestwa Polskiego w Zachodniej Galicji, 1895-1904.* Warsaw, 1967.

Nettle, J. P. *Rosa Luxemburg*, 2 vols. London, 1966.

Pobog-Malinowski, W. *Najnowsza historia polityczna Polski.* Vol. I (1864-1914). 2nd ed. London, 1963.

Prochnik, A. *Ku Polsce socjalistycznej*. Warsaw, 1936.

Res [Feliks Perl]. *Dzieje ruchu socjalistycznego w zaborze rosyjskim do powstania PPS.* 1910; rpt. Warsaw, 1958.

Rostworowski, E. *Ostatni Krol Rzeczypospolitej Polskiej: Geneza i upadek Konstytucji 3° maja.* Warsaw, 1966.

Sabine, G. *A History of Political Theory.* 3rd ed. New York, 1962.

Serejski, M. H. ed. *Historycy o historii: Od Adama Naruszewicza do Stanislawa Ketrzynskiego, 1775-1918.* Warsaw, 1963.

————. *Joachim Lelewel.* Warsaw, 1953.

Sikora, A. "Socjalizm utopijny A. Mickiewicza w Trybune Ludow." (Unpublished pamphlet, Warsaw, 1955.)

Stone, B. "Nationalist and Internationalist Currents in Polish Socialism: The PPS and SDKPiL, 1893-1921." (Unpublished doctoral
dissertation at the University of Chicago, 1965.)

Struve, H. *Filozofia polska w ostatniem dziesiecioleciu, 1894-1904.*
Translated from the German by Kazimierz Krol. Warsaw, 1907.

————. *Historia filozofii w Polsce na tle ogolnego rozwoju zycia
umyslowego.* Warsaw, 1900.

————. *Historia logiki jako teorii poznania w Polsce.* 2nd ed. Warsaw, 1911.

Suchodolski, B. *Polskie tradycje demokratyczne: idea demokracji w
ujeciu myslicieli polskich od St. Staszica do St. Witkiewicza.* Wroclaw, 1946.

Swoboda, H. [Adam Prochnik]. *Pierwsze pietnastolecie Polski niepodleglej, 1918-1933. Zarys dziejow politycznych.* Warsaw, 1933.

Targalski, J.*Szermierz Wolnosci: Boleslaw Limanowski.* Warsaw, 1973.

Tatarkiewicz, W. *Historia Filozofii.* 4th ed. Vol. II. Warsaw, 1949.

Ulashchik, N. *Predposylki krestianskoi reformy 1861 g. v Litve i
Zapadnoi Belorusi.* Moscow, 1965.

Uruski, Count S., A. A. Kosinski and A. Wlodarski. *Rodzina: Herbarz szlachty polskiej.* Vol. IX. Warsaw, 1912.

Venturi, F. *Roots of Revolution: A History of the Populist and Socialist Movements in Nineteenth Century Russia.* Translated by
Francis Haskell, with an Introduction by Isaiah Berlin. London,
1964.

Woodcock, G. *Anarchism: A History of Libertarian Ideas and Movements.* 1962; rpt, Cleveland and New York, 1967.

Zarys historii prasy polskiej. Part II, No. 1, (1864-1900, excluding
Prussian Poland), edited by T. Butkiewicz and A. Mlynarski; Part
II, No. II (1900-1918 and Prussian Poland, 1864-1918) edited by
B. Krzywoblocka and A. Slisz. (Mimeographed monograph at
University of Warsaw, 1959.)

Znaniecki, F. *Modern Nationalities: A Sociological Study.* Urbana,
Ill., 1952.

Zarnowska, A. *Geneza rozlamu w Polskiej Partii Socjalistycznej,
1904-1906.* Warsaw, 1965.

Zechowski, Z. *Socjologia Boleslawa Limanowskiego.* Poznan, 1964.

Zychowski, M. *Boleslaw Limanowski, 1835-1935.* Warsaw, 1971.

C. Bibliographical Works.

"Archiwum CKW PPS," *Dzieje Najnowsze.* Wroclaw, 1947, pp. 360-
63.

Garlicki, A. and others. "Stan badan nad dziejami PPS (1892-1939),"
Z Pola Walki, V, No. 4/20 (1962), pp. 128-60.

Kancewicz, J. "Stan badan nad dziejami PPS i kryteria jego oceny," *Z Pola Walki*, VI, No. 3/23 (1963), pp. 109-16.

Korman, Z. *Materialy do bibliografii drukow socjalistycznych na ziemiach polskich w latach 1866-1918*. Warsaw, 1935.

Moscicki, H. J. "Materialy do bibliografii PPS w latach 1892-1918." Warsaw, 1960. (A mimeographed monograph, based on the contents of the PPS Archives at AZHP.)

Suchodolski, W. *Krytyka, 1899-1914: Bibliografia zawartosci*. Wroclaw, 1953.

1. *Political Ideas and the Enlightenment in the Romanian Principalities, 1750-1831.* By Vlad Georgescu. 1971.

2. *America, Italy and the Birth of Yugoslavia, 1917-1919.* By Dragan R. Zivojinovic. 1972.

3. *Jewish Nobles and Geniuses in Modern Hungary.* By William O. McCagg, Jr. 1972.

4. *Mixail Soloxov in Yugoslavia: Reception and Literary Impact.* By Robert F. Price. 1973.

5. *The Historical and Nationalist Thought of Nicolae Iorga.* By William O. Oldson, 1973.

6. *Guide to Polish Libraries and Archives.* By Richard C. Lewanski. 1974.

7. *Vienna Broadcasts to Slovakia, 1938-1939: A Case Study in Subversion.* By Henry Delfiner. 1974.

8. *The 1917 Revolution in Latvia.* By Andrew Ezergailis. 1974.

9. *The Ukraine in the United Nations Organization: A Study in Soviet Foreign Policy, 1944-1950.* By Konstantin Sawczuk. 1975.

10. *The Bosnian Church: A New Interpretation.* By John V. A. Fine, Jr. 1975.

11. *Intellectual and Social Developments in the Habsburg Empire from Maria Theresa to World War I.* Edited by Stanley B. Winters and Joseph Held. 1975.

12. *Ljudevit Gaj and the Illyrian Movement.* By Elinor Murray Despalatovic. 1975.

13. *Tolerance and Movements of Religious Dissent in Eastern Europe.* Edited by Bela K. Kiraly. 1975.

14. *The Parish Republic: Hlinka's Slovak People's Party, 1939-1945.* By Yeshayahu Jelinek. 1976.

15. *The Russian Annexation of Bessarabia, 1774-1828.* By George F. Jewsbury. 1976.

16. *Modern Hungarian Historiography.* By Steven Bela Vardy. 1976.

17. *Values and Community in Multi-National Yugoslavia.* By Gary K. Bertsch. 1976.

18. *The Greek Socialist Movement and the First World War: The Road to Unity.* By George B. Leon. 1976.

19. *The Radical Left in the Hungarian Revolution of 1848.* By Laszlo Deme. 1976.

20. *Hungary between Wilson and Lenin: The Hungarian Revolution of 1918-1919 and the Big Three.* By Peter Pastor. 1976.

21. *The Crises of France's East-Central European Diplomacy, 1933-1938.* By Anthony J. Komjathy. 1976.

22. *Polish Politics and National Reform, 1775-1788.* By Daniel Stone. 1976.

23. *The Habsburg Empire in World War I.* Robert A. Kann, Bela K. Kiraly, and Paula S. Fichtner, eds. 1977.

24. *The Slovenes and Yugoslavism, 1890-1914.* By Carole Rogel. 1977.

25. *German-Hungarian Relations and the Swabian Problem.* By Thomas Spira. 1977.

26. *The Metamorphosis of a Social Class in Hungary During the Reign of Young Franz Joseph.* By Peter I. Hidas. 1977.

27. *Tax Reform in Eighteenth Century Lombardy.* By Daniel M. Klang. 1977.

28. *Tradition versus Revolution: Russia and the Balkans in 1917.* By Robert H. Johnston. 1977.

29. *Winter into Spring: The Czechoslovak Press and the Reform Movement 1963-1968.* By Frank L. Kaplan. 1977.

30. *The Catholic Church and the Soviet Government, 1939-1949.* By Dennis J. Dunn. 1977.

31. *The Hungarian Labor Service System, 1939-1945.* By Randolph L. Braham. 1977.

32. *Consciousness and History: Nationalist Critics of Greek Society 1897-1914.* By Gerasimos Augustinos. 1977.

33. *Emigration in Polish Social and Political Thought, 1870-1914.* By Benjamin P. Murdzek. 1977.

34. *Serbian Poetry and Milutin Bojić.* By Mihailo Dordevic. 1977.